SpringerWienNewYork

CISM COURSES AND LECTURES

The series presents lecture notes, monographs, edited works and proceedings in the field of Mechanics, Engineering, Computer Science and Applied Mathematics.
Purpose of the series is to make known in the international scientific and technical community results obtained in some of the activities organized by CISM, the International Centre for Mechanical Sciences.

INTERNATIONAL CENTRE FOR MECHANICAL SCIENCES

COURSES AND LECTURES - No. 536

EXPLOITING NONLINEAR BEHAVIOR IN STRUCTURAL DYNAMICS

EDITED BY

DAVID J. WAGG
UNIVERSITY OF BRISTOL, GREAT BRITAIN

LAWRENCE VIRGIN
DUKE UNIVERSITY, DUHAM, NC, USA

SpringerWienNewYork

This volume contains 127 illustrations

SPIN 86067282

All contributions have been typeset by the authors.

ISBN 978-3-7091-1186-4 SpringerWienNewYork

PREFACE

There are many physical phenomena which lead to nonlinear vibration problems. Modern structures are increasingly being built using more sophisticated materials that have a range of nonlinear material properties, some of which can be designed into the system. In some cases there are clear advantages in deliberately including nonlinear effects into the design of a structure. An obvious example are structural dampers. The most effective dampers contain highly nonlinear processes such as friction, fluids and most recently magneto-rheological fluids. Understanding and modelling the behaviour of these nonlinear effects is not a trivial process. However, there has been a dramatic increase in our understanding of nonlinear systems in the past 20 years, which has led to the realisation that beyond just modelling nonlinear effects, engineers can also use them to their advantage.

Nonlinearity arises from a range of phenomena. For example, geometric nonlinearity, including the effects of large deformations, combined stretching or compressing with vibration and nonlinear alignment of structural elements. Another source of nonlinearity is external forces acting on a linear system, such as fluid or magnetic forces. Nonlinear behaviour also comes from constraints in the system, freeplay, backlash, impact and friction.

Control forces can be added to a structural system in order to control the behaviour in some way and make it an adaptive structure. For example to reduce unwanted vibrations, detect damage, harvest energy or to shape change (morph) the structure. However, to create adaptive structures, the structure needs to have some awareness of its condition and/or the environment it is in. This is achieved by having a series of measurement sensors mounted on (or integrated into) the structure. Information from the sensors is then used by the global control system. This is where the smart (or intelligent) behaviour is generated.

This volume is a direct result of a course on "Exploiting Nonlinear Behaviour in Structural Dynamics" held at the International Centre for Mechanical Sciences (CISM) Udine, Italy Sep 13-17, 2010. Each chapter corresponds to a summary of the content of the lectures presented by each of the expert speakers who participated in the course.

Chapter 1 by Virgin & Wagg gives an overview and introduction to nonlinear phenomena in structural dynamics. The analysis of nonlinear effects using state space and bifurcation analysis is introduced, followed by an introduction to nonlinear control techniques. Chapter 2 by Neild covers material on so called approximate methods for analysing nonlinear systems where the level of nonlinearity is assumed to be relatively small. Examples include a device for harvesting mechanical energy, where the nonlinear effects are exploited to increase the useful energy which can be extracted. Chapter 3 by Virgin is devoted to the topic of vibration isolation. In particular, buckled structures and structures with large amounts of geometric nonlinearities are used as nonlinear vibration absorbers, to reduce significant unwanted vibrations in the systems under consideration. Chapter 4 by Shaw focuses attention on the mitigation of undesirable torsional vibration in rotating systems utilizing specifically nonlinear features in the dynamics. Reducing vibration in automotive motors is a problem of considerable practical relevance. Tunable vibration absorbers require careful design with the role of nonlinearities of particular relevance. Chapter 5 by Ribeiro discusses the vibration of nonlinear (beam) structures in which the motion is sufficiently large amplitude that elasto-plastic effects are induced. Both free and forced situations are analyzed including practical (numerical) solution procedures. Ultimately, more accurate modeling of these types of systems will assist in a more complete understanding of nonlinear vibration and its relationship with material failure. Finally in Chapter 6 by Wagg, structural systems with control are considered. These include nonlinear systems with active control, as well as morphing structures, where snap-through mechanisms are exploited as hinges so the structure can change its shape between two stable states.

We would like to thank all those at CISM who helped to make the course and the production of this volume such a enjoyable experience, in particular Mrs P. Agnola, Elsa Venir and Carla Toros who dealt with the administration of the meeting. In addition we would like to thank the Rectors of CISM: G. Maier, J. Salençon, W. Schneider and the Secretary General, B.A. Schrefler. Finally, we would like to thank Prof. Serafini for his assistance with the preparation of this book.

Lawrie Virgin (Duke University, USA)
David J. Wagg (University of Bristol, UK)
June 2011

CONTENTS

Introductory Material

Lawrie Virgin and David Wagg
Duke University and University of Bristol.

Abstract This book is based on a one-week worksop at CISM. In order to fully appreciate the benefits of nonlinearity in certain engineering systems it is important to understand the underlying behavior of linear systems, and this first chapter provides a general overview of linear dynamical systems and then begins to explore the effect of nonlinearities.

1 The Linear Oscillator

In mechanics we are primarily interested in the time evolution of systems governed by odes

$$\dot{x} = F(x, \lambda, t) \qquad x \epsilon R^n, \qquad t \epsilon R, \tag{1}$$

where x is a state vector which describes the evolution of the system under the vector field, F. Given an initial condition x_0 at time $t = 0$ we can seek to solve system 1 to obtain a trajectory, or orbit, along which the solution evolves with time. We will then seek to ascertain the stability of the system, generally as a function of the (control) parameter, λ (Guckenheimer and Holmes (1983)).

The cornerstone of dynamics in a mechanics context is, of course, Newton's second law, and thus sets of second-order, ordinary differential equation are dominant:

$$\ddot{x} + \omega_n^2 x = 0, \tag{2}$$

where an overdot represents a derivative with respect to time, and the system has two initial conditions $x(t = 0), \dot{x}(t = 0)$, from whence the dynamics develops. This is the equation of motion governing the dynamic response of the spring-mass system shown in Figure 1 with $\omega_n = \sqrt{k/m}$ and all other parameters set equal to zero. We can write the solution as

$$x(t) = Ae^{\lambda t}. \tag{3}$$

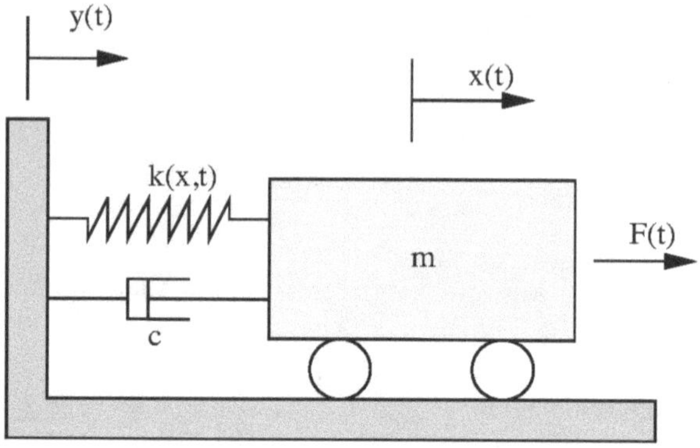

Figure 1. A spring-mass damper.

Placing 3 in to 2 we find that $\lambda = i\omega_n$ and thus the general form of the solution is given by

$$x(t) = ae^{i\omega_n t} + be^{-i\omega_n t}. \tag{4}$$

Alternatively, using Euler's identities we can write:

$$x(t) = c\cos(\omega_n t) + d\sin(\omega_n t). \tag{5}$$

In order to determine a and b, (or c and d), we make use of the initial conditions to get

$$x(t) = x(0)\cos(\omega_n t) + \frac{\dot{x}(0)}{\omega_n}\sin(\omega_n t). \tag{6}$$

This system can be converted in to a pair of coupled, first-order ordinary differential equations (in state variable format) by introducing a new variable

$$y = \dot{x} \tag{7}$$

and substituting in equation 2 gives:

$$\dot{x} = y, \qquad \dot{y} = -\omega_n^2 x \tag{8}$$

and in matrix notation:

$$\begin{bmatrix} \dot{x} \\ \dot{y} \end{bmatrix} = \begin{bmatrix} 0 & 1 \\ -\omega_n^2 & 0 \end{bmatrix} \begin{bmatrix} x \\ y \end{bmatrix}. \tag{9}$$

The solutions to this type of equation are harmonic, with oscillation occurring about the origin (the unique equilibrium position), Inman (1994). The form, and frequency, of the resulting motion is independent of the initial conditions.

2 A Nonlinear Spring

Now suppose we have a spring whereby the applied force and corresponding deflection are related via a cubic expression:

$$F = k(x) = Ax + Bx^3. \tag{10}$$

Adding a small amount of inertia and a little damping we obtain a standard nonlinear oscillator known as Duffing's equation (Virgin (2000)):

$$\ddot{x} + 0.1\dot{x} + Ax + Bx^3 = 0. \tag{11}$$

We still have a spring that initially responds the same way in compression and extension, but now the nonlinearity depends on the signs of A and B. Superposition no longer holds. Suppose we have $A = 1$ and $B > 0$. In this case we have a hardening spring, i.e., it becomes disproportionately stiffer as the deflection increases. This is shown schematically in Figure 2. Also shown is the softening case $A = 1, B < 0$, indicated by the dashed line. Furthermore, if $A = -1$ and $B > 0$ we get a (still symmetrical, and

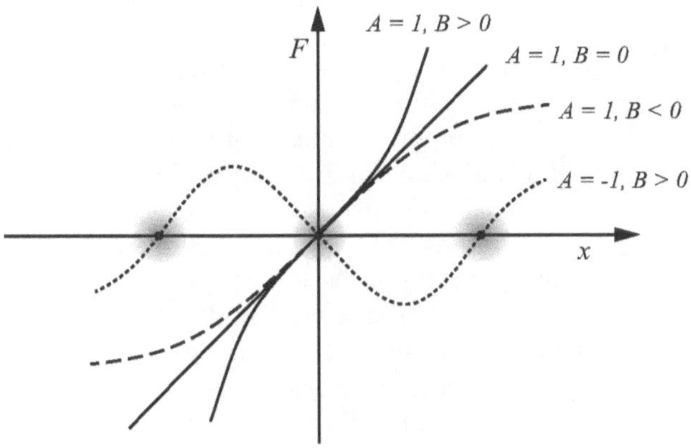

Figure 2. Force-deflection relations for some typical springs.

shown by the dotted line) system in which the origin is now unstable (to be shown later), and the spring, given a typical load will take up one of two available equilibrium positions. The force-deflection characteristic need not be symmetric, and in fact, this will typically be the case under some pre-loading.

When incorporated into the dynamics context (equation 11) we find that the frequency as well as the amplitude depends on the initial conditions. Equation 11 is of course a nonlinear ordinary differential equation and does not submit to standard methods. However, we can gain considerable insight into the behavior of such systems, and we start by ignoring damping. As such, the total mechanical energy is now conserved and using $\ddot{x} = \dot{x}\mathrm{d}\dot{x}/\mathrm{d}x$ we can separate variables, integrate, and write

$$\dot{x} = \pm\sqrt{-Ax^2 - (B/2)x^4 + C}, \tag{12}$$

in which C is determined from the initial conditions. For $A = 1$ and $B = 0$ we obtain ellipses in the phase space corresponding to simple harmonic motion. However, for other combinations of A and B we obtain behavior that may be very different from simple harmonic.

We can also obtain the natural period of motion in a related way. It can be shown that the period is equal to four times the time it takes to move from the maximum amplitude ($\bar{x}$) to zero, and rearranging equation 12 we can evaluate the period of motion, T, (and hence natural frequency) from Jordan and Smith (1977); Stoker (1992)

$$T = 4\int_0^{\bar{x}} \frac{\mathrm{d}x}{\sqrt{-Ax^2 - (B/2)x^4 + C}}. \tag{13}$$

However, this integral is not easy to solve, but whereas the linear oscillator has a natural frequency that is independent of the amplitude of motion, the nonlinear oscillator will have a period that depends on the initial conditions and hence the amplitude of motion. In Figure 2 was shown typical hardening and softening spring systems. The natural frequencies, often referred to as 'backbone curves' corresponding to these two case are shown schematically in Figure 3. These mildly nonlinear cases are centered on an equilibrium position. In some cases, e.g., when $A = 1$ and $B = -1$ and the motion exceeds $x = 1$ the behavior can become unbounded. In general, this type of behavior has to be investigated numerically, and we shall see that these backbone curves have a profound resonance effect when incorporated into the context of (harmonically) forced oscillators.

2.1 Linearization

With no force applied the 'rest length' of the linear spring is the unique position of equilibrium, which we consider without loss of generality to be the origin. For a nonlinear spring, for example, the dotted line in Figure 2

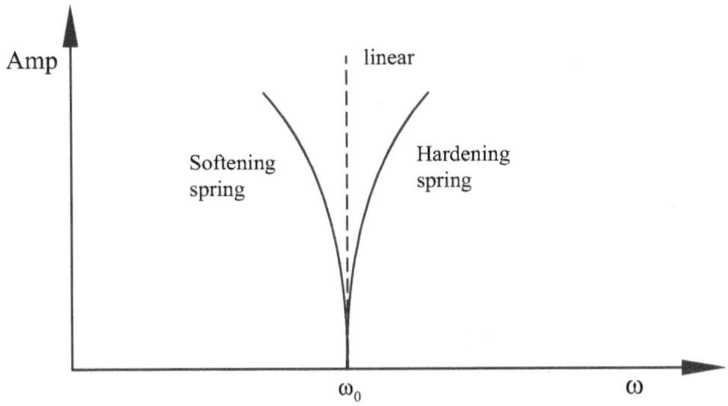

Figure 3. The frequency relation for mildly nonlinear oscillators.

we see two 'remote' equilibrium positions. These happen to be stable: their local slopes are positive and any deviation from the position will be opposed by the (restoring) force. In the vicinity of equilibrium (indicated by the three shaded regions) we see that the force-deflection relation is practically linear. More formally, we can take a Taylor series expansion about equilibria. The equilibria are found from $Ax+Bx^3 = 0$, from which have $x_e = 0$ (sometimes referred to as the trivial equilibrium) but there may also be other roots. For example with $A = -1, B = +1$ we get two more real roots at $x_e = \pm 1$, but if A and B have the same sign then the origin is the only real root. Consider a small perturbation, δ about an equilibrium, x_e:

$$x = x_e + \delta. \tag{14}$$

Placing this into equation 10 and assuming $A = -1, B = +1$ we obtain:

$$F = -x_e - \delta + x_e^3 + 3x_e^2\delta + 3x_e\delta^2 + \delta^3. \tag{15}$$

Since δ is small we neglect terms in δ higher than linear, and $-x + x_e^3 = 0$ to satisfy equilibrium, and thus we have

$$F = \delta(3x_e^2 - 1), \tag{16}$$

which is valid in the vicinity of the equilibria. For $x_e = 0$ we have a locally negative slope and the force tends to move the system further away with deflection. For $x_e = \pm 1$ we have a locally positive slope and a restoring force. If the spring undergoes 'large' deflections then the system becomes progressively more nonlinear.

In terms of dynamic response we can still use the approach of the first part of this chapter, but now the motion local to the equilibrium at the origin is described by

$$\ddot{x} - \omega_n^2 x = 0. \tag{17}$$

The solution again has the form $x(t) = Ae^{\lambda t}$ where $\lambda = \pm\omega_n$, and thus

$$x(t) = ae^{\omega_n t} + be^{-\omega_n t}, \tag{18}$$

and using the definition of hyperbolic functions we also have

$$x(t) = x(0)\cosh\omega_n t + (\dot{x}(0)/\omega_n)\sinh\omega_n t. \tag{19}$$

In this case we do not have a periodic solution: the positive exponential indicates that typically $x \to \infty$ as $t \to \infty$. Hence, this behavior is unstable (Virgin (2007)). However, we also observe that we can choose very specific initial conditions (unlikely but nevertheless important cases), where the trajectory will end up at the origin, i.e., where the positive exponential term is completely suppressed, as well as the case where the negative exponential term in equation 18 dominates for a short time before the trajectory is swept away. For all practical purposes, i.e., arbitrary initial conditions, the motion is clearly unstable.

Thus, the key stability information is contained in (the sign of) λ in equations 3. If it is negative then the motion will decay with time, otherwise it will grow (Thompson and Stewart (1986)). It is convenient to introduce a more general (state) matrix of the form:

$$\begin{bmatrix} \dot{x} \\ \dot{y} \end{bmatrix} = \begin{bmatrix} a & b \\ c & d \end{bmatrix} \begin{bmatrix} x \\ y \end{bmatrix}, \tag{20}$$

and extract the determinant, Δ, and trace, T:

$$\Delta = (ad - bc), \quad T = a + d, \tag{21}$$

and recast stability given the fact that these relate to the system eigenvalues:

$$\lambda_{1,2} = \tfrac{1}{2}\left(T \pm \sqrt{T^2 - 4\Delta}\right), \quad \Delta = \lambda_1\lambda_2, \quad T = \lambda_1 + \lambda_2. \tag{22}$$

Hence for stability (negative eigenvalues) we require T negative and Δ positive (Strogatz (1994)). In the case of the two systems considered earlier both times the trace is zero (Newton's laws provide certain restrictions for typical mechanical systems), but for the system in equation 2 we have a positive Δ and hence stability, and a negative Δ for the system described by equation 17. We will now refine this to take account of damping.

2.2 Damping

Most real systems undergo a form of energy dissipation. With the inevitable presence of damping the question of stability becomes less ambiguous. Typical motion will then consist of a transient followed by some kind of recurrent long-term behavior, e.g. the motion will die out and the mass position is maintained at equilibrium.

Suppose we now allow for some energy dissipation in the form of linear viscous damping, i.e., $c \neq 0$ in Figure 1. The equation of motion is now

$$\ddot{x} + 2\zeta\omega_n\dot{x} + \omega_n^2 x = 0, \tag{23}$$

in which a nondimensional damping ratio, $\zeta = c/(2m\omega_n)$ has been introduced. Solutions to this equation now also depend on the value of ζ. For lightly damped systems we have $\zeta < 1$ and solutions of the form

$$x(t) = e^{-\zeta\omega_n t}\left(\frac{\dot{x}(0) + \zeta\omega_n\dot{x}(0)}{\omega_d}\sin\omega_d t + x(0)\cos\omega_d t\right) \tag{24}$$

where the damped natural frequency ω_d is given by

$$\omega_d = \omega_n\sqrt{1-\zeta^2}. \tag{25}$$

A typical underdamped response ($\zeta = 0.1$) is shown in Figure 4(a) and (b) as a time series and phase portrait. The origin in Figure 4 (b) indicates the position of asymptotically stable equilibrium. The trajectory gradually spirals down to this rest state: we can imagine a family of trajectories forming a flow as time evolves. Since this equilibrium is unique, the whole of the phase space is the attracting set for all initial conditions and disturbances. Damping in this range, e.g., $\zeta \approx 0.1$, is quite typical for mechanical and structural systems.

For a heavily (or overdamped) system $\zeta > 1$, and in this case the form of the solution is

$$x(t) = Ae^{(-\zeta+\sqrt{\zeta^2-1})\omega_n t} + Be^{(-\zeta-\sqrt{\zeta^2-1})\omega_n t} \tag{26}$$

where

$$A = \frac{\dot{x}(0) + (\zeta + \sqrt{\zeta^2-1})\omega_n x(0)}{2\omega_n\sqrt{\zeta^2-1}} \tag{27}$$

and

$$B = \frac{-\dot{x}(0) - (\zeta - \sqrt{\zeta^2-1})\omega_n x(0)}{2\omega_n\sqrt{\zeta^2-1}}. \tag{28}$$

The motion is a generally monotonically decreasing function of time and may take a relatively long time to overcome relatively heavy damping forces

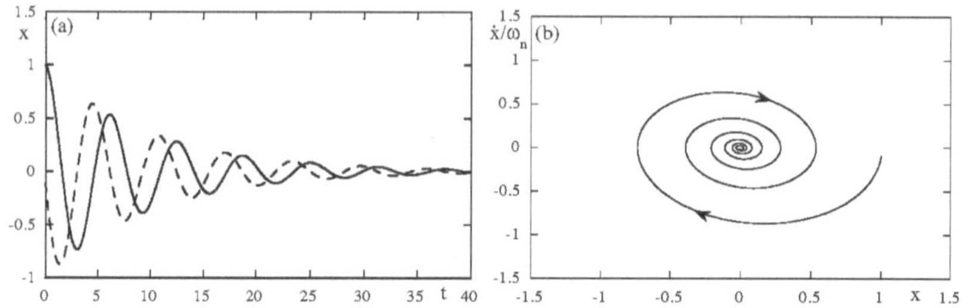

Figure 4. Time series and phase portraits for underdamped (oscillatory) motion. $x(0) = 1.0; \dot{x}(0) = 0.0; \zeta = 0.1$.

on the way to equilibrium. The boundary between these two cases is the critically damped case, i.e., $\zeta = 1$ when the eigenvalues are equal.

Returning to the state variable matrix format of the linear oscillator and adding damping we therefore have

$$\begin{bmatrix} \dot{x} \\ \dot{y} \end{bmatrix} = \begin{bmatrix} 0 & 1 \\ -\omega_n^2 & -2\zeta\omega_n \end{bmatrix} \begin{bmatrix} x \\ y \end{bmatrix}. \tag{29}$$

We can also write the solution in terms of the eigenvalues of the state matrix, i.e., the roots of the characteristic equation

$$\lambda^2 + 2\zeta\omega_n\lambda + \omega_n^2 = 0. \tag{30}$$

Now, the only difference with the expressions given in equation 9 is that the trace, T, becomes $-2\zeta\omega_n$, i.e., negative.

Given the scenario of a system losing stability we can usefully view all the response possibilities of this type of linear system according to the location of the roots in the complex plane. For example, having two complex roots with negative real parts corresponds to an exponentially decaying oscillation. Summarizing these outcomes in terms of the trace and determinant leads to Figure 5. In general we will have a system with positive stiffness and damping and thus a root structure corresponding to the lower right quadrant. Critical damping corresponds to the dashed parabola, and phase portraits and eigenvalues are indicated for various combinations of (T, Δ) and hence the natural frequency and damping.

3 A Nonlinear Damper

Another basic nonlinearity sometimes occurring in mechanical vibration is the appearance of energy dissipation in which the damping force is not

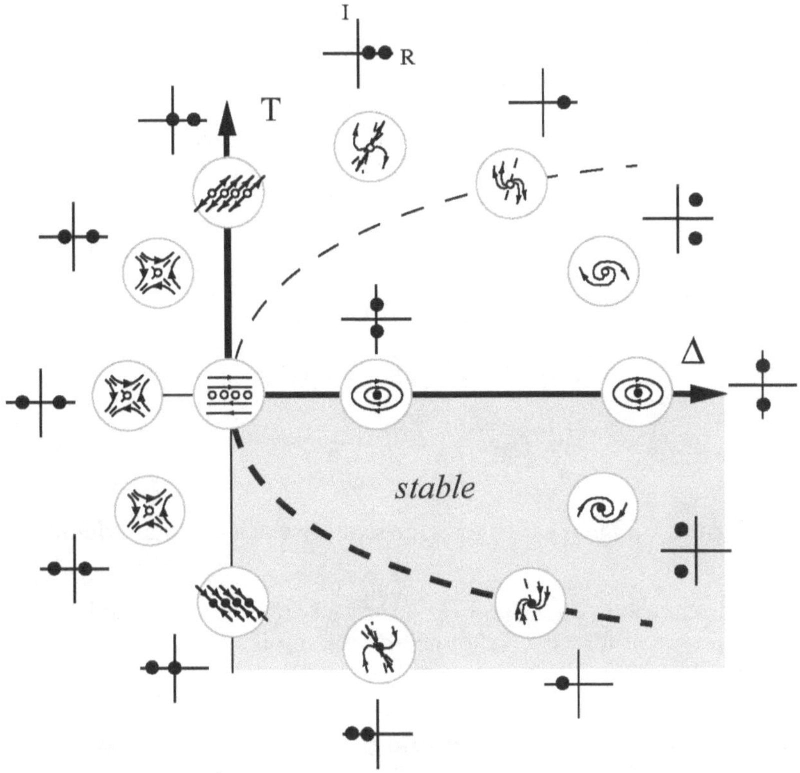

Figure 5. The root structure of a linear oscillator.

necessarily proportional to the velocity of motion. The adoption of linear viscous damping in our spring-mass-damper model is partly motivated by relevant damping processes, for example, the mechanism by which a dashpot (and other related devices) utilizes this type of energy dissipation. However, it is also used because of its relative analytic simplicity, i.e., assumption of linear viscous damping does not violate the rules of linearity, and this also has certain advantages in the study of dynamics in continuous dynamical systems.

3.1 Coulomb Damping

Friction commonly occurs in mechanical systems in which rubbing, or contact, between two dry surfaces causes the dissipation of energy (often in the form of heat). The assumption here is that the damping force is equal to the product of the normal force and a material-dependent coefficient of friction, Thomson (1981). The free response consists of a linear (as opposed to an exponential) decay of motion (with a frequency of oscillation the same as the underlying undamped system), and the mass may come to rest with a slight static offset if the static force of friction is greater than the restoring force of the spring. A typical example is shown in Figure 6. This type of

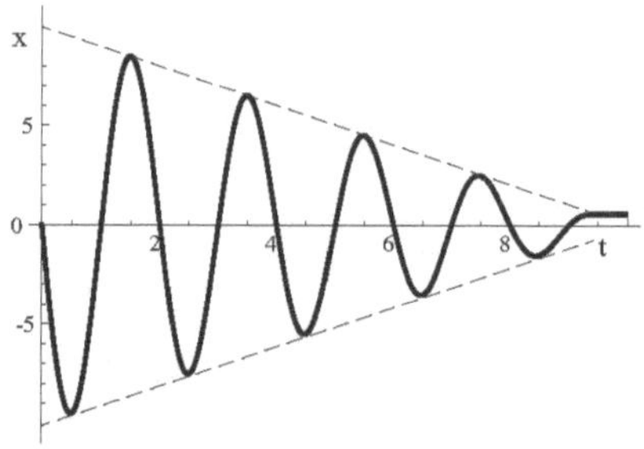

Figure 6. Time series for a mass subject to Coulomb damping.

energy dissipation can lead to a variety of interesting behavior especially in the context of forced vibrations. For example, stick-slip occurs in many mechanical systems.

3.2 Motion-dependent Damping

Another type of nonlinear energy dissipation is the mechanism underlying the appearance of certain types of limit cycle. The classic example is the van der Pol equation (van der Pol (1934)):

$$\ddot{x} - \mu(1 - x^2)\dot{x} + x = 0. \tag{31}$$

The parameter μ has a profound effect on the behavior of this system. For positive μ, we see that if $x^2 > 1$ then the damping term is positive and energy is dissipated. However, again for positive μ the damping becomes negative when $x^2 < 1$. Using the same linearization process as detailed in section 2.1 we consider small perturbations about equilibrium (the origin) which leads to:

$$\begin{bmatrix} \dot{x} \\ \dot{y} \end{bmatrix} = \begin{bmatrix} 0 & 1 \\ -1 & \mu \end{bmatrix} \begin{bmatrix} x \\ y \end{bmatrix}, \tag{32}$$

The eigenvalues of this system are

$$\lambda = \frac{\mu}{2} \pm \frac{1}{2}\sqrt{\mu^2 - 4}. \tag{33}$$

The determinant $\Delta = 1$ and the trace $T = \mu$. Thus, we have real negative roots when $\mu \leq 0$ indicating stable behavior. If $0 < \mu < 2$ the roots are complex with positive real parts indicating an unstable spiral. Thus, we locate these possible responses within Figure 5. This behavior is only valid in the vicinity of equilibrium. Solving equation 31 numerically for various

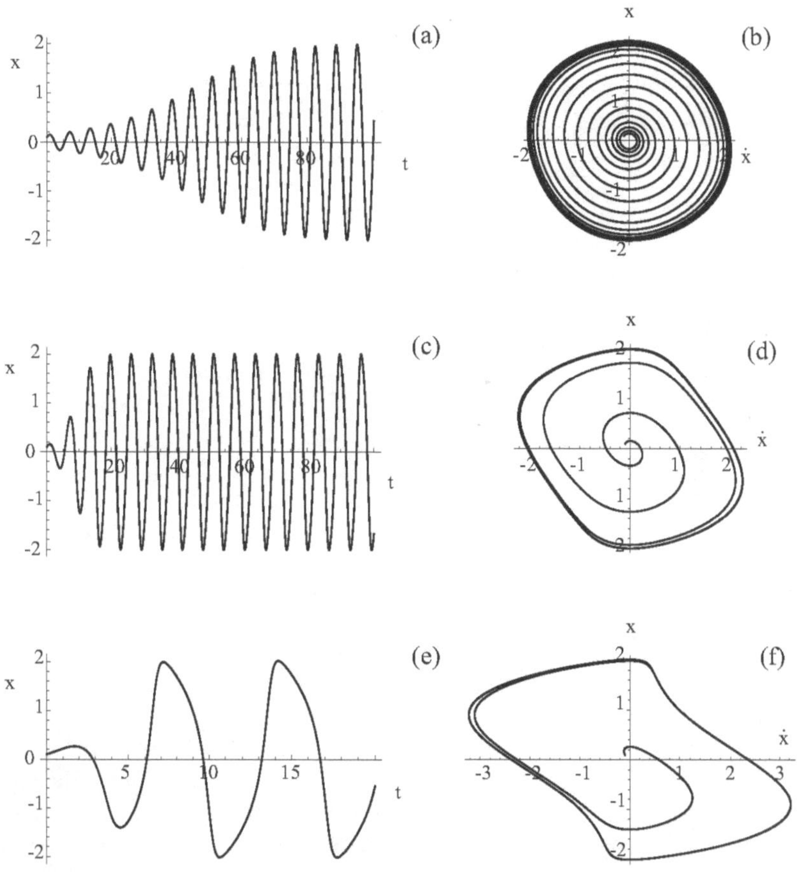

Figure 7. Time series and phase portraits for van der Pol's equation, (a) and (b) $\mu = 0.1$; (c) and (d) $\mu = 0.5$; (e) and (f) $\mu = 1.5$

positive values of μ leads to the results shown in Figure 7 as time series and phase projections. These are sometimes called relaxation oscillations, and since transients are attracted from within and without the oscillation this type of behavior is called a limit cycle oscillation (LCO). A perturbation approach (see chapter 2) and Jordan and Smith (1977), assuming μ is relatively small, can be used to show that the amplitude of the LCO remains close to 2 and that the frequency of oscillation is approximately $\omega = 1 - (1/16)\mu^2$.

We conclude this section by showing a couple of flows in phase space. An unforced Duffing system of the form $\ddot{x} + 0.1\dot{x} - x + x^3 = 0$ is shown for a variety of initial conditions in phase space in Figure 8(a). Likewise, the

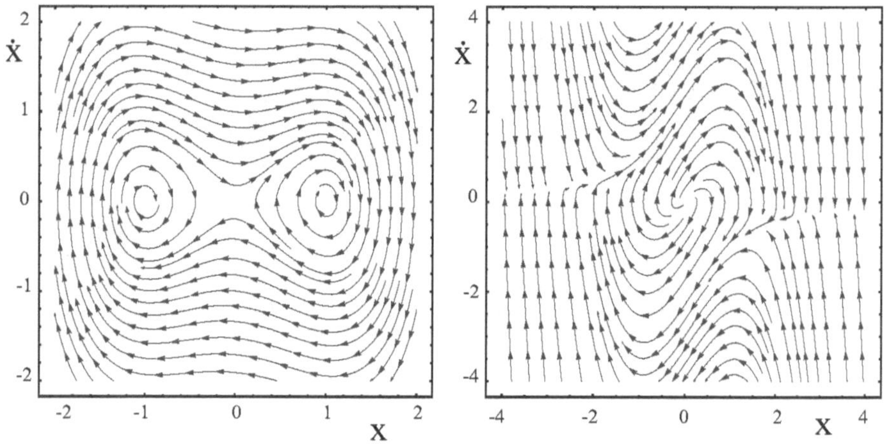

Figure 8. Flows in phase space: (a) Duffing's equation, (b) van der Pol's equation.

behavior of van der Pol's equation of the form $\ddot{x} - 1.5(1 - x^2)\dot{x} + x = 0$ is displayed in part (b). In both cases, for these parameter values, we see an unstable origin. In part (a) we have $(\Delta, T) = (-1, -0.1)$ indicating a saddle point and motion is swept away (and ultimately settles about one of the two stable equilibria at $x_e = \pm 1$), whereas in part (b) we have $(\Delta, T) = (+1, +1.5)$ and motion spirals away from the origin and settles onto the stable periodic orbit.

We thus observe what will typically happen when the stiffness or damping of the system changes, and especially where one of these parameters drops to zero, corresponding to an instability. The important issue here is that linear scenarios occur naturally within the context of nonlinear oscillators. The geometric view afforded by a consideration of the root structure and phase portraits of families of solution about equilibrium points is very useful.

4 Bifurcations

In many practical situations the forces acting on a system change. For example, the spring force in equation 10 might be subject to changing values of A and B, and this of course has a fundamental impact on the nature of solutions. Bifurcation theory (Doedel (1986); Seydel (1994)), classifies the generic ways in which an equilibrium loses its stability. Under the action of a single control parameter, for example, the stiffness or damping in an oscillator, we have already seen how the instability corresponds to an eigenvalue moving into the positive half-plane. The behavior of the linear oscillator provides an informative local view of behavior, but in a practical situation we might expect nonlinear effects to influence, or limit, the response in some way. As a parameter is varied the response of a system changes, and often gradually, but it is the qualitative change in the dynamics that is classified as a bifurcation. The elementary bifurcations are essentially one-dimensional but since the focus here is dynamics, we embed these (four elementary) bifurcations within the context of oscillations (Guckenheimer and Holmes (1983)).

4.1 Bifurcations from a Trivial Equilibrium

There are some systems in which some kind of initial symmetry is present, e.g., Euler buckling (Virgin (2007)). They represent an important class of instability in structural mechanics: super- and sub-critical pitchfork bifurcations. For the super-critical pitchfork bifurcation we can consider the oscillator:

$$\ddot{x} + 0.1\dot{x} + x^3 - \mu x = 0. \tag{34}$$

Again we observe the fundamental $x_e = 0$ solution, which is stable for $\mu < 0$. We immediately see how the example of Duffing's equation described earlier is a specific example with positive μ. At $\mu = 0$ a secondary equilibrium intersects the fundamental and it can be shown that the two (symmetric) non-trivial solutions are stable (see Figure 9(a)). The stability of equilibrium is determined using the (linearization) approach of section 2.1. This corresponds to the classic 'double-well' potential which is also shown superimposed for a specific (positive) value for μ.

The corresponding sub-critical pitchfork bifurcation is given by:

$$\ddot{x} + 0.1\dot{x} - x^3 - \mu x = 0, \tag{35}$$

and is shown in Figure 9(b). In this case, starting from a negative value of μ the trivial equilibrium is again stable but becomes completely unstable at the critical point, i.e., there is no local stable equilibrium to gradually

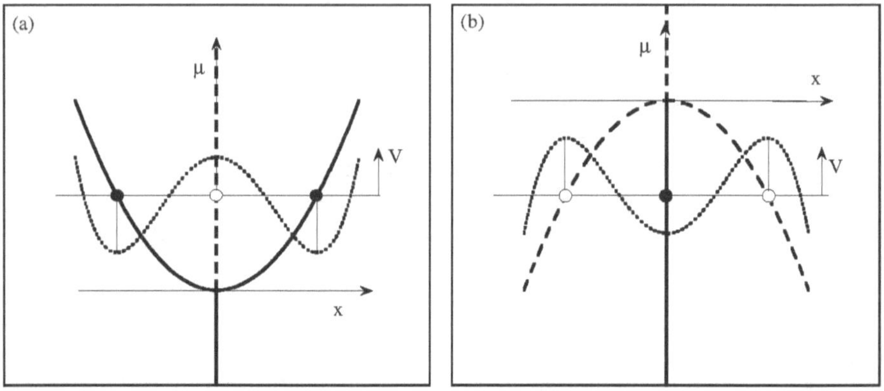

Figure 9. (a) A super-critical pitchfork bifurcation, (b) A sub-critical pitchfork bifurcation.

move onto. Furthermore, as the critical point is approached, the adjacent saddle points (associated with the unstable equilibria) start to erode the size of allowable perturbations. Although these two bifurcations have the same stable trivial equilibrium and critical point they have quite different consequences if encountered in practice. Hence, they are sometimes characterized as 'safe' or 'unsafe' according to whether a local post-critical stable equilibrium is available.

Another elementary bifurcation is the transcritical, or asymmetric, bifurcation:

$$\ddot{x} + 0.1\dot{x} + x^2 - \mu x = 0, \tag{36}$$

and illustrated in Figure 10(a). Here, a fundamental (trivial) equilibrium for negative μ loses stability as μ passes the through the origin. The other equilibrium becomes stable at this point and deflection occurs in the positive x direction.

Figure 10(b) shows the final example of an elementary bifurcation. It is also perhaps the most fundamental, since later we will show that the symmetry of the bifurcations already described is unlikely to be exactly observed in practice. The saddle-node bifurcation is characterized by the control parameter μ and coordinate x linked quadratically:

$$\ddot{x} + 0.1\dot{x} + x^2 - \mu = 0. \tag{37}$$

Equilibrium corresponds to the rest state and thus

$$x_e = \pm\sqrt{\mu}. \tag{38}$$

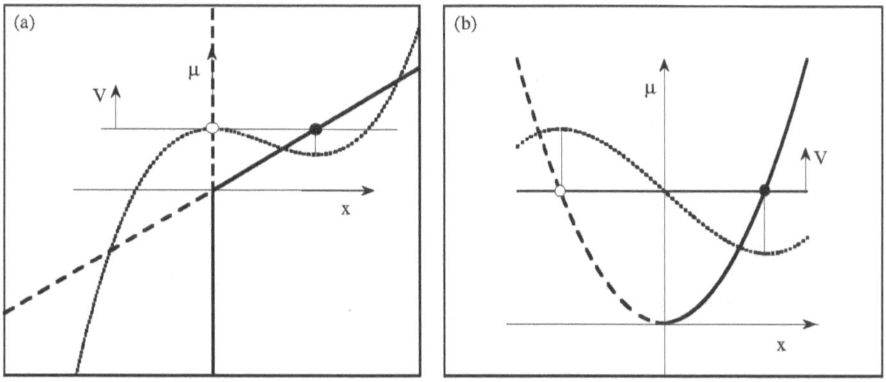

Figure 10. (a) A transcritical bifurcation, (b) A saddle-node bifurcation.

and we see either two co-existing solutions (one stable and the other stable) or no (real) solutions, depending on the sign of μ. The fundamental path is nonlinear, rather than the trivial initial path exhibited by the other bifucations.

We conclude this section by relating these situations to the changing potential energy (Bazant and Cedolin (1991)). For example, the potential energy associated with the saddle-node can be written as

$$V = \frac{x^3}{3} - \mu x + C, \tag{39}$$

and equilibrium from

$$\frac{\mathrm{d}V}{\mathrm{d}x} = x^2 - \mu = 0. \tag{40}$$

The sign of the curvature of the potential energy governs stability:

$$\frac{\mathrm{d}^2V}{\mathrm{d}x^2} = 2x, \tag{41}$$

which is evaluated about equilibrium. When $x_e = \sqrt{\mu}$ the second derivative of the potential energy function is positive indicating that this is a minimum and hence is stable. The opposite conclusion can be drawn from the other equilibrium branch thus confirming the results of the stability properties based on the decay or growth of local perturbations. As the value of μ is reduced the two equilibria come together (the frequency of small oscillations will decrease and effective damping increases) as the potential surface flattens out. Just prior to coalescence the stable equilibrium can be thought of as a node, and the unstable equilibrium is a saddle. Hence their approach (at the critical point) is called a saddle-node bifurcation. No equilibria exist for negative μ and trajectories would simply be swept away. This instability is also sometimes referred to as a fold or limit point. The potential energy is shown in Figures 9 and 10 (shown dotted) for a given value of the control parameter.

4.2 Initial Imperfections

Initial geometric imperfections or load eccentricities that tend to break the symmetry may have a relatively profound effect on stability (Virgin (2007)). We shall consider this type of effect and its influence on the sub-critical pitchfork. Incorporating a small offset causes equation 35 to be altered to

$$\ddot{x} + 0.1\dot{x} - x^3 - \mu x + \epsilon = 0, \tag{42}$$

where ϵ is a small parameter which breaks the symmetry. Figure 11 shows how the instability transition is changed. We see that for large negative μ

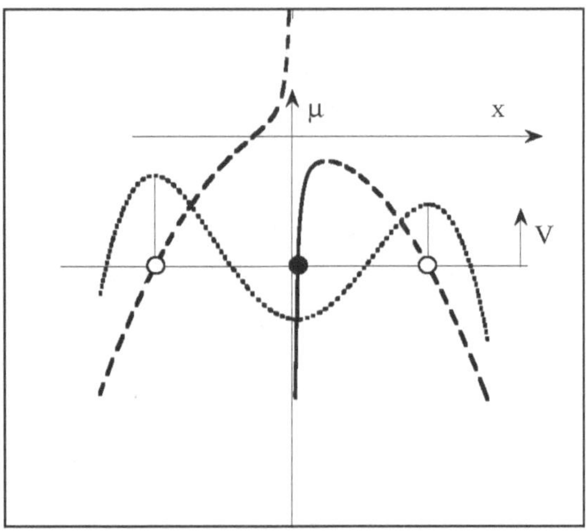

Figure 11. A perturbed sub-critical pitchfork bifurcation.

we have an equilibrium slightly offset from $x = 0$, and this grows as μ approaches the underlying critical value for the perfect geometry, but then falls off and the system completely loses stability. There is also a complementary (remote) solution for negative x but this wouldn't ordinarily be accessed as μ is monotonically increased (and it is unstable in any event). However, the fundamental solution does posses a critical point, and this is actually a saddle-node bifurcation. We also note the small tilt in the potential energy function. This behavior is termed 'imperfection-sensitive' (Thompson and Hunt (1973)) since the maximum value of the control parameter diminishes with the magnitude of the initial imperfection. The saddle-node and super-critical bifurcations are not imperfection sensitive.

4.3 Hopf Bifurcation

The other way in which an equilibrium can lose its stability under the operation of a single control parameter is the Hopf bifurcation (Thompson and Stewart (1986)). We have already seen this in van der Pol's equation, in which a complex conjugate pair of eigenvalues changes from having negative real parts to positive. That is, given a positive value of the determinant, the trace becomes positive (see figure 5). This instability is inherently dynamic, and is the main mechanism by which limit cycle oscillations occur. This also occurs in both the sub- and super-critical forms.

Figure 12 shows some typical transitions through these elementary bifurcations, in which the control is made a linear (ramp) function of time. Part (a) is a sub-critical pitchfork bifurcation in which the control parameter evolves with time according to $\mu = 0.01t - 1$ and thus the (quasi-static) critical point is reached after 100 time units. A slight delay in the realization is observed since the system remains somewhat in the vicinity of the unstable equilibrium after the critical point, before losing stability completely. The initial conditions for this case are $x(0) = 0.2, \dot{x}(0) = 0.0$. In part (b) is shown the corresponding super-critical case where the post-critical path follows one of the two available non-trivial (but stable) equilibrium paths. Part (c) is the saddle-node. Since there is no trivial equilibrium in this case the simulation was initiated at $x(0) = 1.2$, i.e., not far from equilibrium at $x_e = 1$ when $\mu = -1$. Finally part (d) illustrates a realization of a super-critical Hopf bifurcation. In this case $\mu = 1 - 0.5t$ and thus the quasi-static critical point is reached after approximately 20 time units, and again a delay is observed. Also seen in this figure is the lengthening of the period for larger t (and hence μ) anticipated from the initial post-critical approximation $\omega = 1 - (1/16)\mu^2$, as well as the motion becoming gradually less sinusoidal.

5 Forced (Linear) Oscillators

This section will focus on externally-excited systems, i.e., where $F(t) \neq 0$ in Figure 1. An important class of forcing function is harmonic excitation:

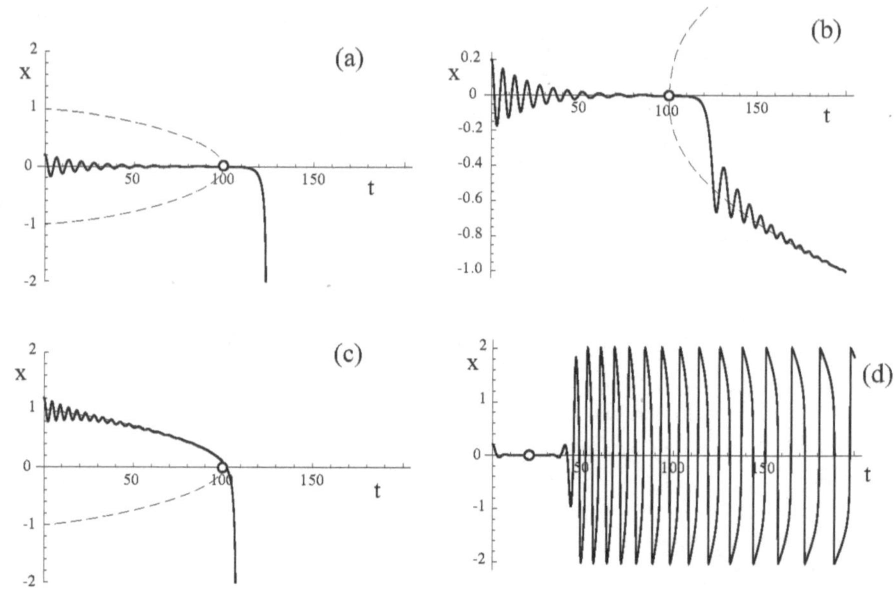

Figure 12. Examples of transitions through generic instabilities, (a) sub-critical pitchfork, (b) super-critical pitchfork, (c) saddle-node, (d) Hopf.

$F(t) = F_0 \sin \omega t$, or $y(t) = Y_0 \sin \omega t$, where this latter expression relates to a base movement that transmits motion to the mass via the support system (Thomson (1981); Inman (1994)). This latter situation is an important practical aspect of vibration and underlies the concept of vibration isolation to be considered from a nonlinear perspective in chapter 3.

For the case when the force is applied directly to the mass we have a governing equation of motion of the form

$$m\ddot{x} + c\dot{x} + kx = F_0 \sin \omega t, \tag{43}$$

or in nondimensional terms

$$\ddot{x} + 2\zeta\omega_n\dot{x} + \omega_n^2 x = f_0 \sin \omega t, \tag{44}$$

where $f_0 = F_0/m$. The solution of equation 44 consists of the summation of two parts: a homogeneous solution, obtained from the free vibration (obtained in the previous section); and the particular solution, which is related primarily to the forcing. Its general solution has the form:

$$x(t) = X_1 e^{-\zeta\omega_n t}\sin(\sqrt{1-\zeta^2}\omega_n t + \phi_1) + \frac{f_0}{k}\frac{\sin(\omega t - \phi)}{\sqrt{[1-(\omega/\omega_n)^2]^2 + [2\zeta\omega/\omega_n]^2}}, \tag{45}$$

where trigonometric identities have been used to combine the harmonic terms from equation 24, and X_1 and ϕ_1 depend on the initial conditions.

The first (transient) part of the solution decays with time leaving the second part as the steady-state oscillation. Some sample responses are shown in Figure 13 in which the (lightly damped) system is started from rest at three different forcing frequencies. Parts (a) and (b) show that for a forcing frequency, $\omega = 0.3$, which is less than the system natural frequency, $\omega_n = 1.0$, the transient is relatively mild compared with the steady-state response and is quickly attracted to the harmonic oscillation. When the forcing frequency is equal to the natural frequency, as in parts (c) and (d) resonance occurs, i.e., a significant magnification effect (the denominator in the second term in equation 45 becomes small for $\omega \approx \omega_n$). Note the much larger amplitude of the response. In parts (e) and (f) the forcing frequency is increased to a value of 1.6, and now the transient solution is on the same order of magnitude as the steady-state, and the steady-state amplitude is back down to a lower level. Thus we observe that both parts of the solution depend quite strongly on the frequency ratio. The rate, and hence duration, of the transient decay is primarily a function of the damping. In all these cases the final steady-state motion is independent of the initial conditions (the choice of the origin in Figure 13 is arbitrary). This will not necessarily be the case for nonlinear systems, and indeed transients may be repelled by an unstable solution, as for example one of the cases shown in Figure 2.

It is useful to summarize how the maximum amplitude of the (steady-state) response ($A = x_{max}k/F_0$) varies with the frequency ratio Ω, where $\Omega = \omega/\omega_n$. The normalized amplitude of response can also be written as $A = x_{max}\omega_n^2/f_0$ (Inman (1994)). The response scales linearly with the forcing amplitude f_0. Figure 14 (a) shows a typical amplitude response diagram for four different damping values. The phenomenon of resonance is apparent, i.e., a significant amplitude magnification when the forcing frequency is close to the natural frequency (i.e., $\Omega \approx 1$). In fact we see for zero damping a growth to infinite amplitudes. The resonant peak is thus very sensitive to damping ($A_{res} \approx 1/2\zeta$, for light damping), and since many of the nonlinearities of interest are related to larger amplitude motion, we might anticipate interesting behavior in the vicinity of resonantly forced, lightly damped systems.

When the system is subject to $y(t) = Y\sin\omega t$ (a displacement applied

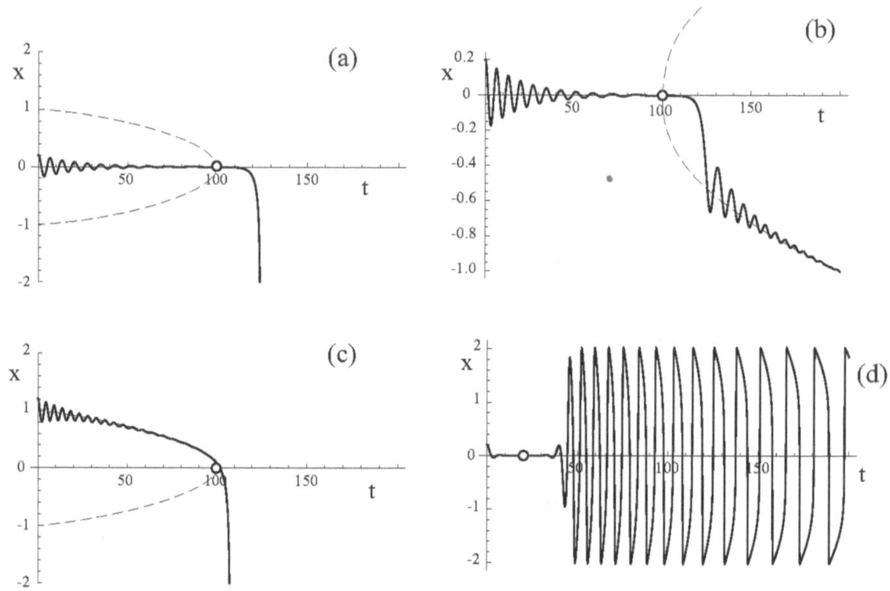

Figure 12. Examples of transitions through generic instabilities, (a) sub-critical pitchfork, (b) super-critical pitchfork, (c) saddle-node, (d) Hopf.

$F(t) = F_0 \sin \omega t$, or $y(t) = Y_0 \sin \omega t$, where this latter expression relates to a base movement that transmits motion to the mass via the support system (Thomson (1981); Inman (1994)). This latter situation is an important practical aspect of vibration and underlies the concept of vibration isolation to be considered from a nonlinear perspective in chapter 3.

For the case when the force is applied directly to the mass we have a governing equation of motion of the form

$$m\ddot{x} + c\dot{x} + kx = F_0 \sin \omega t, \tag{43}$$

or in nondimensional terms

$$\ddot{x} + 2\zeta\omega_n\dot{x} + \omega_n^2 x = f_0 \sin \omega t, \tag{44}$$

where $f_0 = F_0/m$. The solution of equation 44 consists of the summation of two parts: a homogeneous solution, obtained from the free vibration (obtained in the previous section); and the particular solution, which is

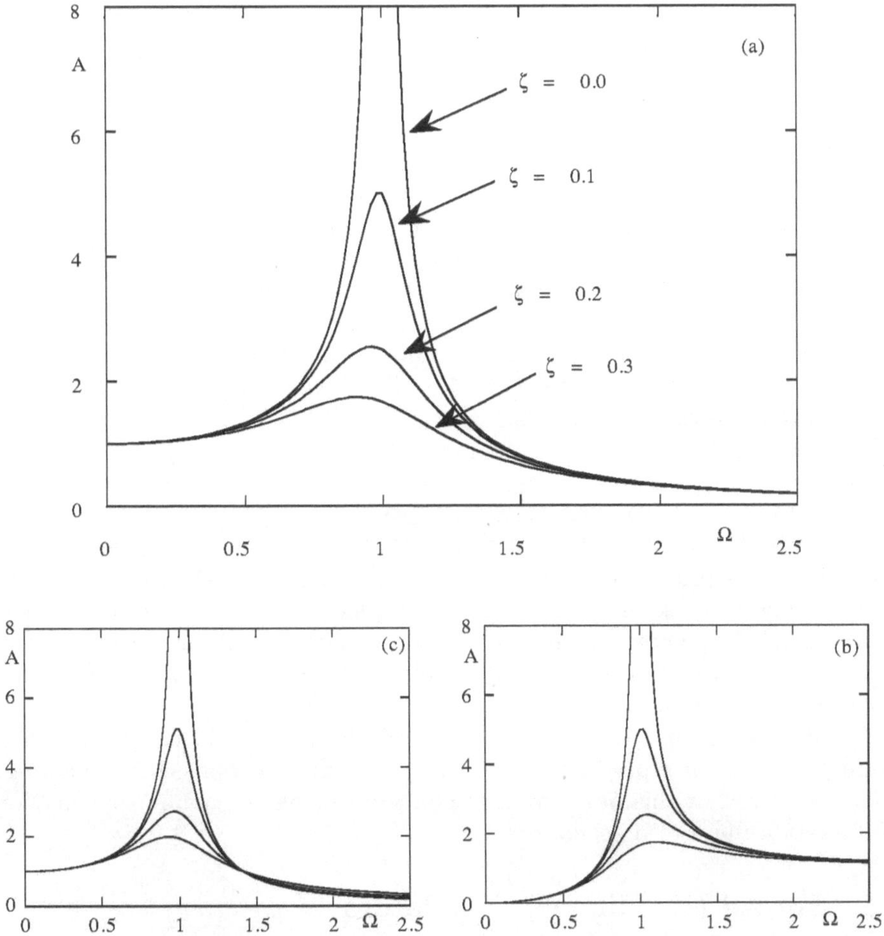

Figure 14. Amplitude response diagrams for linear oscillators, (a) direct mass excitation, (b) support motion (relative response), (c) support motion (absolute response). The same damping values as used in (a).

to the supporting frame in Figure 1) we obtain a related resonant effect. If the position of the mass is measured relative to the moving base, then the governing equation of motion takes the form

$$m\ddot{z} + c\dot{z} + kz = mY\omega^2 \sin(\omega t + \phi), \tag{46}$$

where $z(t) = x(t) - y(t)$. Thus, the forcing frequency enters into the forcing amplitude, which is also proportional to the amplitude of the base motion. This is familiar from rotating machinery with mass eccentricity in shafts Thomson (1981). The steady-state amplitude is given by the second part of equation 45 but with an additional ω^2 in the numerator, and the response amplitude is adjusted to take account of the fact that the force now arises via a transmitted base movement ($A = |Z/Y|$, where Z is the amplitude of z). This is shown in Figure 14(b).

A third form of resonant response (Figure 14(c)) is also found for base-excited systems where the absolute motion of the mass is measured (here $A = |X/Y|$, where Y and X are the input and response amplitudes respectively (Thomson (1981)). We note that the resonant condition can be considered as practically the same in all three cases. In this third case the amplitude response includes an interesting independence of damping when the forcing frequency is close to $\Omega = \sqrt{2}$. At high excitation frequencies the mass is practically stationary, and this is often a desirable characteristic for vibration isolation.

There are many other aspects of forced vibration of linear systems that are important in a practical sense, e.g., random vibrations, shock loading, etc. However, at this point we move on to focus on the nonlinear behavior of periodically forced oscillators.

6 Forced (Nonlinear) Oscillators

The forced response of the Duffing system in its three-dimensional phase space allows for the full spectrum of nonlinear behavior (Thompson and Stewart (1986)). We will restrict ourselves to harmonic excitation and start by considering the simplest case in which the magnitude of excitation is relatively small, such that the system responds in a mildly nonlinear manner. Adding a harmonic forcing to Duffing's equation 11 gives

$$\ddot{x} + 2\zeta\dot{x} + Ax + Bx^3 = F\sin(\omega t + \phi). \tag{47}$$

Equation 47 is by no means easy to solve in the general case. However, we clearly expect to see an approximately linear response for small amplitude motion (or when $B = 0$).

For small F and typical lightly damped systems (e.g., $\beta = 0.1$), we can use a variety of approximate analytic methods to obtain solutions, and these are the subject of chapter 2. We focus at this point on the typical qualitative behavior and specifically how this relates to resonance (but with some important differences from the linear case). The softening and hardening spring cases introduced in section 2 causes a bending over of the resonance response curve and thus the appearance of co-existing solutions. Examples are shown in Figure 15. For relatively heavy damping we expect to

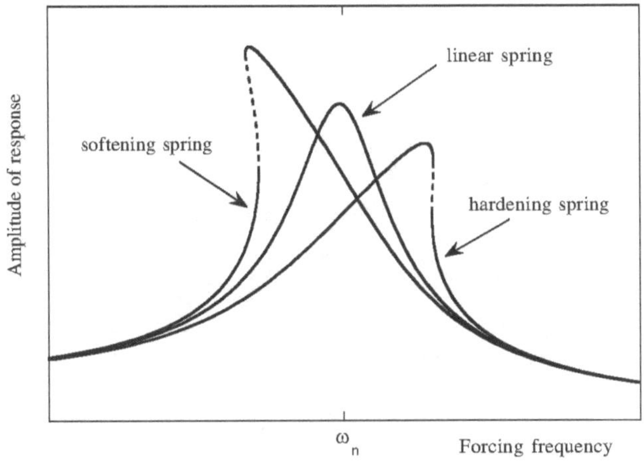

Figure 15. Schematic amplitude response diagrams for linear and nonlinear spring restoring forces. Dashed lines indicate an unstable branch.

see the single-valued case. However, under light damping (and moderately large forcing magnitude), a region of hysteresis is encountered. Thus, under slowly changing forcing frequency a jump up or down occurs (depending on the direction of sweep). These jumps occur via the cyclic analogue of the saddle-node bifurcation and are associated with a characteristic multiplier (CM) of the system penetrating the unit circle in the complex plane at $+1$ (associated with a vertical tangency in the response) Strogatz (1994). This is one of the three generic mechanisms of instability under the action of a single control parameter and will be revisited a little later (see Figure 17 and Guckenheimer and Holmes (1983)).

However, the stability of the steady-state solutions must in general be determined using some kind of additional analysis based on behavior of small perturbations, Jordan and Smith (1977), e.g., Floquet theory, and will also be considered in more detail in chapter 2. Figure 16 also shows some resonance response curves for Duffing's equation in which the damping and forcing magnitude are varied. Here, in part (a), the dashed line corresponds to the free (undamped) vibration case. The continuous curves correspond to different levels of damping (for a fixed level of forcing). With $c = 0$ we

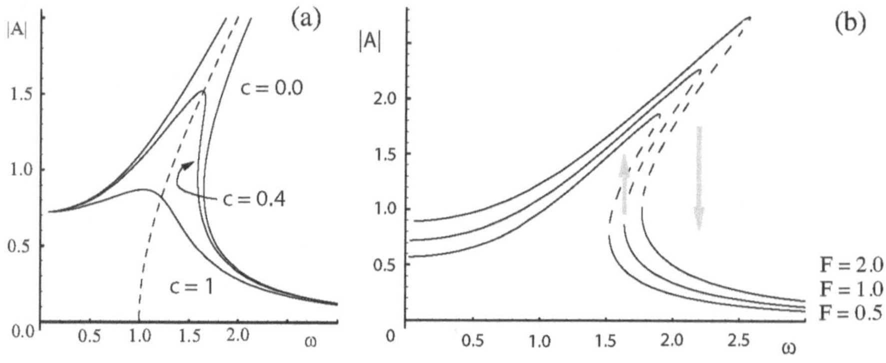

Figure 16. Resonance response curve based on a harmonic balance solution of Duffing's equation with $\alpha = \beta = F = 1$. (a) Various damping levels for fixed forcing magnitude ($F = 1$), (b) Various forcing magnitudes for a fixed level of damping ($c = 0.2$).

get the outer curve. When the damping is increased to a level of $c = 0.4$ we obtain a response that has a maximum amplitude close to $A = 1.5$ that occurs in the vicinity of $\omega = 1.6$. For the more heavily damped case ($c = 1$) the response hardly exhibits resonance at all and is similar to the linear response. This can still be considered underdamped ($\zeta = 0.5$) according to the linear description in an earlier part of this chapter.

6.1 Bifurcation of Maps

We finish this section with a brief consideration of the stability of cycles (Jordan and Smith (1977); Guckenheimer and Holmes (1983)). The amplitude jumps observed in Figure 16 are the cyclic equivalent of the saddle-node bifurcation described earlier. The analytic determination of the stability of periodic orbits is complex, and details are covered in chapter 2. As we have seen, nonlinear oscillators may suffer bifurcations in their periodic behavior as a system parameter is changed. Analogous to the behavior of small perturbations in the vicinity of equilibria, we can consider the behavior of small perturbations about a periodic orbit. Again linearization is a key concept and use will be made of Poincaré sampling and fixed points of maps. It can be shown that the behavior of small perturbations is typically governed by a variational equation of the form

$$\ddot{x} + G(t)x = 0, \tag{48}$$

where $x(t)$ represents motion about an oscillation and $G(t+T) = G(t)$. This equation does not, in general, submit to closed-form analytical solutions. However, Floquet theory has been developed to determine the general behavior of solutions to this equation. It is related to a discrete (Poincaré) map, the stability of which depends on the system eigenvalues, the characteristic multipliers (CM's) of the system. Rather than requiring negative real parts for stability, we now require eigenvalues with magnitude less than one. In this event, any disturbances from the periodic orbit can be shown to decay back onto the original orbit, and thus we have a periodic attractor. As a system parameter is changed we may have the possibility of an eigenvalue exceeding one in magnitude.

The three typical ways in which multipliers leave the unit circle are shown in Figure 17, with R and I signifying real and imaginary, respectively. We note that the Neimark bifurcation is less commonly encountered in practice.

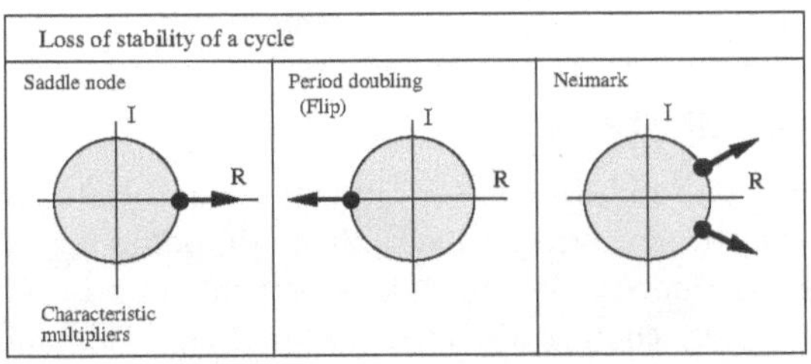

Figure 17. The generic routes to instability (in terms of CM's) for system under the action of a single control parameter.

6.2 Subharmonic Behavior

The response shown in the previous section is relatively close to primary resonance. Another commonly encountered behavior in nonlinear systems is a subharmonic, i.e., a period-n oscillation which takes n forcing periods to complete a full cycle. This may also occur over a variety of parameter values, and a typical output of a numerical simulation of equation 47 ($n = 5$), is shown in Figure 18 as a time series (a) and phase projection (b) after transients have been allowed to decay. For other parameters various types

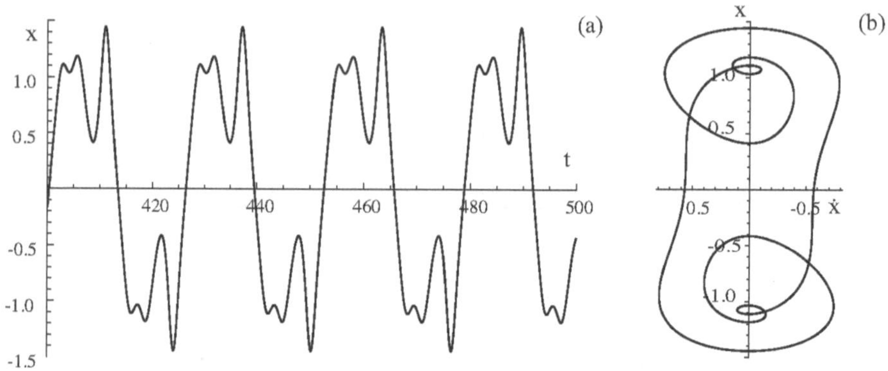

Figure 18. A typical subharmonic response from Duffing's equation with $A = -1, B = 1, \zeta = 0.15, F = 1, \omega = 1.2, \phi = 0$.

of subharmonics occur. Also, it is possible to obtain responses that repeat n times within a single forcing cycle and these are called superharmonics. Subharmonics are also often associated with peaks in the resonance curve at one-half the frequency of the main resonant peak for $n = 2$, etc.

6.3 Quasi-periodic Behavior

In nonlinear dynamics it is also not uncommon to encounter quasi-periodic behavior. In similarity to subharmonic behavior, quasi-periodicity occurs when two or more frequencies are present, but in this case they are incommensurate, i.e., their ratio is a non-integer. The important consequence is that a time series series will not appear repetitive, and a typical example is shown in Figure 19. The somewhat random-like response is particularly apparent in the phase portrait shown in part (b). The behavior of both subharmonic and quasi-periodic motion can be usefully visualized as shown in Figure 20. In the case of a subharmonic response, the orbit closes after a finite number of traverses around the surface of the torus. For the quasi-periodic response, the trajectories keeps wandering around the surface and it slowly fills out.

The Poincaré section is a stroboscopic sampling technique that reveals the periodic nature of a response, facilitates stability considerations, and is especially useful in the characterization of chaos. If we have an n-dimensional phase space, we can extract data at regular intervals to obtain an $n-1$-dimensional discrete map. This is shown schematically in Figure 21. In this example the sampling is achieved by extracting the penetrations of a

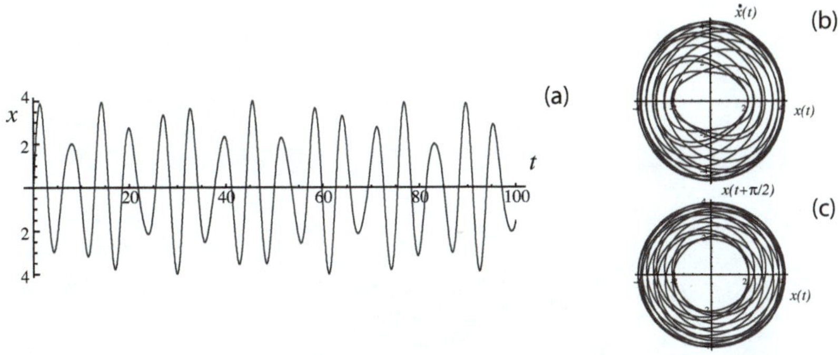

Figure 19. (a) A time series of a quasi-periodic response, (b) a conventional phase projection, (c) a reconstruction using time-delay coordinates.

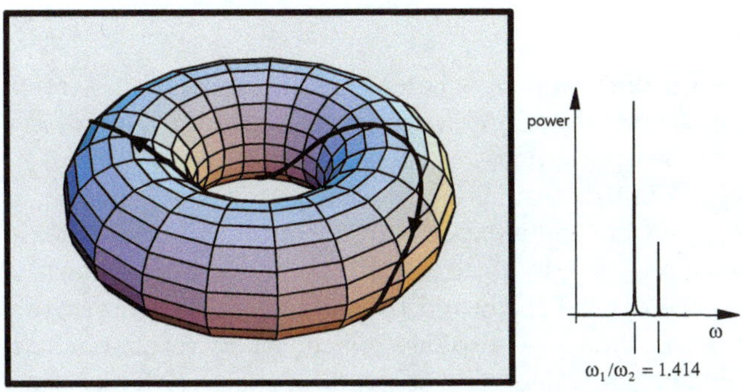

Figure 20. (a) In a response with two frequencies the trajectories can be envisioned as wrapping around the surface of a torus, (b) a frequency spectrum.

surface of section. For periodically forced systems it is natural to choose a given forcing phase to trigger the Poincaré section, since a periodic response will then result in a repeating point (after transients have been allowed to decay). Taking a Poincaré section of the data from Figure 18 results in

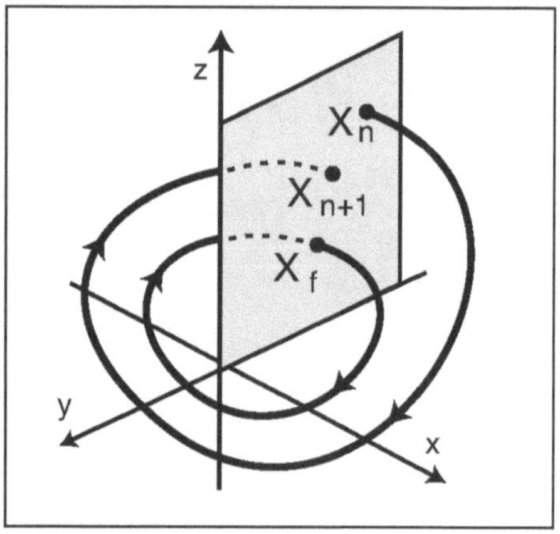

Figure 21. A Poincaré section reduces a continuous flow to a discrete map by examining a surface of section.

a repeating set of 5 data points, whereas taking a Poincaré section of the data from Figure 19 results in points that eventually form a closed curve, a characteristic of quasi-periodicity.

Also shown in Figure 19(c) is an example of using time-lag, or time-delay, embedding. Often, in an experimental context for example, it is only practical to measure a single state variable. It has been shown that the original phase space can be recovered by plotting this single state variable versus itself at later times. The phase space, when reconstructed in this way, is topologically equivalent to the original system. This also has the added advantage of obviating the need to perform numerical differentiation (e.g., when extracting velocity from a time series of position), which has the effect of adding noise.

Another diagnostic tool that is very useful in nonlinear dynamics is the frequency spectrum (Newland (1984)). This has a firm basis in linear signal processing of course (based on Fourier analysis), but can shed light on the differences between various types of periodic and non-periodic behavior. A sine wave is represented as a single delta function located at the single frequency of the signal. White noise has a broadband spectrum, etc. A subharmonic response is reflected in a finite number of peaks, as is quasi-periodic behavior, despite the incommensurate relation between frequencies. An example of a frequency spectrum for the time series shown in Figure 19 is shown in Figure 20(b).

6.4 Chaos

One of the most interesting features of nonlinear dynamics is chaos. That is, a low-order, deterministic dynamical system is capable of exhibiting random-like behavior that exhibits an extreme sensitivity to initial conditions. Chaos is characterized by a broadband power spectra, i.e., motion in which a multitude of frequencies actively participate. However, although a similar response might be obtained from a very high-order system or one in which noise were present, these types of responses are examples of low-order deterministic chaos. This is a feature of nonlinear dynamical systems that has received considerable attention recently.

Again, using Duffing's equation we now change the parameters, such that we obtain

$$\ddot{x} + 0.3\dot{x} - x + x^3 = 0.5 \sin 1.2t. \tag{49}$$

Numerically integrating equation (49) leads to the results shown in Figure

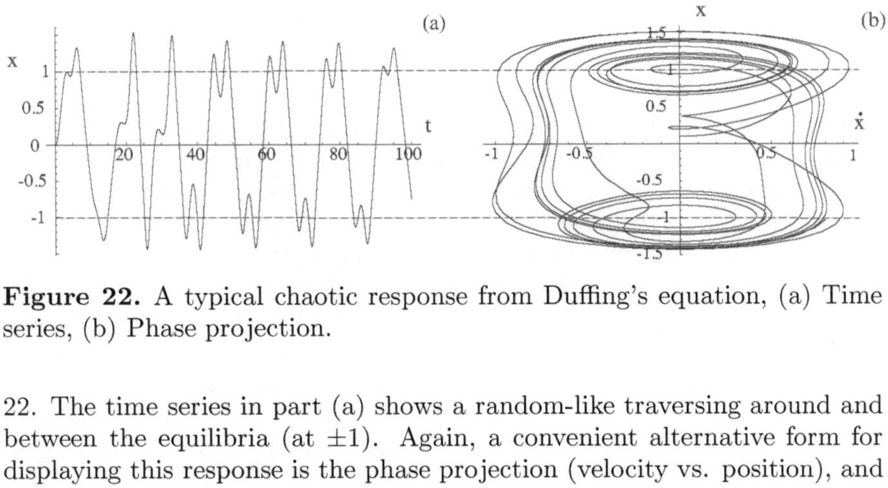

Figure 22. A typical chaotic response from Duffing's equation, (a) Time series, (b) Phase projection.

22. The time series in part (a) shows a random-like traversing around and between the equilibria (at ± 1). Again, a convenient alternative form for displaying this response is the phase projection (velocity vs. position), and this is shown for the same data in part (b).

However, despite the apparent randomness of this response, it is deterministic with significant underlying structure. The corresponding Poincaré section is shown in Figure 23, with an interesting sequence of points mapped due to the folding and stretching evolution of the chaotic attractor. This is persistent behavior and perturbations will decay as any transients are

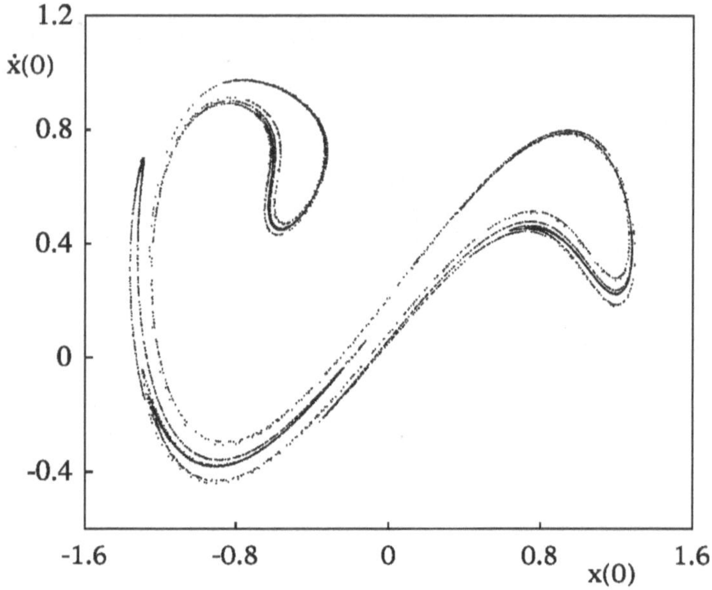

Figure 23. A Poincaré section revealing the chaotic attractor corresponding to the response shown in Figure 22.

attracted back onto this strange shape (about 10,000 points are plotted here). This fine structure shows some fractal characteristics and displays an extreme sensitivity to initial conditions. Many numerical tools have been developed to shed light on chaos. The broadband nature of the frequency spectrum has been mentioned (Newland (1984)), but the complex geometry of the attractor can also be described in terms of dimension, see Strogatz (1994) (and this is where certain fractal features are apparent). For a chaotic response the sensitivity to initial conditions is extreme. There is a local exponential divergence of adjacent points on a trajectory and this feature is described by a positive Lyapunov exponent (LE) Ott (1993).

In a system with multiple attractors, each attractor will possess its own basin of attraction (for a linear system with a unique attractor, all initial conditions end up on the steady state). Hence there is some dependence on initial conditions, but sometimes there appear fractal basin boundaries, such that under certain ranges of initial condition it is practically impossible to predict where the trajectory will end up. Figure 24 illustrates just such a situation in which, even though the attractors are periodic, the basins are intermingled in a very complex manner. The black and white regions correspond to those initial conditions (basins of attraction) that lead to periodic motion in the vicinity of the +1 and -1 equilibrium positions, respectively Virgin (2007). The fractal nature of these basin boundaries remain no

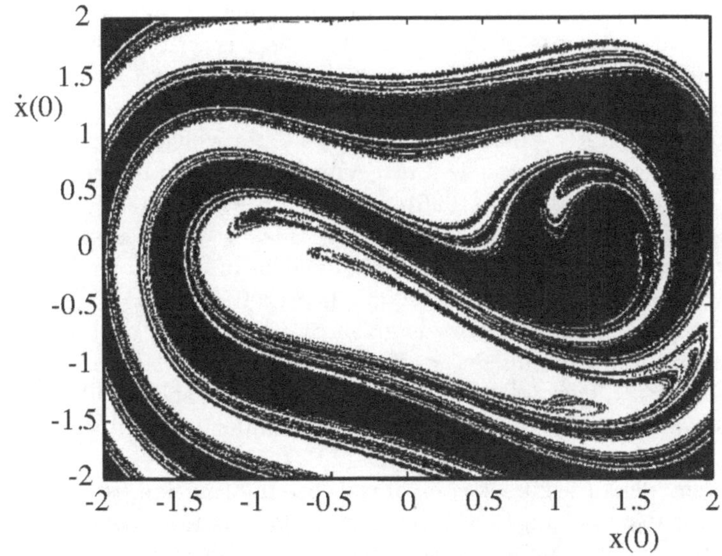

Figure 24. Fractal basin boundaries based on the numerical simulation of Duffing's equation from a fine grid of initial conditions. $\zeta = 0.168, A = -0.5, B = 0.5, \omega = 1, F = 0.15$.

matter how fine the grid, and of course, in an experimental context there is always a degree of imprecision. Thus, we see that sensitivity to initial conditions in terms of basin boundaries may occur even when steady-state chaos is not present, since only periodic solutions are present in Figure 24. The is kind of sensitivity may also occur in terms of the parameter space.

Often the broad characteristics of chaotic attractors are quite similar, and certain universal behavior has been observed. For example, chaos can often occur after a sequence of period-doubling bifurcations (Feigenbaum (1978)), i.e., an accumulating sequence of super-critical pitchfork bifurcations. Chaos can also occur as a result of other standard sequences including intermittency, and quasi-periodicity (Strogatz (1994)).

7 Control of Nonlinear Vibrations

An increasing number of applications in structural dynamics require the application of some form of control. The most usual form of control for low frequency vibration applications is feedback control. The areas of "dynamics" and "control" have tended to grow as separate subjects which can make it difficult for the interested reader who is familiar with one discipline to understand the concepts of the other. This text is written from a nonlinear dynamics viewpoint, and for those who already have a good grounding in the concepts of nonlinear dynamics the following introductory literature may be useful. Firstly, there are a series of texts which discuss the application of control techniques to *linear* vibration problems (Inman (2006); Beards (1981); Fuller et al. (1996); Moheimani et al. (2003)). There are also texts which discuss the (predominately linear) vibration and control of smart structures, see Clark et al. (1998); Srinivasan and McFarland (2001); Preumont (1997); Worden et al. (2003); Leo (2007); Vepa (2010). A good overview of linear control theory is given by Goodwin et al. (2000), and a good introduction to nonlinear control can be found in Khalil (1992). The concepts discussed here follow the approach of Wagg and Neild (2009) who apply both linear and nonlinear control techniques to nonlinear vibration problems.

The basic idea of feedback control is to use information from the system response in order to change the input and achieve an improved or otherwise desirable behaviour. The response of the system is obtained from measurements taken using sensors to record system variables such as displacements, accelerations or forces. These measurements are then fed back and used to update the control forces via a control algorithm. The control algorithm is designed to give an improved system performance based on the requirements of the task at hand.

A block diagram of a typical feedback control system is shown in Fig. 25. This type of block diagram shows the logic of the feedback system, from an input demand signal to a system response output. The input demand is also known as a setpoint or reference signal, and this is specified by the control designer as the desired system output. In the example shown here, the output from the system is subtracted from the demand signal to form an error signal, $e(t) = r(t) - x(t)$, which is used by the controller in order to give an improved response. It is important to note that the feedback is negative, and as a general rule positive feedback will cause the system to become unstable.

In control terminology, the system to be controlled is usually referred to as the plant. If there is just one input demand and one response output

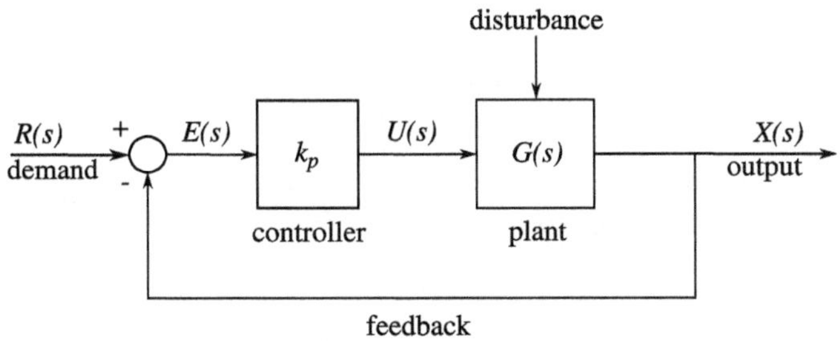

Figure 25. Feedback control block diagram.

from the plant, the control system is said to be a single-input, single-output (SISO) system. If there is more than one input or output, the control system is said to be multiple-input, multiple-output. (MIMO).

Adding feedback introduces additional complexity to the stability of the combined controller and system behaviour. Feedback controllers can destabilise a system in certain situations, and the design and application of these types of system is a large subject in its own right. The problem is usually split into two parts. Firstly design the system to be stable in a perfect (no noise) environment, and second design the stable system to be as robust as possible to noise and other disturbances (shown in Fig. 25 acting on the plant) and to uncertainty which may occur under operating conditions, this is the robustness problem.

7.1 Feedback control of linear systems

As an example, consider the single-degree-of-freedom oscillator shown in Fig. 1, when there is a single control input, u, such that the equation of motion is given by

$$m\ddot{x} + c\dot{x} + kx = F = pu. \tag{50}$$

Here x is the displacement of the mass m, with damping constant c, stiffness coefficient k and p is a scalar constant which can be thought of as a control gain (where parameters are chosen with suitable units). Note that dividing by m, with $u = 0$, gives Eqn. (23), from which an analysis of the type of damping can be carried out. Here we assume that the oscillator is underdamped, $\zeta \ll 1$. Control system analysis of linear systems is often undertaken using Laplace transform techniques. For example, we can analyse the steady state response by taking the Laplace transform of Eq. (50), which, assuming zero initial conditions gives

$$X(s) = \frac{p}{(ms^2 + cs + k)} U(s) = G(s)U(s), \tag{51}$$

where s is the Laplace variable, and $G(s)$ is the transfer function for the single-degree-of-freedom oscillator given in Eq. (50). Note we use the convention that a capital letter denotes the Laplace transform of the variable i.e. $x(t) \to X(s)$.

Assuming that the control task is to make the displacement $x(t)$ follow a predetermined reference signal $r(t)$ such that $x(t) \to r(t)$ in the steady state, the error is defined as $e(t) = r(t) - x(t)$. Now the error can be used as a feedback, so that when $e \neq 0$, some control effort is applied to the system as shown in Fig. 25.

Using the Laplace transform variables shown in Fig. 25, $U(s) = k_p E(s) = k_p(R(s) - X(s))$, and by using the logic of the block diagram in Fig. 25 (assuming no disturbance), it can be shown that the steady state relationship for the closed-loop feedback system is

$$\frac{X(s)}{R(s)} = \frac{G(s)k_p}{1 + k_p G(s)}, \tag{52}$$

where k_p is known as a proportional control gain. The process of choosing the best k_p is the control design, and for the design process the Laplace parameter s is related to a generalised frequency parameter, such that $s = \mathrm{i}\omega$. Then the relationship between the input function $R(\mathrm{i}\omega)$ and the output $X(\mathrm{i}\omega)$ is governed by $G(\mathrm{i}\omega)k_p/(1 + k_p G(\mathrm{i}\omega))$, which is otherwise known as the closed-loop transfer function, which is denoted as $L(\mathrm{i}\omega)$.

Note from Fig. 25 that the system has negative feedback. However, assuming the signals are sinusoidal, if the output signal becomes so far phase shifted, while maintaining a high enough amplitude, it will have the same effect as positive feedback and the system will become unstable. For this to happen the amplitude of the closed-loop transfer function must be $|L(\mathrm{i}\omega)| \geq 1$ and the phase $\arg(L(\mathrm{i}\omega)) \geq -\pi$. Notice that only half a wavelength is required because the negative feedback inverts the signal to look like the input signal after only π phase lag. These conditions are known as the Nyquist stability criterion, and the use of this criterion is a fundamental design technique for linear control systems, see Goodwin et al. (2000) for details. For any selected k_p value, the frequency can be varied from $0 \leq \omega \leq \omega_{max}$ to see if $L(\mathrm{i}\omega)$ remains stable.

Now let's consider an example of how the stability analysis would work for the controlled oscillator given by Eq. (50). We need to find the stability of the closed-loop transfer function, and we will take parameter values when mass $m = 1$kg, stiffness $k = 100$N/m, damping is $c = 2.0$kg/s and $p =$ 1N/V. First, using the mass and stiffness values, the natural frequency is $\omega_n = \sqrt{100/1} = 10$rads/s. Then $G(s) = 1/(s^2 + 2s + 100)$, which has *poles* of $-1 \pm \mathrm{i}9.95$. The poles are in the left-hand side of the complex plane which means that the uncontrolled, open-loop, system is stable.

For this example, the poles for the controller system the closed-loop poles are found from the poles of

$$L(s) = \frac{k_p}{s^2 + 2s + 100 + k_p}, \tag{53}$$

which gives poles of $s_{1,2} = -1 \pm 0.5\sqrt{4 - 4(100 + k_p)}$, which for positive k_p are always complex and in the left-hand plane and therefore stable. Note also that the effect of k_p is analogous to adding stiffness to the oscillator. Substituting $s = \mathrm{i}\omega$ gives

$$L(\mathrm{i}\omega) = \frac{k_p}{100 + k_p - \omega^2 + \mathrm{i}2\omega}. \tag{54}$$

Multiplying top and bottom of Eq. (54) by $(100 + k_p - \omega^2 - \mathrm{i}2\omega)$ allows the real and imaginary parts to be found, from which

$$|L(\mathrm{i}\omega)| = \frac{k_p}{\sqrt{(100 + k_p - \omega^2)^2 + (2\omega)^2)}}, \tag{55}$$

and

$$\arg(L(\mathrm{i}\omega)) = -\arctan\left(\frac{2\omega}{(100 + k_p - \omega^2)}\right). \tag{56}$$

The Nyquist stability criterion defines the ω values at which instability occurs as when $|L(\mathrm{i}\omega)| = 1$, so for example computing Eq. (55) when $k_p = 50$ gives two points at which $|L(\mathrm{i}\omega)| = 1$. Of these, the value closest to the instability point is $\omega \approx 13.8$rad/s, from which $\arg(L(\mathrm{i}13.8)) \approx -2.47$ (using Eq. (56) minus π). So in this example the system is stable by a margin of $-\pi - (-2.47) \approx -0.6$ radians or 34 degrees. This stability margin is called the phase margin for the system. It can usually be inferred that the larger the stability margin, the more robust the system will be to disturbances and modelling uncertainty. The Bode plot showing the phase margin computed using Matlab function 'margin' is shown in Figure 26.

Now consider the example of a multi-degree-of-freedom linear system, where control forces have been added to the equation of motion so that

$$M\ddot{\mathbf{x}} + C\dot{\mathbf{x}} + K\mathbf{x} = \mathbf{F}_E + \mathbf{F}_C,$$

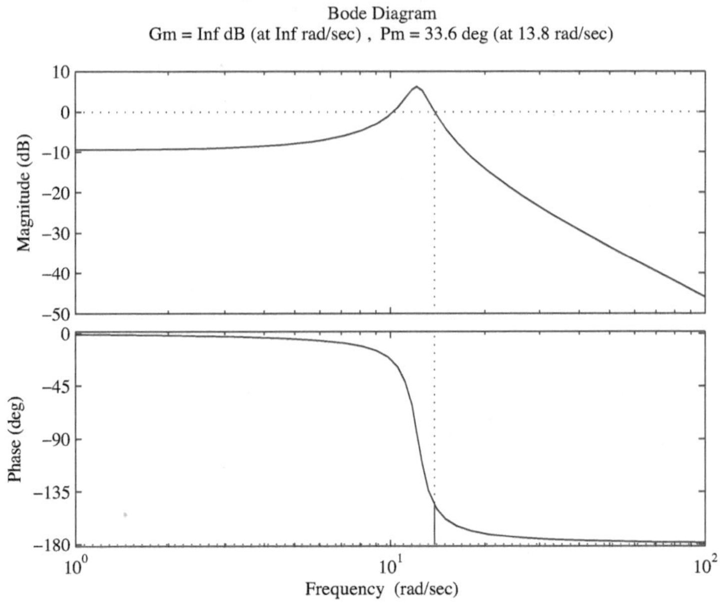

Figure 26. Bode plot of the complex frequency response function of Eq. (53) with $k_p = 50$ using the Matlab margin command. Note that the phase margin is indicated as the solid portion of the vertical line.

where $\mathbf{x}$ represents the displacements, M is an $N \times N$ mass matrix, C is the $N \times N$ damping matrix and K is the $N \times N$ stiffness matrix, $\mathbf{F}_E$ is the dynamic forcing vector and $\mathbf{F}_C$ is the vector of control forces. N is the number of degrees-of-freedom. To carry out state space analysis we put the differential equations into first-order form, and to do this we let $x = [\mathbf{x}^T, \dot{\mathbf{x}}^T]^T$, $F_E = 0$ (unforced) and $\mathbf{F}_C = P\mathbf{u}$ such that

$$\dot{x} = Ax + Bu, \tag{57}$$

where

$$A = \begin{bmatrix} 0 & I \\ M^{-1}C & M^{-1}K \end{bmatrix}, \quad B = \begin{bmatrix} 0 \\ M^{-1}P \end{bmatrix},$$

and where $u = \{u_1, u_2, ..., u_N\}^T$ is the vector of control signals u_i, I is the identity matrix and P is a constant matrix representing the control mechanism/hardware. Equation (57) represents the state space form of the controlled system. It is typically written with an output equation $y =$

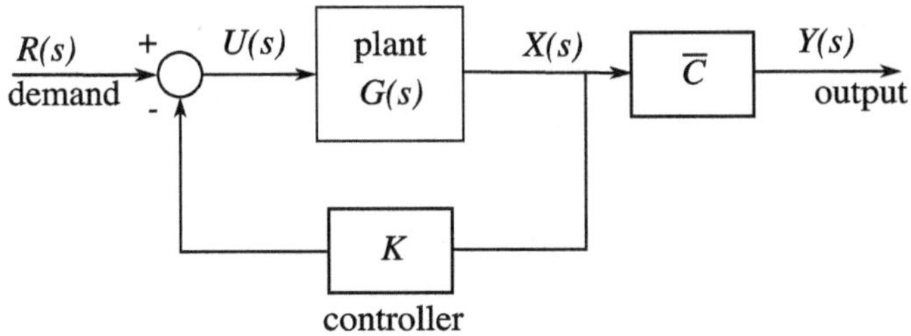

Figure 27. Schematic diagram of state feedback controller for a linear system.

$\bar{C}x$, which represents the case when the state vector, x, cannot be directly observed, and instead y are the observations from the sensors. Note that to avoid confusion with the damping matrix, $\bar{C}$ is used as the control output matrix representing the relationship between x and y.

Now using the Laplace transform approach on Eqn. 57 (assuming zero initial conditions) we obtain

$$sX(s) = AX(s) + BU(s), \quad \rightsquigarrow \quad X(s) = (sI - A)^{-1}BU(s) \tag{58}$$

and taking the Laplace transform of the output relationship gives $Y(s) = CX(s)$ so that

$$Y(s) = \bar{C}(sI - A)^{-1}BU(s) = L(s)U(s) \tag{59}$$

where $L(s)$ represents the dynamics of the closed loop system.

There are multiple ways to choose the control signal, $U(s)$, here we show a state feedback approach, shown schematically in Fig. 27. In this case $U(s) = R(s) - KX(s)$ so that

$$Y(s) = \bar{C}(sI - A)^{-1}BU(s) = \bar{C}(sI - A)^{-1}B(R(s) - KX(s)), \tag{60}$$

from which

$$Y(s) = \bar{C}(sI - A + BK)^{-1}BR(s). \tag{61}$$

Now the control design can be used to choose values of the feedback gain matrix K to give the desired response. One of the most common ways of doing this is to use pole placement, or in other words to choose a set of poles which would make the system suitably stable. For example Ackermans formula can be used

$$K = [0, 0, ..., 1]M_c^{-1}\phi_d(A), \tag{62}$$

where $M_c = [B \mid AB \mid .... A^{n-1}B]$, and ϕ_d is the characteristic equation for the closed loop poles at $s = A$.

Let's take the example where

$$\dot{x} = \begin{bmatrix} 0 & 1 \\ -10 & -2 \end{bmatrix} x + \begin{bmatrix} 0 \\ 7 \end{bmatrix} u. \tag{63}$$

Then

$$\det(sI - A) = \begin{vmatrix} s & -1 \\ 10 & s+2 \end{vmatrix} = s(s+2) + 10 = 0 \tag{64}$$

and

$$s^2 + 2s + 10 = 0 \quad \rightsquigarrow \quad s = -1 \pm 3\mathbf{i}, \tag{65}$$

which means that the open loop system in this case is stable.

Now we define $u = r - Kx$ so that

$$u = r - [k_1 \quad k_2] \begin{bmatrix} x_1 \\ x_2 \end{bmatrix}. \tag{66}$$

Now let us assume that we are required to place the poles at $s_{1,2} = -5, -6$. Then

$$(s+5)(s+6) = s^2 + 11s + 30 = \phi_d(s). \tag{67}$$

Now

$$M_c = [B|AB] = \left[\begin{bmatrix} 0 \\ 7 \end{bmatrix} \middle| \begin{bmatrix} 0 & 1 \\ -10 & -2 \end{bmatrix} \begin{bmatrix} 0 \\ 7 \end{bmatrix}\right] = \begin{bmatrix} 0 & 7 \\ 7 & -14 \end{bmatrix}. \tag{68}$$

From which

$$K = [0\ 1] = \begin{bmatrix} 0 & 7 \\ 7 & -14 \end{bmatrix}^{-1} \left(\left[\begin{bmatrix} 0 & 1 \\ -10 & -2 \end{bmatrix}^2 + 11 \begin{bmatrix} 0 & 1 \\ -10 & -2 \end{bmatrix} \right] + 30I \right)$$

$$= [2.8571 \quad 1.2857]. \tag{69}$$

This type of computation can be carried out using the 'place' or 'acker' commands in Matlab. When considering nonlinearity the frequency response becomes even more complex to such an extent that nonlinear control design requires a different approach.

7.2 Feedback control of nonlinear systems

Nonlinear control systems can be written in the general form of

$$\dot{x} = f(x) + g(x)u, \tag{70}$$

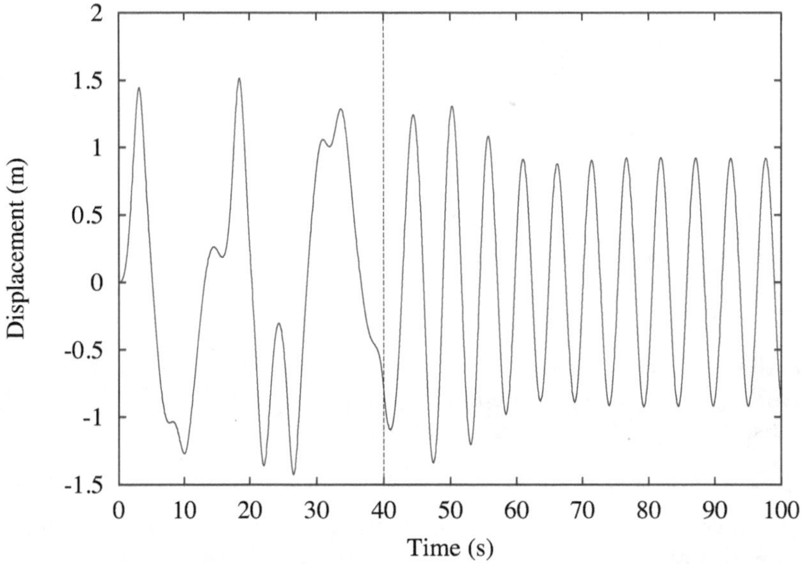

Figure 28. Simulation of feedback linearization control of the Duffing oscillator given in Eqn. 49. The control $u = (x_1^3 - k_p x_1)/p$ is turned on at time $t = 40$s. Parameter values $k_p = 2$ and $p = 4$. Initial conditions $(x_1(0), x_2(0)) = (0, 0)$.

where f is the nonlinear system function (as in Eqn. 1 with lower case f), g is the nonlinear controller function and x is the state vector $x = [\mathbf{x}^T, \dot{\mathbf{x}}^T]^T$. This is a nonlinear version of Eq. (57). In the case where system states cannot be observed directly, $y = h(x)$ is used, where h is the output function which may or may not be nonlinear.

For some systems, the nonlinear functions can be approximated as linear, such that $f(x) \approx Ax$ and $g(x)u \approx Bu$ and $h(x) \approx \bar{C}x$. In this situation the nonlinear system can be approximated by Eq. (70), for a very limited range of system parameters, for example close to an equilibrium.

For systems where nonlinearities cannot be ignored there are a range of possible techniques including (but not exclusively); gain scheduling; optimal control; adaptive control; sliding mode control. See Isidori (1995); Krstić et al. (1995); Slotine and Li (1991); Åström and Wittenmark (1995); Khalil (1992) for an in depth discussion of nonlinear control. Here we consider an introductory example to demonstrate some of the main features of these problems.

One of the simplest ways of dealing with nonlinear systems is to use the control signal to cancel the nonlinear part of the system and effectively turn it back into a linear system. Consider the Duffing oscillator (similar to that defined in Eqn. (47)) with a control input which can be written in first-order matrix form as

$$\begin{bmatrix} \dot{x}_1 \\ \dot{x}_2 \end{bmatrix} = \begin{bmatrix} 0 & 1 \\ -\omega_n^2 & -2\zeta\omega_n \end{bmatrix} \begin{bmatrix} x_1 \\ x_2 \end{bmatrix} + \begin{bmatrix} 0 \\ -\gamma x_1^3 \end{bmatrix} + \begin{bmatrix} 0 \\ p \end{bmatrix} u(t). \quad (71)$$

In vector form this becomes

$$\dot{x} = Ax + \mathcal{N}(x) + Bu, \quad (72)$$

where the nonlinear function, $f(x)$, has been split into a linear part, Ax, and a vector of nonlinear terms, denoted by $\mathcal{N}(x)$ such that $Ax + \mathcal{N}(x) = f(x)$. Note also that $B = g(x)$ and $x = [x_1, x_2]^T$ in this example.

It can be seen by inspection of Eq. (71), that if one sets $u = \gamma x_1^3/p$ then $\mathcal{N}(x) + Bu \to 0$ as $t \to \infty$, ignoring the possibility of transient effects destabilising the system. The system then reduces to $\dot{x} = Ax$, which is linear, and, providing A has stable eigenvalues, it is also stable.

However, although this has removed the nonlinearity, it has applied no additional control to the linear part of the system. To apply proportional control, one could set $u = (\gamma x_1^3 - k_p x_1)/p$, which would remove the nonlinearity and apply proportional control to the resulting linear system, as now $\mathcal{N}(x) + Bu \to k_p x_1$ as $t \to \infty$, providing in this case that the closed-loop proportional controller is stable. However, during the transient phase, there will be both linear feedback and nonlinear effects present, which will make the assessment of stability and robustness difficult.

A numerical simulation of the Duffing example is shown in Fig. 28. Here the Duffing system given by Eqn. 49 is forced with a sinusoidal forcing, and has a chaotic response without control. This can be seen in the displacement response from time zero to $t = 40$ in Fig. 28. After $t = 40$ when the control is turned on, the response rapidly becomes linear. Note that in this case, the underlying linear system is unstable, and therefore the k_p value is selected to stablize the linearized system.

Rather than subtracting off the nonlinearity, an alternative approach is to try to establish a linear relationship between the output and input to the system. In general, the output from the nonlinear system is represented as a vector $y = h(x)$. In practice this will be one or more measurements of displacement, velocity or acceleration which may or may not be a nonlinear function of the system states. For example, assume that the output from the Duffing system in Eq. (71) is the velocity, x_2 so that $y = x_2$. The input to the Duffing system is the control signal, u. If a linear relationship can be

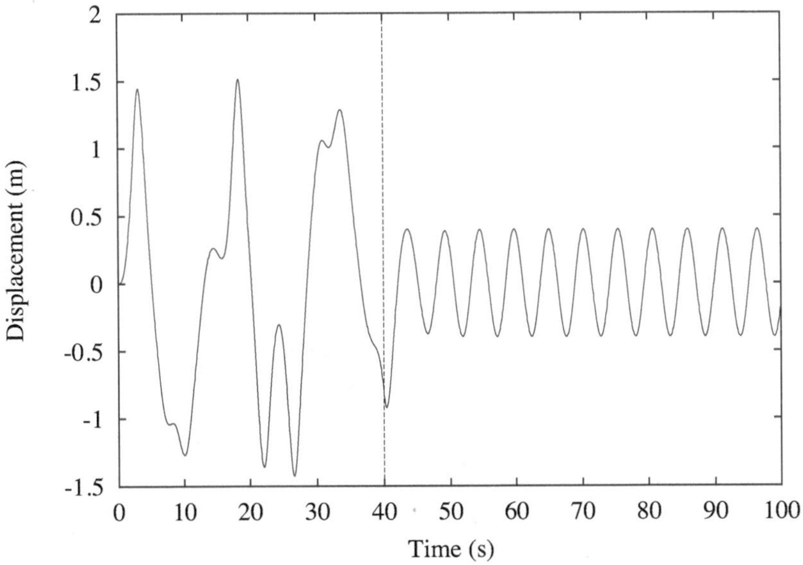

Figure 29. Simulation of input-output linearization control of the Duffing oscillator given in Eqn. 49. The control $\frac{1}{p}[v(t) - x_1 + 0.3x_2 + x_1^3]$ is turned on at time $t = 40$s, where $v(t) = -2x - \dot{x}$, with $p = 4$ and initial conditions $(x_1(0), x_2(0)) = (0, 0)$.

established between y and u, then the system will have been input-output linearized.

Now consider an example where we are required to design a feedback linearization controller for the Duffing system defined by Eq. (71), assuming that the output is the velocity, x_2. To tackle this problem we first take the output and differentiate with respect to time, t, to give $\dot{y} = \dot{x}_2$. From Eq. (71), the expression for $\dot{x}_2$ can be used to write

$$\dot{y} = \dot{x}_2 = -\omega_n^2 x_1 - 2\zeta\omega_n x_2 - \alpha x_1^3 + pu(t). \tag{73}$$

If the control signal is chosen as

$$u = \frac{1}{p}[v(t) + \omega_n^2 x_1 + 2\zeta\omega_n x_2 + \alpha x_1^3], \tag{74}$$

then the effect is to subtract off all the terms and replace them with a new control signal $v(t)$, giving an input-output relationship of the form $\dot{y} = v(t)$.

Then a linear expression can be chosen for $v(t)$ to give the required control effect.

A numerical simulation of this input-output linearization approach is shown in Fig. 29. After $t = 40$ when the input-output linearization control is turned on, the response rapidly becomes linear. Note that as before, the underlying linear system is unstable, and therefore the $v(t)$ function is selected to stablize the linearized system. Note however that $v(t)$ also allows us to increase the damping in the controlled system.

It should be noted in this example that although only x_2 is the measured output, both states are required to form the control signal. Compared with the more basic feedback linearization discussed above, this approach has not just cancelled the nonlinear dynamics, the linear part of the $\dot{x}_2$ dynamics has been removed as well and replaced with $v(t)$. The whole process can be formalised by using *Lie derivatives*, and will be described in Chap. 6.

The usefulness of feedback linearization is that once the system has been linearized addition linear control tasks can be included using well known techniques. Typically the control tasks of interest are to remove unwanted vibration and/or get the system to follow some predefined reference signal. This and other nonlinear control techniques are discussed in more detail in Chap. 6.

7.3 Adaptive control of a scalar linear system

In Chap. 6 we will consider adaptive control techniques for nonlinear systems, and as a final part of this introductory discussion of control techniques we outline an adaptive approach for a simple scalar linear system. The type of adaptive control we will consider is that which uses a model reference. So we have a (scalar) plant

$$\dot{x} = -ax + bu \tag{75}$$

which we assume has the output $y = x$, and a reference model

$$\dot{x}_m = -a_m x_m + b_m r \tag{76}$$

where we assume the output $y_m = x_m$ and r is the reference signal. We assume that the reference model is stable, so $a_m > 0$ and we also assume that $b_m > 0$. A schematic diagram of this type of control is shown in Fig. 30.

It is usual (see for example Åström and Wittenmark (1995)) to use a control signal which consists of a feedback and feedforward component

$$u = kx + k_r r \tag{77}$$

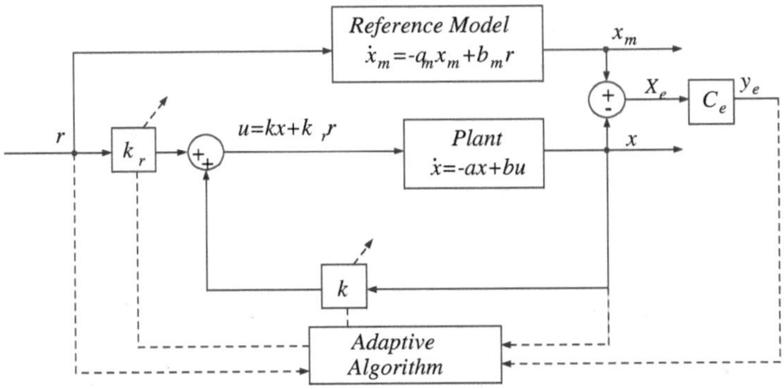

Figure 30. Model reference adaptive control system.

where k is the feedback gain, and k_r is the feedforward gain. Substituting equation 77 into equation 75 gives

$$\dot{x} = (bk - a)x + bk_r r. \tag{78}$$

If $(bk - a) = -a_m$ and $bk_r = b_m$, the systems match perfectly. These conditions are called the Erzberger conditions. We can now write

$$\begin{aligned} k &= \frac{a - a_m}{b} = k^*, \\ k_r &= \frac{b_m}{b} = k_r^*. \end{aligned} \tag{79}$$

We define the error as the difference between the model and plant $x_e = x_m - x \Rightarrow \dot{x}_e = \dot{x}_m - \dot{x}$. Then if we subtract equation 78 from 76 and add $a_m x - a_m x$ to the result of the subtraction we obtain

$$\dot{x}_e = -a_m x_e + (a - a_m - bk)x + (b_m - bk_r)r. \tag{80}$$

From equation 79 we see that $a - a_m = bk^*$ and $b_m = bk_r^*$ which enables us to write

$$\begin{aligned} b(k^* - k) &= a_m - a - bk, \\ b(k_r^* - k_r) &= k_m - bk_r r. \end{aligned} \tag{81}$$

Then we define the parameter errors as the difference between the Erzberger value and the actual value $\phi = (k^* - k)$ and $\psi = (k_r^* - k_r)$, then equation 80 can be written as

$$\dot{x}_e = -a_m x_e + b(\phi x + \psi r). \tag{82}$$

Note also that $\phi = -\dot{k}$ and $\psi = -\dot{k}_r$, because k^* and k_r^* are constants. Equation 82 is now a first order differential equation in terms of the error between the the plant and the reference model.

For general model reference adaptive control, the adaptive gains are commonly defined as

$$k = \alpha \int_0^t y_e x(\tau)\mathrm{d}\tau + \beta y_e x(t), \quad k_r = \alpha \int_0^t y_e r(\tau)\mathrm{d}\tau + \beta y_e r(t). \tag{83}$$

where α and β are control weightings representing the adaptive effort, and $y_e = C_e x_e$ where C_e can be chosen to ensure the stability of the feed forward block Landau (1979). Eq.83 has been shown to satisfy hyperstability theory (i.e. the Popov Criteria), see Åström and Wittenmark (1995); Popov (1973) for a detailed derivations.

By using the fact the k^* is a constant, $\dot{\phi} = -\dot{k}$, and by using Eq. 83 we can obtain expressions for $\dot{\phi}$ and $\dot{\psi}$. These expressions can be combined with the equation for the error dynamics, so that we can write

$$\begin{aligned} \dot{x}_e &= -a_m x_e + b(\phi x + \psi r) \\ \dot{\phi} &= -\alpha C_e x_e x - \beta C_e(\dot{x}_e x + x_e \dot{x}) \\ \dot{\psi} &= -\alpha C_e x_e r - \beta C_e(\dot{x}_e r + x_e \dot{r}) \end{aligned} \tag{84}$$

In Eq.84, x is implicitly defined in x_e, so we must substitute via the relation $x = x_m - x_e$, Khalil (1992), which gives

$$\begin{aligned} \dot{x}_e &= -a_m x_e + b[\phi(x_m - x_e) + \psi r] \\ \dot{\phi} &= -\alpha C_e x_e(x_m - x_e) - \beta C_e[\dot{x}_e(x_m - 2x_e) + x_e \dot{x}_m] \\ \dot{\psi} &= -\alpha C_e x_e r - \beta C_e(\dot{x}_e r + x_e \dot{r}) \end{aligned} \tag{85}$$

A further substitution for $\dot{x}_e$ in the expressions for $\dot{\phi}$ and $\dot{\psi}$ can be made to give

$$\begin{aligned} \dot{x}_e &= -a_m x_e + b[\phi(x_m - x_e) + \psi r] \\ \dot{\phi} &= -\alpha C_e x_e(x_m - x_e) - \beta C_e[(-a_m x_e + b[\phi(x_m - x_e) + \psi r])(x_m - 2x_e) + x_e \dot{x}_m] \\ \dot{\psi} &= -\alpha C_e x_e r - \beta C_e((-a_m x_e + b[\phi(x_m - x_e) + \psi r])r + x_e \dot{r}) \end{aligned} \tag{86}$$

This substitution means that we have a dynamical system of the form $\dot{\xi} = f(\xi, t)$. The parameters a_m, b, α, β and C_e are constant and r, $\dot{r}$, x_m, $\dot{x}_m$ are time varying signals.

The numerical example of a step input for this system is shown in Fig.31. Here we have computed the response of the system, Eq.86, to the unit step input of the form

$$\begin{array}{lll} t < t_0 & r = 0, & \dot{r} = 0 \\ t > t_0 & r = 1, & \dot{r} = 0 \end{array} . \tag{87}$$

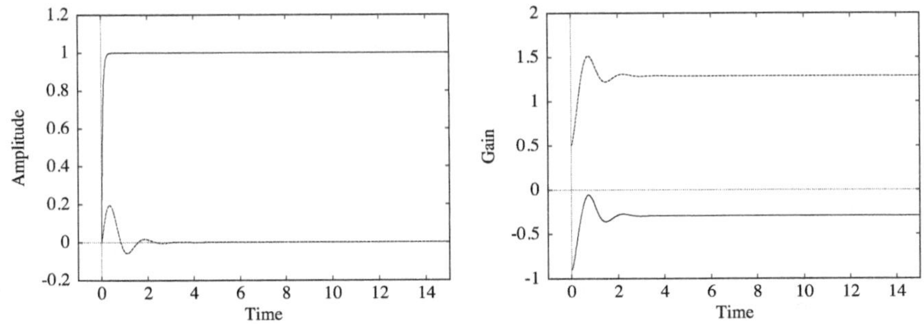

(a) (b)

Figure 31. Example of unity step input. Parameters are $\alpha = 10$, $\beta = 1$, $t_s = 1.0$, $a_m = 1$, $b = 1$. Initial conditions $x_e(0) = 0.0$, $k(0) = 0.9$, $k_r(0) = 0.5$. (a) The solid line shows x_m and the dashed line x_e. (b) The solid line shows k and the dashed line k_r.

where we take $t_0 = 0$. In Fig.31 (a) we show the output from the reference model, x_m, and the error signal x_e. The error signal tends to zero within five seconds, so that the steady state condition is reached $x = x_m = r = 1$. The adaptive gains are shown in Fig.31 (b), and both settle to steady state values within five seconds.

For this example we find that we can identify a set of steady state gain values which satisfy the control problem and define an exact matching manifold, Γ, which gives a set of gain values $\{k, k_r : x = x_m \text{ as } t \to \infty\}$. By examining Eqs. 85 and 86, we note that if the system parameters are identical, $a_m = a$ and $b_m = b$, and the system has reached steady state such that $x = x_m = r = 1$, then

$$k + k_r = 1 \tag{88}$$

provides exact matching. This infinite set of steady state gain values forms the set Γ which is the exact matching set for this example, with the Erzberger gain at $k = 0, k_r = 1$.

Considering the local dynamics of any of the infinite number of equilibrium points in Γ we find that each point has three eigenvalues, $\lambda_1 = 0$ and

$$\lambda_{2,3} = -\frac{1}{2}(a_m + b\tilde{\phi} + \beta C_e b(x_m^2 + r^2)) \pm \frac{1}{2}D, \tag{89}$$

where $D^2 = (a_m + b\tilde{\phi} + \beta C_e b(x_m^2 + r^2))^2 - 4C_e b[\alpha(x_m^2 + r^2) + \beta(x_m \dot{x}_m + r\dot{r})]$.

We should emphasize at this point that λ_1 is time invariant, and indicates neutral (i.e. non-asymptotic) stability in the eigen-direction which is given

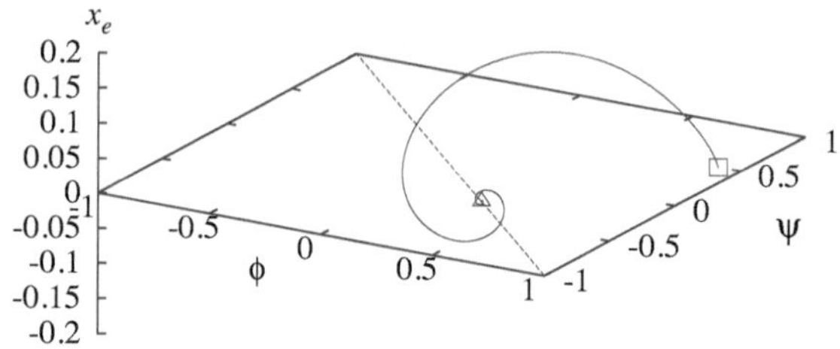

Figure 32. Evolution of the system states onto the gain set Γ. The solid line shows the evolution of the gains in the example shown in Fig.31. The dashed line represents the gain set Γ. The start point of the trajectory is marked with a square, □, and the end point with a triangle, △.

by

$$e_1 = [0, -x_m/r, 1]^T. \tag{90}$$

From Eq.90 with $x_m = r = 1$, the steady state eigenvector, e_1 is given by, $e_1 = \{0, 1, -1\}$. If we transform the set Γ from k, k_r coordinates into the system space, Σ, where $\Sigma = \{\mathbb{R} \times \mathbb{R} \times \mathbb{R} : (x_e, \phi, \psi)\}$, then $\Gamma = \{x_e = 0 : \phi = -\psi\}$, with the Erzberger gain at $x_e = \phi = \psi = 0$. Therefore the eigenvector e_1 is tangent to the set Γ for all values of $\phi = -\psi$.

In Fig. 32 we show the adaption of the system onto the exact matching set, Γ, in the reduced state space, Σ. Here the solid line shows the evolution of the trajectory from the example in Fig.31, where Γ is shown as a dashed line. In Fig.33 we show the the projections of the trajectory in Fig.32 onto the ϕ, ψ plane and the x_e, ϕ plane. In Fig.33 (a) the solid line denotes the system trajectory projected into the ϕ, ψ plane, the dashed line is Γ and the triangle represents the steady state value. We see that as $x_m \rightarrow r$ a diagonal line occurs in Fig.33 (a). This indicates that movement in the e_1 direction occurs when $x_m \neq r$ — i.e. in this case the initial adaption. In Fig.33 (b) we see that the system trajectory is a spiral which settles onto Γ in ϕ, x_e space.

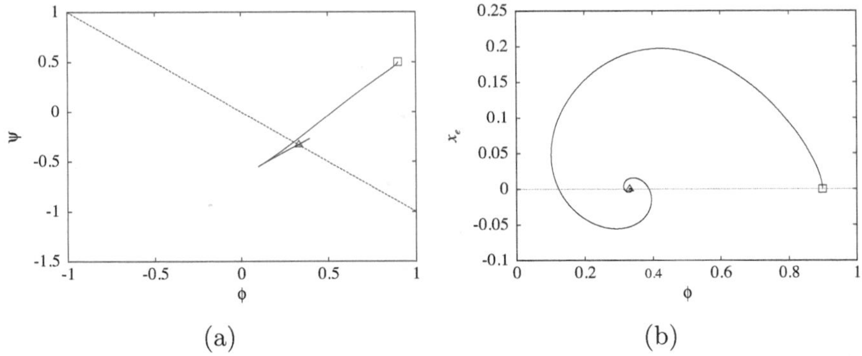

Figure 33. Example trajectory in phase space. The controller parameters are $a_m = 1$, $b = 1$, $C = 4/t_s$, $\alpha = 10$, $\beta = 1$ and settling time $t_s = 1.0$. Initial conditions $(x_e(t_0), \phi(t_0), \psi(t_0)) = (0.0, 0.9, 0.5)$. (a) Shows ϕ, ψ, (b)Shows x_e, ϕ. The start point of the trajectory is marked with a square, $\square$, and the end point with a triangle, $\triangle$.

In the steady state $x \to r = 1$ and $\dot{x} \to \dot{r} = 0$. For the case when $x_e, \phi, \psi \to 0$, such that the steady state gains match the Erzberger gains, the system goes to equilibrium point $\tilde{\xi}_0$, then $D^2 \to [(a_m + 2\beta C_e b)^2 - 8C_e b\alpha]$, and the non-zero eigenvalues become

$$\lambda_{2,3} = -\frac{1}{2}(a_m + 2\beta C_e b) \pm \frac{1}{2}\sqrt{(a_m + 2\beta C_e b)^2 - 8C_e b\alpha}. \tag{91}$$

Because of the invariance of λ_1 the fixed point will always be neutrally stable along e_1. We can see from Eq.91 that if $(a_m + 2\beta C_e b)^2 > 8bC_e\alpha$, then λ_2 and λ_3 are real, and for $(a_m + 2\beta C_e b) > 0$ and $8C_e b\alpha > 0$, asymptotically stable. In addition, if $(a_m + 2\beta C_e b)^2 < 8bC_e\alpha$, then the eigenvalues are complex, and $\text{Re}\lambda_2 = \text{Re}\lambda_3 = -(a_m + 2\beta C_e b)/2$, so as before this is stable if $(a_m + 2\beta C_e b) > 0$ and $8C_e b\alpha > 0$. Thus we can say that if

$$(a_m + 2\beta C_e b) > 0, \quad 8C_e b\alpha > 0 \tag{92}$$

the steady state motion will be locally asymptotically stable in the e_2 and e_3 directions. Thus for $\tilde{\xi}_0$ to lose local stability either $(a_m + 2\beta C_e b) = 0$ or $8C_e b\alpha = 0$ — conditions entirely dependent on constant system parameters.

For any other equilibrium point $\tilde{\xi} \neq \tilde{\xi}_0$ Eq.92 becomes

$$(a_m + b\tilde{\phi} + 2\beta C_e b) > 0, \quad 8C_e b\alpha > 0 \tag{93}$$

For this case local stability can be lost if $(a_m + b\tilde{\phi} + 2\beta C_e b) = 0$ or $8C_e b\alpha = 0$ — conditions which now include the arbitrary equilibrium point value $\tilde{\phi}$.

Note that although we could select the system parameters so that the point $\tilde{\xi}_0$ was locally stable via Eq.92, this point will always be neutrally stable in the e_1 direction. This means that we cannot guarantee to stay on $\tilde{\xi}_0$ without some additional condition. Note that for excitation signal signals other than a step, the system dynamics are to move along Γ towards $\tilde{\xi}_0$. An example of this is shown in Fig. 34, where we have plotted simulations with a square wave demand signal of amplitude ± 1 and frequency 1 Hz. Two cases are shown in Fig. 34, the first (Fig. 34 (a) and (b)) with initial conditions which lead to an initial intersection with Γ below ξ_0. In the second case (Fig. 34 (c) and (d)), the initial conditions lead to a first intersection with Γ above ξ_0. In both cases the additional excitation at each step progressively moves the system trajectory towards the Erzberger equilibrium point at ξ_0.

It is possible that a set of arbitrary fixed points $\tilde{\xi} \neq \tilde{\xi}_0$, are local unstable because the $\tilde{\phi}$ value means that condition Eq. 93 is not satisfied. It is easy to see that for a set of finite parameter values, then in general the set of $\tilde{\phi}$ values which violate the stability condition given by equation 93 will not be empty. We split Γ into two subsets, $\Gamma_u = \{x_e = 0, \tilde{\phi} = -\tilde{\psi} : \tilde{\phi} < -(a_m + 2\beta Cb)/b\}$ and the complement Γ_s — the part of Γ which is locally stable. However, if we define the set of initial conditions which lead to unstable steady state $\tilde{\phi}$ values as $\mathcal{U} = \{(x_e(0), \phi(0), \psi(0)) \in \Sigma : \tilde{\phi} < -(a_m + 2\beta Cb)/b\}$, then it is possible that $\mathcal{U} = \emptyset$. Numerical simulations of the global dynamics indicate that system trajectories starting in the region near Γ_u evolve towards the locally stable part of the manifold, Γ_s. An example is shown in Fig. 35, where the initial conditions $(x_e(t_0), \phi(t_0), \psi(t_0)) = (0.0, -10.0, 10.0)$ have been used. In Fig. 35 (a) the system trajectory can be seen to move significantly along Γ, finally settling on to a point well inside the Γ_s region. The eigenvalue plot shown in Fig. 35 (c) indicates how $\lambda_1 = 0$, and λ_2 and λ_3 are initially unstable.

We will return to adaptive systems in Chap. 6 where we will consider how adaptive control techniques can be applied to feedback linearization methods for nonlinear systems. As a final comment, it should be noted that the type of linear MRAC discussed here has robustness problems, and the interested reader can find more information in Neild et al. (2008).

Bibliography

K. J. Åström and B. Wittenmark. *Adaptive Control.* Addison Wesley: Menlo Park, CA, 1995.

Z.P. Bazant and L. Cedolin. *Stability of Structures.* Oxford University Press, 1991.

C. F. Beards. *Vibration analysis and control system dynamics.* Ellis Hor-

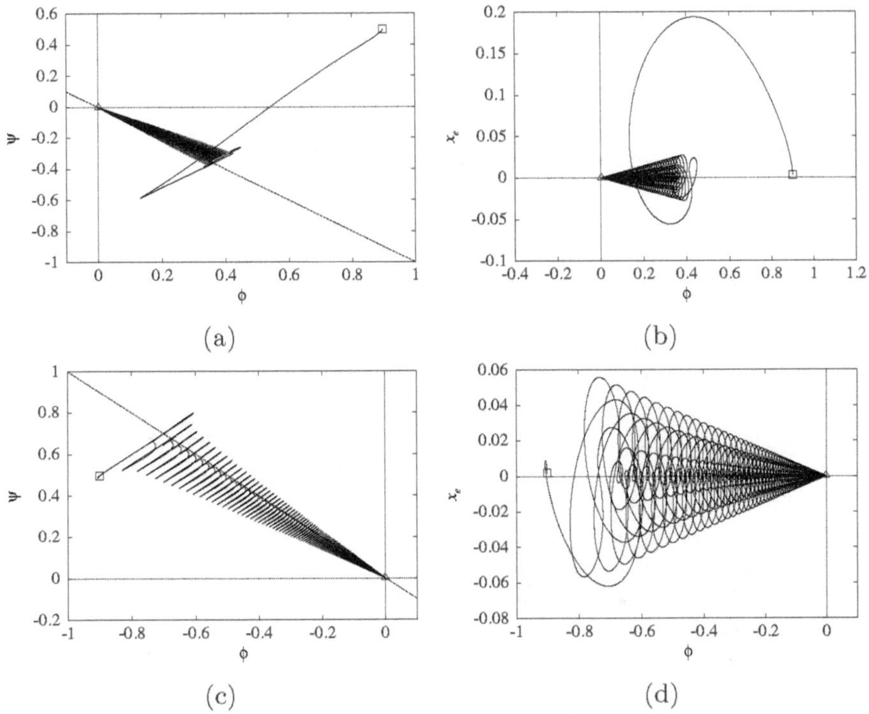

Figure 34. Square wave excitation showing evolution of system to the equilibrium at ξ_0. Parameters; $a_m = 1$, $b = 1$, $C = 4/t_s$, $\alpha = 10$, $\beta = 1$ and settling time $t_s = 1.0$. (a) and (b) $(x_e(t_0), \phi(t_0), \psi(t_0)) = (0.0, 0.9, 0.5)$. (c) and (d) $(x_e(t_0), \phi(t_0), \psi(t_0)) = (0.0, -0.9, 0.5)$. The start point of the simulation is marked with a □, and the end point with a △.

wood: Chichester, 1981.

R. L. Clark, W. R. Saunders, and G. P. Gibbs. *Adaptive structures; dynamics and control.* John Wiley: New York, 1998.

E.J. Doedel. *AUTO - Software for continuation and bifurcation problems in ordinary differential equations.* California Institute of Technology, 1986.

M.J. Feigenbaum. Quantitative universality for a class of nonlinear transformations. *Journal of Statistical Physics*, 19:25–32, 1978.

C. R. Fuller, S. J. Elliot, and P. A. Nelson. *Active control of vibration.* Academic Press, 1996.

G. C. Goodwin, S. F. Graebe, and M. E. Salgado. *Control System Design.* Pearson, 2000.

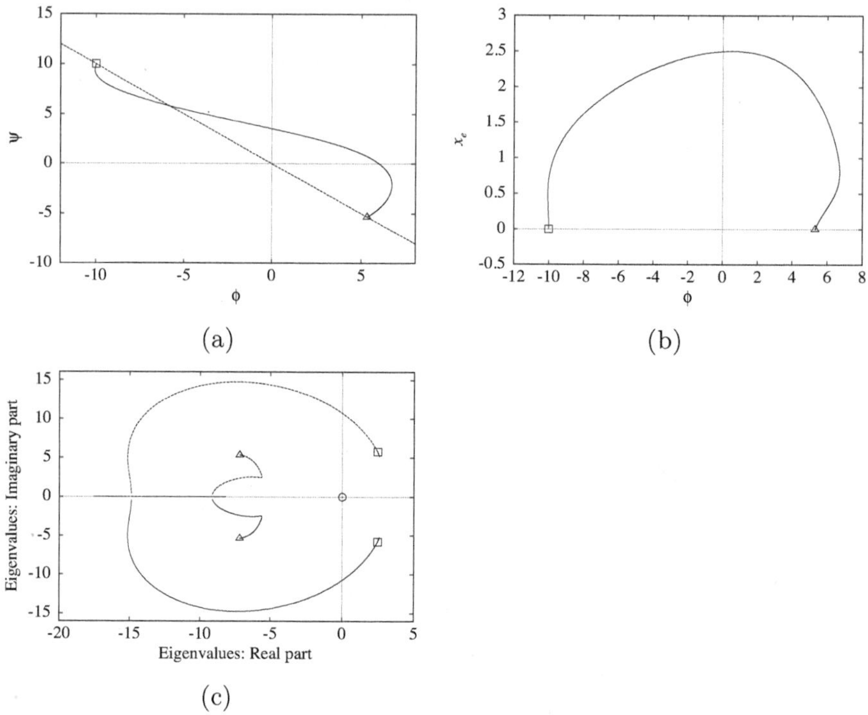

Figure 35. Initial conditions on Γ_u showing system trajectory evolving to Γ_s. Parameters; $a_m = 1$, $b = 1$, $C = 4/t_s$, $\alpha = 10$, $\beta = 1$ and settling time $t_s = 1.0$. $(x_e(t_0), \phi(t_0), \psi(t_0)) = (0.0, -10.0, 10.0)$. (a) and (b) projections of Σ; (c) eigenvalues showing how λ_2 and λ_3 are initially unstable, e_1 denoted by a circle $\circ$. The start point of the simulation is marked with a $\square$, and the end point with a $\triangle$.

J. Guckenheimer and P.J. Holmes. *Nonlinear Oscillations, Dynamical Systems, and Bifurcations of Vector Fields.* Springer-Verlag, 1983.

D. J. Inman. *Vibration with control.* Wiley: Chichester, 2006.

D.J. Inman. *Engineering Vibration.* Prentice Hall, 1994.

A. Isidori. *Nonlinear Control Systems.* Springer, 1995.

D.W. Jordan and P. Smith. *Nonlinear Ordinary Differential Equations.* Clarendon Press, 1977.

H. K. Khalil. *Nonlinear Systems.* Macmillan: New York, 1992.

Miroslav Krstić, Ioannis Kanellakopoulos, and Petar Kokotović. *Nonlinear and adaptive control design.* John Wiley, 1995.

Y. D. Landau. *Adaptive control:The model reference approach.* Marcel Dekker: New York, 1979.

D. J. Leo. *Smart material systems.* Wiley: New York, 2007.

S. O. R. Moheimani, D. Halim, and A. J. Fleming. *Spatial control of vibration.* World Scientific, 2003.

S. A. Neild, L. Yang, and D. J. Wagg. Modified model reference adaptive control approach for systems with noise or unmodelled dynamics. *Proceedings of the IMechE, Part I: Journal of Systems and Control Engineering*, 222:197–208, 2008.

D.A. Newland. *An Introduction to Random Vibrations and Spectral Analysis.* Longman, 1984.

E. Ott. *Chaos in Dynamical Systems.* Cambridge University Press, 1993.

V. M. Popov. *Hyperstability of control systems.* Springer, 1973.

A. Preumont. *Vibration control of active structures.* Kluwer Academic Publishers: Dordrecht, 1997.

R. Seydel. *Practical Bifurcation and Stability Analysis.* Springer, 1994.

J-J. E. Slotine and W Li. *Applied nonlinear control.* Prentice Hall: Englewood Cliffs, NJ, 1991.

A. V. Srinivasan and D. M. McFarland. *Smart structures.* Cambridge: New York, 2001.

J.J. Stoker. *Nonlinear Vibrations in Mechanical and Electrical Systems.* Wiley Classics Library, 1992. Originally published in 1950.

S.H. Strogatz. *Nonlinear Dynamics and Chaos.* Addison-Wesley, 1994.

JMT Thompson and GW Hunt. *A General Theory of Elastic Stability.* Wiley, 1973.

J.M.T. Thompson and H.B. Stewart. *Nonlinear Dynamics and Chaos.* Wiley, 1986.

W.T. Thomson. *Theory of Vibration with Applications.* Prentice-Hall, 1981.

B. van der Pol. The nonlinear theory of electric oscillations. *Proceedings of the Institute of Radio Engineers*, 22:1051–1086, 1934.

R. Vepa. *Dynamics of Smart Structures.* Wiley, 2010.

L.N. Virgin. *Introduction to Experimental Nonlinear Dynamics.* Cambridge University Press, 2000.
L.N. Virgin. *Vibration of Axially Loaded Structures.* Cambridge University Press, Cambridge, UK, 2007.
D. J Wagg and S. A. Neild. *Nonlinear Vibration with Control.* Springer-Verlag, 2009.
K. Worden, W. A. Bullough, and J. Haywood. *Smart Technologies.* World Scientific: Singapore, 2003.

Approximate Methods for Analysing Nonlinear Structures

Simon A. Neild[†*]

[†] Department of Mechanical Engineering, University of Bristol, Bristol, UK.

Abstract

The dynamics of the majority of nonlinear structures cannot be solved exactly. In this chapter, approximate methods for solving the equations of motion of weakly nonlinear structures are presented. Common types of nonlinear response behaviour are identified using an example structure. Perturbation techniques and the method of second-order normal forms are then discussed and used to analyse three applications in which the nonlinear behaviour is exploited.

1 Introduction

The presence of nonlinearity in a dynamic system is often viewed as undesirable and historically the emphasis has been to design systems ensuring that nonlinear behaviour regimes are avoided. However, with ever more ambitious design envelopes it is becoming more common for structures to behave nonlinearly. This nonlinearity may be a consequence of pushing existing design solutions beyond their linear limit. Increasingly however, it is because of engineers deliberately designing a structure to have nonlinear dynamic properties. The motivaton being to exploit some characteristic of the nonlinear dynamic response to maximise the structure's performance envelope. One example of this is the design of passive vibration suppression systems for buildings. It has been reported that nonlinear devices, such as tuned-mass dampers with nonlinear stiffness characteristics, can operate effectively over far wider frequency ranges than the equivalent linear devices (see, for example, Soong and Dargush, 1997 or Reed *et al.*, 1998).

Alternatively, the desire for more efficient structural solutions can result in structures operating in nonlinear regimes. For example, large deflections of more flexible structures can result in geometric nonlinearites and the use

*The author would like to acknowledge the contributions from Alicia Gonzalez-Buelga, Siming Liu and Xie Zhenfang and the helpful discussions with Anthony Croxford and David Wagg.

of novel materials can introduce nonlinear stress-strain behaviour. Beyond this, “intelligent” structures which react to their surroundings can potentially be designed. These structures could, for example, sense motion and actively suppress their vibration response using arrays of piezo-electric actuators or change their shape in response to their surroundings. Intelligence, in the form of structural health monitoring, can also be added to a structure. The structure could monitor its own structural integrity and potentially even self-repair or reconfigure themselves in response to the prognosis.

To understand and optimise the nonlinear behaviour of such structures, it is helpful to be able to study them either numerically or analytically. To find steady-state solutions, numerical time-stepping studies can be time-consuming especially if damping is light or if, due to the presence of nonlinearities, there are multiple possible solutions. In this second case many simulation runs with different initial conditions would be needed at each parameter set to ensure that all the multiple solutions have been identified. Alternatively, approximate analytical solutions to the nonlinear dynamics can often be found, particularly if the nonlinearity is relatively small, as is often the case in structural dynamics. It is some of these techniques that will be discussed in this chapter. More detailed analysis of nonlinear systems can be achieved through numerical continuation. Once a valid solution to the system dynamics has been identified, the continuation technique numerically tracks the solution as system parameters change allowing a picture of the full system response to be built up, see for example Krauskopf *et al.* (2007).

In Section 2 the dynamic response of an example nonlinear structure, an inclined cable, is discussed. Using this example, typical dynamic response features introduced by the nonlinearity are discussed. These features are the possibility of multiple stable solutions for a single parameter set, sudden jumps in the response amplitudes as the excitation amplitude or frequency is changed, the generation of harmonics of the forcing frequency in the response and the parametric excitation of modes of vibration that are not directly influenced by the external forcing. Two analytical approaches for studying the response of weakly nonlinear systems, perturbation methods and normal forms, are then discussed in Sections 3 and 4 respectively. In Sections 5–7, example applications that involve the exploitation of nonlinear dynamics are analysed using these analytical methods. In Section 5 the Duffing oscillator, which has applications in vibration-based energy harvesting and vibration suppression, is analysed. In Section 6 the generation of harmonics of the forcing frequency due to the presence of material nonlinearity is exploited for non-destructive monitoring of a structure subjected to acoustic excitation. In this section a combination of two perturbation techniques, the regular perturbation theory and multiple scales analysis, is used to examine the generation of a response at the second harmonic of the excitation frequency as the excitation propagates through the structure. Finally, in Section 7, the use of parametric excitation to increase modal damping in a target mode is examined using the normal forms technique.

1.1 Equations of Motion

Before considering the dynamics of nonlinear structures it is helpful to define some notation. In this chapter both one- and multi-degree-of-freedom systems will be examined. The general form of nonlinear equation of motion for a one-degree-of-freedom system used here is

$$\ddot{x} + 2\zeta\omega_n\dot{x} + \omega_n^2 x + \gamma_x(x, \dot{x}, p) = p = P\cos(\Omega t + \phi).$$

The nonlinearity is captured by the γ_x function and the external forcing is p. When discussing the normal form technique, in Section 4, an N-degree-of-freedom matrix version of the equation of motion will be used.

In considering the response of systems with small nonlinear terms, it is useful to define three frequencies;

ω_n	the natural frequency of the linear undamped system
Ω	the excitation (or forcing) frequency
ω_r	the dominant frequency of the response

The frequency of the response is equal to the nonlinear natural frequency if the system is unforced. If the system is forced close to resonance then the response frequency matches the forcing frequency, so $\omega_r = \Omega$ in this case.

Finally, in the discussion of the perturbation and normal form techniques, a small parameter, ε, will be used. This parameter will be added to small terms such as the nonlinear term and often the damping term by writing for example $\gamma = \varepsilon\hat{\gamma}$. The ε parameter can be viewed as a book-keeping aid (Nayfeh, 1993) which allows the tracking of small terms such that when, during algebraic manipulations, they are combined with other small terms (resulting in ε^2 terms) they can be ignored or at least be easily identified as very small terms.

2 Nonlinear Behaviour

An example of a nonlinear structure will now be considered. Using this structure some common forms of nonlinear dynamic response behaviour will be identified. In later sections these behaviour types will be examined in more detail.

2.1 A Nonlinear Structure

The example structure is a taut inclined cable excited through vertical motion of the lower support, as shown in Figure 1. The application is a cable-stayed bridge with vertical deck motion. The case where the cable is excited via the support motion at frequencies close to the second in-plane natural frequency will be considered. To understand this motion three modes must be modelled; the second in-plane mode and the first and second out-of-plane modes (Gonzalez-Buelga *et al.*, 2008). Note that in- and out-of-plane motion refers to motion in the vertical $x - z$ and the sideways $x - y$ planes respectively.

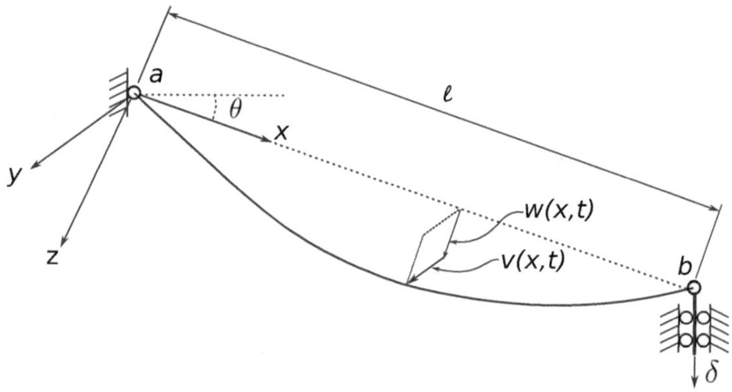

Figure 1. An inclined cable subject to vertical support motion, δ, that can result in in-plane and out-of-plane motion.

To show where the nonlinearity arises in this structure, a few key aspects of the derivation of the modal equations of motion are now discussed briefly. The full methodology behind these equations is described by Warnitchai *et al.* (1995) and a detailed description of their derivation is given in Wagg and Neild (2010). The partial differential equations governing the motion of the cable are found by considering the forces on a small element of cable and are given by

$$(T_s + T_d)\frac{\partial^2 v_d}{\partial x^2} = \rho A \frac{\partial^2 v_d}{\partial t^2}, \tag{1}$$

$$T_d \frac{\partial^2 w_s}{\partial x^2} + (T_s + T_d)\frac{\partial^2 w_d}{\partial x^2} = \rho A \frac{\partial^2 w_d}{\partial t^2}, \tag{2}$$

for the out-of- and in-plane directions respectively. Here the subscripts d and s indicate dynamic and static terms and the cable tension T is defined as the tension component acting along the x-axis. Note that in the static state gravity causes the cable to sag (albeit only by a small amount as the cable is taut) resulting in a deflected shape defined by w_s.

There are two main sources of nonlinearity. Firstly, the dynamic tension is a function of cable deflection. This function can be calculated with reference to Figure 2. The dynamic tension is proportional to the dynamic strain $T_d = AE\varepsilon_d$, which itself may be written in terms of the element lengths

$$T_d = AE\frac{\Delta s_d - \Delta s_s}{\Delta s_s} \quad \rightsquigarrow \quad T_d = AE\frac{\frac{\mathrm{d}s_d}{\mathrm{d}x} - \frac{\mathrm{d}s_s}{\mathrm{d}x}}{\frac{\mathrm{d}s_s}{\mathrm{d}x}}, \tag{3}$$

where Δs_s and Δs_d are the static and dynamic element lengths respectively. In the second equation the limit as $\Delta x \to 0$ has been applied (such that $\Delta s_s/\Delta x \to \mathrm{d}s_s/\mathrm{d}x$ etc). With reference to Figure 2, the static element length may be written as

$$\Delta s_s = \sqrt{\Delta x^2 + \Delta w_s^2} \quad \rightsquigarrow \quad \frac{\mathrm{d}s_s}{\mathrm{d}x} = \sqrt{1 + \left(\frac{\mathrm{d}w_s}{\mathrm{d}x}\right)^2}. \tag{4}$$

Likewise, the dynamic deflection may be written as

$$\frac{\mathrm{d}s_d}{\mathrm{d}x} = \sqrt{\left(1 + \frac{\mathrm{d}u_d}{\mathrm{d}x}\right)^2 + \left(\frac{\mathrm{d}v_d}{\mathrm{d}x}\right)^2 + \left(\frac{\mathrm{d}w_s}{\mathrm{d}x} + \frac{\mathrm{d}w_d}{\mathrm{d}x}\right)^2}, \tag{5}$$

where the deflections are defined in Figure 2. Substituting these expressions into Equation 3 and then using the Taylor series expansions $\sqrt{1+\Delta} = 1 + \Delta/2 - \Delta^2/8 + \ldots$ and $^{-1}\!\!\sqrt{1+\Delta} = 1 - \Delta/2 + \ldots$ for the numerator and denominator terms respectively gives

$$T_d = AE\left[\frac{\partial u_d}{\partial x} + \frac{\mathrm{d}w_s}{\mathrm{d}x}\frac{\partial w_d}{\partial x} + \frac{1}{2}\left(\frac{\partial v_d}{\partial x}\right)^2 + \frac{1}{2}\left(\frac{\partial w_d}{\partial x}\right)^2 + \mathcal{O}\left(\frac{\partial}{\partial x}^3\right)\right]. \tag{6}$$

The second main source of nonlinearity arises from the dynamic motion of the lower support and affects just the in-plane modes. In-plane, the dynamic motion is split into two components to account for the support motion. One component is the dynamic response of the cable, taking the cable to be simply supported at both ends. The other component ensures that the true boundary conditions are met and so accounts for the motion of the lower support. This is achieved by considering how the cable deflection and tension changes as the support moves quasi-statically, i.e without considering inertial effects (which are accounted for in the dynamic response

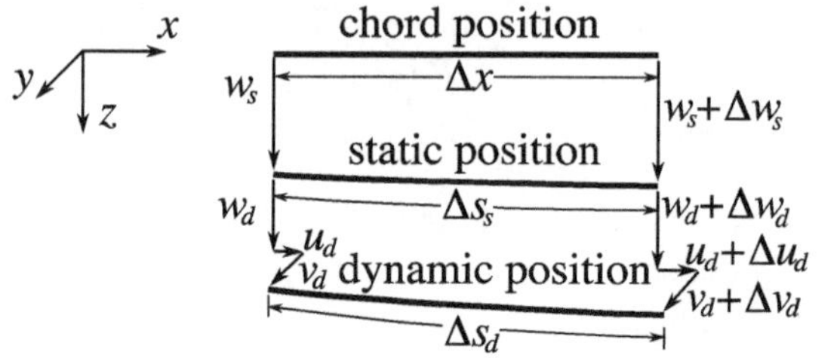

Figure 2. Deflection of a small element of cable away from the chord position due to static and dynamic loading. Reproduced from Wagg and Neild (2010).

component). The second in-plane mode is not affected by this source of nonlinearity and so we do not discuss it in detail here (instead see, for example, Wagg and Neild, 2010).

Equations 1 and 2 along with the expression for the dynamic tension, Equation 6, can now be used to generate modal equations of motion by applying the Galerkin technique. The mode shapes for a linearised version of the system are first identified by assuming that both the modal and quasi-static motions of the cable are small in comparison with the static sag. From these assumptions it follows that the dynamic tension is small in comparison with the static tension. Hence the out-of- and in-plane equations of motion, Equations 1 and 2, may be written as

$$T_s\frac{\partial^2 v_d}{\partial x^2} = \rho A\frac{\partial^2 v_d}{\partial t^2}, \quad T_d\frac{\partial^2 w_s}{\partial x^2} + T_s\frac{\partial^2 w_d}{\partial x^2} = \rho A\frac{\partial^2 w_d}{\partial t^2}, \tag{7}$$

where the dynamic tension, T_d, is given by

$$T_d = AE\left[\frac{\partial u_d}{\partial x} + \frac{\mathrm{d}w_s}{\mathrm{d}x}\frac{\partial w_d}{\partial x}\right]. \tag{8}$$

These equations are linear and can be solved to find the linear mode shapes and corresponding natural frequencies by using the separation of variable technique, with the dynamic displacements in the out-of- and in-plane directions being represented in the form $v_d = \sum_n \psi_n(x)y_n(t)$ and $w_d = \sum_n \phi_n(x)z_n(t)$, where ψ and ϕ are the mode shapes in the two planes.

The nonlinear version of these equations, in which the assumptions are relaxed such that the dynamic deflections are not taken to be small compared to the static deflections, are considered next. This is done by applying the Galerkin technique based on the linear mode shapes found using the separation of variable technique. Since these mode shapes are not the true mode shapes for the nonlinear system, the resulting modal equations contain cross-coupling terms.

The resulting equations of motion for the first two in-plane modes and the second out-of-plane mode, in the case where the support excitation is close to the resonance of the second in-plane mode, are given by

$$\begin{aligned} \ddot{y}_1 + 2\zeta_{y1}\omega_1\dot{y}_1 + \omega_1^2 y_1 + W_{11}y_1^3 + W_{12}y_1\left(y_2^2 + z_2^2\right) + N_1\delta y_1 &= 0, \\ \ddot{y}_2 + 2\zeta_{y2}\omega_2\dot{y}_2 + \omega_2^2 y_2 + W_{21}y_2 y_1^2 + W_{22}y_2\left(y_2^2 + z_2^2\right) + N_2\delta y_2 &= 0, \\ \ddot{z}_2 + 2\zeta_{z2}\omega_2\dot{z}_2 + \omega_2^2 z_2 + W_{21}z_2 y_1^2 + W_{22}z_2\left(y_2^2 + z_2^2\right) + N_2\delta z_2 &= B\ddot{\delta}. \end{aligned} \quad (9)$$

In these equations modal damping, ζ, has been introduced and two types of nonlinearity exist: cubic terms which have coefficients W_{ab} and parametric excitation terms which have coefficients N_c. The equations for the coefficients B, W_{ab} and N_c are not important for our discussion but can be found in Gonzalez-Buelga *et al.* (2008). It can be seen that the linear undamped natural frequency for each second modes is the same, namely ω_2. The natural frequency for the first in-plane mode is half of the second mode natural frequencies, i.e. $\omega_1 = \omega_2/2$. These natural frequency relationships are important as they govern the nonlinear interaction between the modes.

2.2 Nonlinear Response Behaviour

Considering the equations of motion for the three cable modes, Equation 9, it can be seen that for both the out-of-plane modes, a trivial solution to the modal equations of motion exists, namely $y_1 = 0$ and $y_2 = 0$. However, the second in-plane mode is directly excited by the support excitation through the $B\ddot{\delta}$ term so a zero-solution does not exist for this mode. Note that the other in-plane modes also contain direct excitation terms, but, as the support motion is close to the second mode, the only significant response will be in the second mode. With $y_1 = 0$ and $y_2 = 0$ the system reduces to

$$\ddot{z}_2 + 2\zeta_{z2}\omega_2\dot{z}_2 + \omega_2^2 z_2 + W_{22}z_2^3 + N_2\delta z_2 = B\ddot{\delta}. \quad (10)$$

Figure 3 shows the amplitude of response of this mode, Z_2, at the forcing frequency for a range of frequencies around the second natural frequency. The data was generated from a time-stepping simulation using the Matlab ode45 routine. The points are the steady-state responses as the excitation frequency is (a) stepped from below the natural frequency to above it, shown as dots, and (b) stepped back down to below the natural frequency, shown as circles. It can be seen that the resonance curve bends over to the right. This is because the natural frequency of the nonlinear equation of motion is a function of the amplitude of response. This relationship will be discussed for a simpler but related system in Section 2.3. A consequence of this relationship is that, as the frequency increases or decreases across the resonance peak, a jump in the amplitude of oscillation occurs. For increasing and decreasing frequencies this jump is from point A to B and from point C to D respectively. Between these jumps, the simulations highlight two steady-state amplitudes that might be observed, one on the upper solution curve from D to A and one on the lower curve from C to B. There is also a third solution that is unstable and so is not observed in the simulation. The curve for this third solution joins point C to point A. This type of unstable solution is studied in more detail in Section 5.

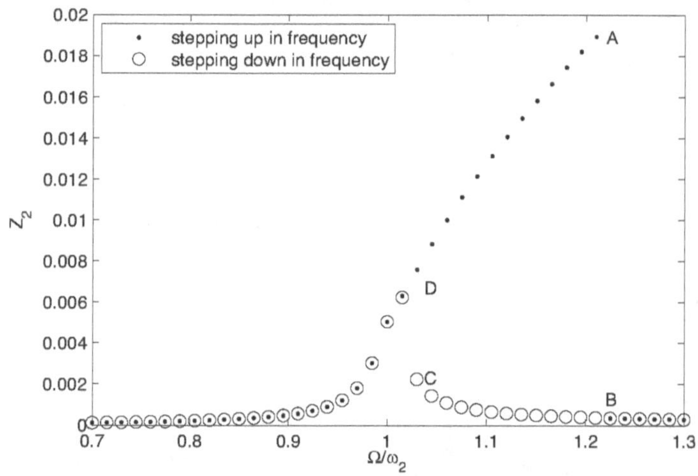

Figure 3. Steady-state response amplitude, Z_2, of the second in-plane mode, z_2, at the forcing frequency for a range of excitation frequencies and an excitation amplitude of 0.4mm. The simulation uses the same cable parameters as Gonzalez-Buelga *et al.* (2008): a 0.8mm diameter, 1.98 long, steel cable of mass 0.67kg/m inclined at 20^0 to the horizonal with a damping ratio of $\zeta = 0.2\%$ and a static tension of 205N.

Across the frequency range between points C and B in Figure 3, the multiple solutions can also be seen by considering a constant excitation frequency and varying the excitation amplitude. This is shown in Figure 4, where jumps occur as the excitation amplitude is increased from points A to B and as the excitation amplitude reduces from points C to D. The phenomena of multiple solutions is examined in Section 5.

As well as an amplitude-dependent natural frequency and multiple solu-

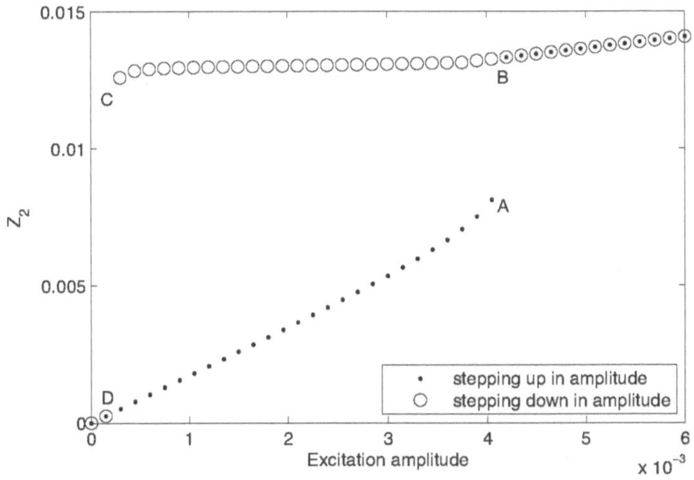

Figure 4. Steady-state response amplitude, Z_2, of the second in-plane mode, z_2, for a range of excitation amplitudes and a excitation frequency of $\Omega = 1.1\omega_2$.

tions, the nonlinear terms present in the pure z_2 response (i.e. with no other modes present) produces a response that contains harmonics of the forcing frequency. This is shown in Figure 5, where the frequency content of the steady-state response is shown for the case where the excitation frequency matches the second natural frequency and the amplitude of excitation is 0.4mm. In addition to the dominant response at the excitation frequency, there is a response at frequencies $0, 2\Omega$ and 3Ω and a much smaller response at 4Ω. One application of the generation of harmonics caused by the presence of a nonlinearity is in structural health monitoring. An example of this is studied in Section 6.

Finally, the response of the out-of-plane modes is considered. In the previous simulations it has been assumed that the two out-of-plane modes do not respond to the excitation. The zero-response solution is a valid solution to both of the out-of-plane equations of motion (see Equation 9). However, it is not always a stable solution due to parametric forcing terms (see, for example, Cartmell, 1990, for a discussion on parametric vibration). These are nonlinear terms where the forcing is multiplied by the modal co-ordinate and are present here in the form $N_a \delta y_a$. There are also auto-parametric forcing terms consisting of nonlinear terms in which the modal co-ordinate is multiplied by a different modal co-ordinate, for example $W_{12} y_1 z_2^2$ in the

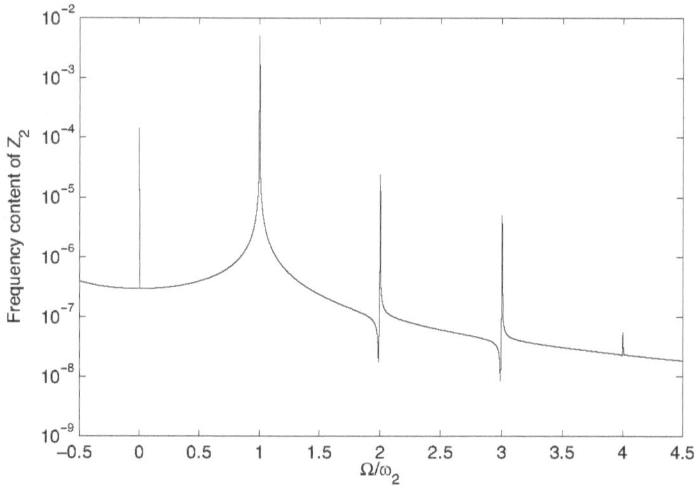

Figure 5. Frequency content of the steady-state response of the second in-plane mode for the case where the excitation amplitude is 0.4mm and the excitation frequency is $\Omega = \omega_2$. Note that, for consistency with the other Figures, the frequency content is in terms of sinusoidal amplitudes rather than the standard complex exponential amplitudes used in Fourier transforms.

y_1 equation of motion. At low amplitudes of excitation the zero steady-state response solution for the cable system is stable. However, as the excitation amplitude increases one or both of these zero-response solutions can become locally unstable. If the zero solution is locally unstable and the mode is subjected to any unmodelled excitation, such as a gust loading or some sway motion of the deck, the modal response will settle at a non-zero steady-state response solution. This is demonstrated in Figure 6, where the system is left to settle with no out-of-plane excitation and then at 50 s a pulse is applied to the out-of-plane modes. In Figure 6(a), the out-of-plane modal response decays away to the zero solution, which is stable. In Figure 6(b) the excitation amplitude has been increased to the point where the zero-response solution is unstable for the first out-of-plane mode. The result is a non-zero steady-state response after the pulse has been applied.

This parametric excitation of cable modes using these equations of motion is discussed in detail in Gonzalez-Buelga *et al.* (2008) and Macdonald *et al.* (2010a). Based on these studies, design guidelines to avoid parametric excitation have been proposed (Macdonald *et al.*, 2010b). These

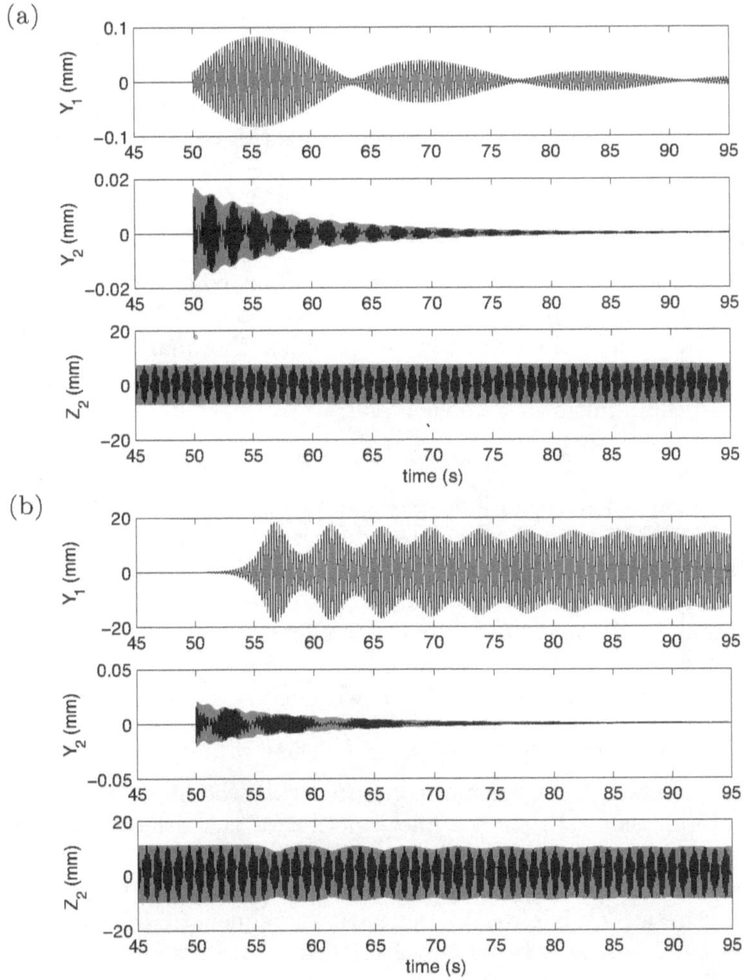

Figure 6. Response of the three modes for the case where the excitation frequency is $\Omega = 0.98\omega_2$ and the amplitude of excitation is (a) 2 mm and (b) 4 mm.

studies use an approximation technique termed *averaging* (discussed in, for example, Verhulst, 1989, Bakri *et al.*, 2004 and Wagg and Neild, 2010), but the equations have also been studied using numerical continuation. This allows the prediction of the response amplitudes for the various modes in the more complex multi-mode response regions (Marsico *et al.*, submitted). Parametric excitation will be discussed in Section 7.

2.3 Extension of Linear Analysis

To understand further some of the behaviour exhibited by the second in-plane cable mode, a slightly simpler but related equation of motion, the Duffing oscillator equation, will now be analysed. The analysis will be based on standard linear techniques. So, first, consider a sinusoidally-excited one-degree-of-freedom linear system such as

$$\ddot{x} + 2\zeta\omega_n\dot{x} + \omega_n^2 x = P\cos(\Omega t). \tag{11}$$

To solve the equation of motion, it is usual to make a trial solution of the form $x = X\cos(\Omega t - \phi)$, where the response frequency ω_r (the frequency at which the system responds to the excitation) matches the forcing frequency Ω and ϕ is the phase lag of the system at frequency Ω. Alternatively, by shifting the time origin we can write the forcing as $p = P\cos(\Omega t + \phi)$ and the response as $x = X\cos(\Omega t)$, where time has undergone the shift $t \to t + \phi/\Omega$.

Now consider the Duffing oscillator, with the time-shifted forcing

$$\ddot{x} + 2\zeta\omega_n\dot{x} + \omega_n^2 x + \gamma_x(x, \dot{x}, p) = P\cos(\Omega t + \phi), \quad \gamma_x = \alpha x^3,$$

where the nonlinear term is small compared with the linear stiffness term. Since the nonlinear term is a cubic function of the response amplitude, the effect of the nonlinear term is most significant near resonance, where the response amplitude is large. Therefore we shall study the case where the forcing frequency is close to the natural frequency. Initially, as with linear systems, a trial solution to the steady-state response of the form $x = X\cos(\omega_r t)$ is considered. Since the forcing is close to resonance, the dominant response frequency, ω_r, is taken to be at the forcing frequency. Substituting this trial solution into the equation of motion gives

$$(\omega_n^2 - \Omega^2)X\cos(\Omega t) - 2\zeta\omega_n\Omega X\sin(\Omega t) + \alpha X^3\cos^3(\Omega t) = P\cos(\Omega t + \phi).$$

It can be seen that using a time-shift such that the response is a pure cosine, at the expense of introducing a sinusoidal component to the forcing, is algebraically beneficial since the nonlinear term requires the response to be cubed. Using the trigonometric expansion

$$\cos^3(\omega_r t) = \frac{1}{4}\left[3\cos(\omega_r t) + \cos(3\omega_r t)\right]$$

gives

$$(\omega_n^2 - \Omega^2)X\cos(\Omega t) - 2\zeta\omega_n\Omega X\sin(\Omega t) + \frac{\alpha}{4}X^3\left[3\cos(\Omega t) + \cos(3\Omega t)\right] \\ = P\cos(\Omega t + \phi).$$

The $\cos(3\Omega t)$ term means that this equation cannot be satisfied exactly without setting $X = 0$ and hence $P = 0$. To satisfy the equation approximately, the coefficients of the $\cos(\Omega t)$ and $\sin(\Omega t)$ terms are equated (while ignoring the $\cos(3\Omega t)$ term). This gives

$$\begin{aligned} \cos(\Omega t) \text{ terms:} \quad & (\omega_n^2 - \Omega^2)X + \tfrac{3\alpha}{4}X^3 \approx P\cos(\phi), \\ \sin(\Omega t) \text{ terms:} \quad & 2\zeta\omega_n\Omega X \approx P\sin(\phi). \end{aligned}$$

The phase, ϕ, can be eliminated by squaring and adding the expressions, giving

$$X^2\left[\left(\omega_n^2 - \Omega^2 + \frac{3\alpha}{4}X^2\right)^2 + 4\zeta^2\omega_n^2\Omega^2\right] \approx P^2, \tag{12}$$

which leads to

$$\frac{X}{P} \approx \frac{1}{\sqrt{(\omega_n^2 - \Omega^2 + \frac{3\alpha}{4}X^2)^2 + 4\zeta^2\omega_n^2\Omega^2}}.$$

The phase can be found from the ratio of the sine and cosine balance equations, which gives

$$\phi = \arctan\left(\frac{2\zeta\omega_n\Omega}{\omega_n^2 - \Omega^2 + \frac{3\alpha}{4}X^2}\right).$$

Hence for this nonlinear system the phase is also dependent on the amplitude of response.

The process adopted in this example is the most basic form of *harmonic balance*. This technique involves assuming a harmonic (sine and cosine wave) solution for the steady-state response. Then, having substituted the assumed solution into the governing equation, the second part of the process is to *balance* the coefficients of the harmonic terms that are included in the trial solution. This does not give an exact solution to the equation of motion, since the harmonic terms that are absent from the trial solution but do appear in the equation of motion are not balanced. In the Duffing oscillator, for example, the $\cos(3\omega_r t)$ term remained unbalanced. The general form of the trial solution with a primary response at frequency ω_r is

$$x = a_0 + \sum a_n\cos(n\omega_r t) + b_n\sin(n\omega_r t) \quad n = 1, 2, 3...,$$

where a_n and b_n are coefficients. The accuracy of the harmonic balance approximation is dependent on the number of these terms that are included in the trial solution. Often the trial solution is limited merely to terms at frequency ω_r as was the case in the example.

To plot the response curve predicted using the harmonic balance technique, Equation 12 can be rewritten as a quadratic in Ω^2, which gives

$$(\Omega^2)^2 + (\Omega^2)\left(4\zeta^2\omega_n^2 - 2\omega_n^2 - \frac{3\alpha}{2}X^2\right) + \left(\omega_n^2 + \frac{3\alpha}{4}X^2\right)^2 - \left(\frac{P}{X}\right)^2 = 0.$$

The two solutions for the quadratic in Ω^2 can be found for a range of values of X. Real positive values of Ω^2 are valid solutions and make up the response curve. Figure 7(a) shows the prediction of the response curve using the harmonic balance technique compared to time-stepping simulation results. The corresponding prediction of the phase ϕ is shown in Figure 7(b).

As with the cable example, it can be seen that the resonance curve bends over. This is due to an amplitude-dependent natural frequency, which can be derived by considering the unforced, undamped, version of the Duffing oscillator

$$\ddot{x} + \omega_n^2 x + \alpha x^3 = 0.$$

Making the substitution $x = X\cos(\omega_r t)$, where the response frequency is now the nonlinear natural frequency, and balancing the $\cos(\omega_r t)$ terms, while again ignoring the $\cos(3\omega_r t)$ term, yields

$$(\omega_n^2 - \omega_r^2)X + \frac{3\alpha X^3}{4} \approx 0 \quad \rightsquigarrow \omega_r \approx \omega_n\sqrt{1 + \frac{3\alpha X^2}{4\omega_n^2}}. \tag{13}$$

This equation gives an approximate value for the response frequency as a function of the response amplitude, which defines the **backbone curve**, plotted as a dashed line in Figure 8(a). Note that $\omega_r = \omega_n$ when $X = 0$. As a comparison, resonance peaks for the forced system are also plotted, using a range of forcing amplitudes in conjunction with equation 12. Again for comparison, the linear equivalent system (where $\alpha = 0$) is shown in Figure 8(b).

The response will also have content at three times the forcing frequency, as is indicated by the presence of the $\cos(3\Omega t)$ term that was previously ignored. In the case of the in-plane cable equation, it was seen that significant harmonics exist at 0Ω, 2Ω and 3Ω (see Figure 5). The reason for this is that in addition to a cubic nonlinear term, $W_{22}z_2^3$, which results in the 3Ω harmonic, there is a parametric term $N_2\delta z_2$. Upon substitution of a response trial solution at the forcing frequency, this parametric term contains

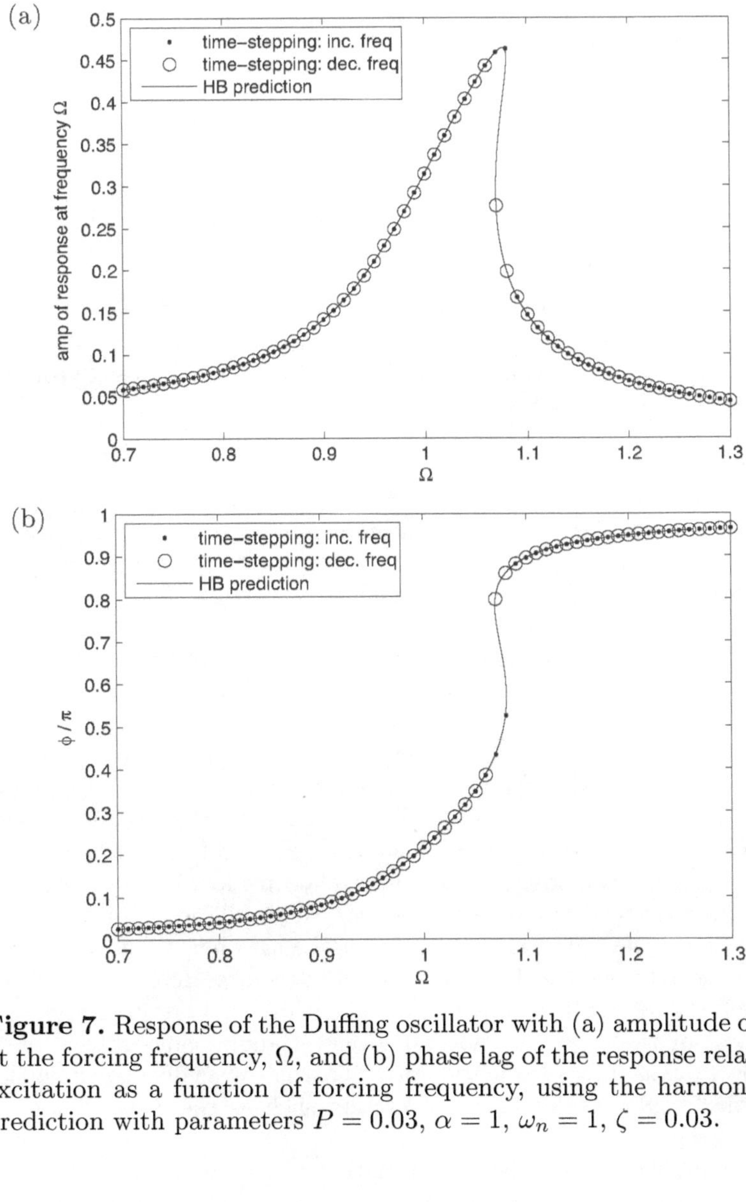

Figure 7. Response of the Duffing oscillator with (a) amplitude of response at the forcing frequency, Ω, and (b) phase lag of the response relative to the excitation as a function of forcing frequency, using the harmonic balance prediction with parameters $P = 0.03$, $\alpha = 1$, $\omega_n = 1$, $\zeta = 0.03$.

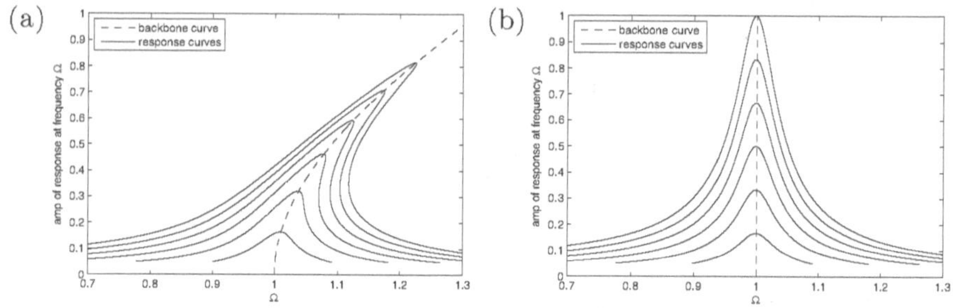

Figure 8. Amplitude of response at the forcing frequency, Ω, as a function of forcing frequency for the range of forcing amplitudes from $P = 0.01$ to $P = 0.06$ along with the backbone curves, for (a) the Duffing oscillator and (b) a linear oscillator with $\alpha = 0$.

$\cos^2(\Omega t) = (1 + \cos(2\Omega t))/2$ type sinusoidal relationships which result in harmonic content at 0Ω and 2Ω. The harmonic at 4Ω and higher frequencies (not shown in the figure) are over 100 times smaller. They are due to mixing between the response and the larger harmonics and can be observed if a higher order trial solution is used. Response content at harmonics of the forcing frequency are examined further using the *normal forms* technique in Sections 5 and 7.

In this analysis, the response frequency is assumed to be at the forcing frequency. This is a realistic assumption for the case where the forcing is around the natural frequency, as the Figures show. For most frequencies away from resonance, the assumption is also reasonable as the nonlinear term will be insignificant due to the cubic relationship with the response, which is small away from resonance. However, there will be regions in which this assumption is less plausible, an example being when the forcing is at a frequency close to one-third of the natural frequency. Here, the $\cos(3\Omega t)$ term is likely to be significant as its frequency will be close to resonance. For this type of response, the normal form technique can be used as it handles the forcing in a systematic way. This is especially beneficial for more complex nonlinear equations of motion such as the cable equations considered earlier.

Finally, note that no transient information is available when using the harmonic balance technique. The multiple scales technique gives transient information, which is useful as it provides a convenient way of examining the stability of the solutions. This will be demonstrated in Section 5 where the unstable steady-state solutions for the Duffing oscillator will be identified.

3 Perturbation Techniques

The perturbation method for analysing weakly nonlinear systems is based on the idea that the response consists of a series of terms of reducing significance, or increasing order of small parameter ε. The response is therefore represented as a power series in ε describing increasingly less significant components of the response

$$x = x_0 + \varepsilon x_1 + \varepsilon^2 x_2 + \varepsilon^3 x_3 \dots \tag{14}$$

It is usual for the most significant term in the series, x_0, to represent the response of the linearised system, ie the system if all the nonlinear terms were removed. The less significant terms x_i, for $i = 1, 2, \dots$, then describes perturbations away from the linear response with reducing significance as i increases as indicated by the ε^i multiplier.

In this section two approaches will be discussed. The first is *regular perturbation theory* in which Equation (14) is substituted into the equation of motion. It will be demonstrated that this technique can produce erroneous results in some circumstances, but it can also work well under the right conditions and will be used in Section 6. Secondly, the *multiple scales* technique will be examined. It will be shown that this technique, which utilises the observation that often a response consists of terms that change rapidly with time and others that change slowly, overcomes the shortcomings of the regular perturbation technique.

The multiple scales technique is powerful since it provides transient as well as steady-state response information, so can be used to assess the stability of steady-state solutions. In addition, it can provide information regarding the response of the harmonics of the forcing frequency. These features will be demonstrated when the multiple scales technique is used in the first case study on energy harvesting in Section 5. The second case study on detecting nonlinear material properties, Section 6, will use the perturbation technique, followed by a form of multiple scales analysis, to predict the amplitude of harmonic response the material generates due to its nonlinear stress-strain relationship.

3.1 Regular perturbation theory

Regular perturbation theory involves making the substitution for the response trial solution given in Equation (14) into the equation of motion. As with the trial solution, the equation of motion includes ε multipliers to indicate terms with reduced significance. For an unforced single-degree-of-freedom nonlinear system this results in

$$\ddot{x} + 2\varepsilon\hat{\zeta}\omega_n\dot{x} + \omega_n^2 x + \varepsilon\hat{\gamma}_x(x, \dot{x}) = 0,$$

where the damping and the nonlinear terms have been taken to be small, hence $\gamma_x = \varepsilon\hat{\gamma}_x$ and $\zeta = \varepsilon\hat{\zeta}$.

Making the trial solution substitution gives

$$\ddot{x}_0+\varepsilon\ddot{x}_1+2\varepsilon\hat{\zeta}\omega_n\dot{x}_0+\omega_n^2(x_0+\varepsilon x_1)+\varepsilon\hat{\gamma}_x(x_0+\varepsilon x_1+\ldots,\dot{x}_0+\varepsilon\dot{x}_1+\ldots)+\mathcal{O}(\varepsilon^2)=0. \tag{15}$$

The nonlinear term can be simplified using a Taylor series expansion, noting that the ε^1 terms are small, giving

$$\varepsilon\hat{\gamma}_x(x_0+\varepsilon x_1+\ldots,\dot{x}_0+\varepsilon\dot{x}_1+\ldots)=\varepsilon\hat{\gamma}_x(x_0,\dot{x}_0)+\mathcal{O}(\varepsilon^2)$$

Now Equation 15 can be satisfied up to order ε^1 by balancing the ε^0 and the ε^1 terms, giving

$$\begin{aligned} \varepsilon^0: &\qquad \ddot{x}_0+\omega_n^2x_0=0,\\ \varepsilon^1: &\qquad \ddot{x}_1+\omega_n^2x_1=-2\hat{\zeta}\omega_n\dot{x}_0-\hat{\gamma}_x(x_0,\dot{x}_0). \end{aligned}$$

Higher order terms of ε could also be balanced but, for a technique to be useful, it is desirable to get a reasonably accurate system response estimate using a low number of terms in the power series expansion.

The order ε^0 equation represents the linear undamped response. Note that if the damping was taken to be order ε^0 then this equation would represent the linear damped response. The order ε^1 equation provides information regarding the first perturbation from the linear response, x_1. The response of x_1 is governed by the x_0 terms which can be viewed as forcing terms in this equation. Importantly, this equation is also linear in x_1 and hence can be solved straightforwardly once the x_0 forcing terms on the right-hand side have been calculated.

To demonstrate the potential difficulty with this technique, the vibration response of a lightly-damped, unforced, linear oscillator will be considered. The initial conditions are taken to be that the system is at rest with displacement $x(0)$ at time zero. The equation of motion is given by

$$\ddot{x}+2\varepsilon\hat{\zeta}\omega_n\dot{x}+\omega_n^2x=0.$$

Applying the substitution power series trial solution for the response, Equation 14, and balancing the ε^0 and ε^1 terms gives

$$\begin{aligned} \varepsilon^0: &\qquad \ddot{x}_0+\omega_n^2x_0 &= 0,\\ \varepsilon^1: &\qquad \ddot{x}_1+\omega_n^2x_1 &= -2\hat{\zeta}\omega_n\dot{x}_0. \end{aligned}$$

To solve these equations, the first step is to write the linearised response as $x_0 = x(0)\cos(\omega_n t)$, noting that the initial displacement $x(0) = x_0(0) +$

$\varepsilon x_1(0)$ is valid for all ε such that $x_0(0) = x(0)$ and $x_1(0) = 0$. Secondly, in the ε^1 equation, the x_0 solution to the ε^0 equation can be substituted into the forcing-like term, giving

$$\ddot{x}_1 + \omega_n^2 x_1 = 2\hat{\zeta}\omega_n^2 x(0)\sin(\omega_n t).$$

Inspecting the right-hand side, the trial solution $x_1 = A\cos(\omega_n t) + B\sin(\omega_n t)$ might be selected. However, when substituted into the left-hand side of the differential equation, both these terms cancel out. Therefore, the substitution $x_1 = At\cos(\omega_n t) + Bt\sin(\omega_n t)$ must be made. This results in the response for x_1 as

$$x_1 = -x(0)\hat{\zeta}\omega_n t\cos(\omega_n t).$$

Hence the full response (to order ε^1) is

$$x = x_0 + \varepsilon x_1 = x(0)\,(1 - \zeta\omega_n t)\cos(\omega_n t),$$

where ε has been eliminated using $\varepsilon\hat{\zeta} = \zeta$. This predicted response can be compared to the actual response

$$x = x(0)\mathrm{e}^{-\zeta\omega_n t}\cos(\omega_n\sqrt{1-\zeta^2}t).$$

It can be seen that there is a slight error in the natural frequency but, more significantly, that the exponential decay term in the accurate solution has been replaced with the first two terms of its Taylor series expansion in the perturbation solution. This is acceptable at low values of $\zeta\omega_n t$ but, however small ζ is, as time increases this approximation will break down. In summary, this analysis has generated *secular* terms, which are terms that are unbounded with time – in this case $t\cos(\omega_n t)$.

There are two timescales occurring in this solution, namely a fast timescale due to the oscillations and a slow timescale due to the exponential decay. When using the regular perturbation technique the slow timescale is poorly described, with the exponential decay being represented as $1 - \zeta\omega_n t$. This leads to the question: how can we obtain a good approximation for the different timescale present in the response? The answer is often to use multiple scales analysis.

Further discussion of regular perturbation theory is given in Verhulst (1989) and Strogatz (2000).

3.2 Multiple scales method

As was seen in the perturbation analysis discussion, the dynamic responses often consist of terms that are functions of different timescales.

Typically these are fast oscillations with slowly changing amplitude envelopes such as the exponential decay envelope. In multiple scales this is reflected by writing the trial solution in the form

$$x = X_c(\varepsilon t)\cos(\omega_r t) + X_s(\varepsilon t)\sin(\omega_r t). \tag{16}$$

Here ε is used to indicate that the amplitude terms X_c and X_s vary slowly over time. This type of solution could also be expressed in the amplitude-phase form $x = X(\varepsilon t)\sin(\omega_r t + \phi(\varepsilon t))$ (this representation is often easier for free vibration, but can be algebraically intensive for forced vibration).

The two timescales can be labelled separately, fast-time over which oscillations occur $t_f = \omega_r t$ and slow-time over which the amplitudes evolve $t_s = \varepsilon t$. Both terms on the right-hand side of Equation 16 now take the form $f_s(t_s)f_f(t_f)$, and can be differentiated using the chain rule to give

$$\frac{\mathrm{d}}{\mathrm{d}t}\{f_s(t_s)f_f(t_f)\} = \omega_r f_s \frac{\mathrm{d}f_f}{\mathrm{d}t_f} + \varepsilon \frac{\mathrm{d}f_s}{\mathrm{d}t_s} f_f.$$

It is perhaps easier to view t_f and t_s as independent variables, such that the derivative with respect to t can be expressed as

$$\frac{\mathrm{d}}{\mathrm{d}t}\{x(t_s, t_f)\} = \omega_r \frac{\partial x}{\partial t_f} + \varepsilon \frac{\partial x}{\partial t_s},$$

which gives the same result. The second derivative with respect to time is then

$$\frac{\mathrm{d}^2 x}{\mathrm{d}t^2} = \omega_r^2 \frac{\partial^2 x}{\partial t_f^2} + 2\omega_r \varepsilon \frac{\partial^2 x}{\partial t_f \partial t_s} + \varepsilon^2 \frac{\partial^2 x}{\partial t_s^2}.$$

In addition to this separation of the response into slow and fast timescales, the power series representation of the solution, Equation 14, is used to give

$$\begin{aligned}
x &= x_0 + \varepsilon x_1 + \mathcal{O}(\varepsilon^2),\\
\frac{\mathrm{d}x}{\mathrm{d}t} &= \omega_r \frac{\partial x_0}{\partial t_f} + \varepsilon\left(\omega_r \frac{\partial x_1}{\partial t_f} + \frac{\partial x_0}{\partial t_s}\right) + \mathcal{O}(\varepsilon^2),\\
\frac{\mathrm{d}^2 x}{\mathrm{d}t^2} &= \omega_r^2 \frac{\partial^2 x_0}{\partial t_f^2} + \varepsilon\left(\omega_r^2 \frac{\partial^2 x_1}{\partial t_f^2} + 2\omega_r \frac{\partial^2 x_0}{\partial t_f \partial t_s}\right) + \mathcal{O}(\varepsilon^2).
\end{aligned} \tag{17}$$

These expressions can be used to analyse free and forced vibration.

Free Vibration: Consider the unforced nonlinear system

$$\ddot{x} + 2\varepsilon\hat{\zeta}\omega_n\dot{x} + \omega_n^2 x + \varepsilon\hat{\gamma}_x(x, \dot{x}) = 0,$$

where the nonlinearity and damping are taken to be small. Since the damping is small, the linear response has a natural frequency of ω_n and therefore the fast-time is set to $t_f = \omega_r t = \omega_n t$. Substituting the expressions for the derivatives of x, Equation 17, into the equation of motion gives

$$\left(\omega_n^2 \frac{\partial^2 x_0}{\partial t_f^2} + \varepsilon\left(\omega_n^2 \frac{\partial^2 x_1}{\partial t_f^2} + 2\omega_n \frac{\partial^2 x_0}{\partial t_f \partial t_s}\right)\right) + 2\varepsilon\hat{\zeta}\omega_n\left(\omega_n \frac{\partial x_0}{\partial t_f}\right) + \omega_n^2\left(x_0 + \varepsilon x_1\right) + \varepsilon\hat{\gamma}_x\left(x_1, \omega_n \frac{\partial x_0}{\partial t_f}\right) + \mathcal{O}(\varepsilon^2) = 0.$$

Here the nonlinear function has been subject to a Taylor series expansion yielding

$$\varepsilon\hat{\gamma}_x\left(x_1 + \varepsilon x_0 + \mathcal{O}(\varepsilon^2), \omega_n \frac{\partial x_0}{\partial t_f} + \varepsilon\left(\omega_n \frac{\partial x_1}{\partial t_f} + \frac{\partial x_0}{\partial t_s}\right) + \mathcal{O}(\varepsilon^2)\right) = \varepsilon\hat{\gamma}_x\left(x_1, \omega_n \frac{\partial x_0}{\partial t_f}\right) + \mathcal{O}(\varepsilon^2).$$

Now the order ε^0 and ε^1 terms are balanced to give

$$\varepsilon^0: \quad \omega_n^2 \frac{\partial^2 x_0}{\partial t_f^2} + \omega_n^2 x_0 = 0, \tag{18}$$

$$\varepsilon^1: \quad \omega_n^2 \frac{\partial^2 x_1}{\partial t_f^2} + \omega_n^2 x_1 = -2\omega_n \frac{\partial^2 x_0}{\partial t_f \partial t_s} - 2\hat{\zeta}\omega_n^2 \frac{\partial x_0}{\partial t_f} - \hat{\gamma}_x(x_0, \omega_n \frac{\partial x_0}{\partial t_f}). \tag{19}$$

Note that higher orders of ε can also be balanced for higher accuracy solutions. However, in practice, this is rarely done as these terms are very small.

As with the regular perturbation technique, both these equations are linear in terms of x_0 and x_1 respectively and, in the case of the second equation, the x_0 terms can be viewed as forcing terms acting on x_1. A solution to Equation 18 is

$$x_0 = X_{0c}(t_s)\cos(t_f) + X_{0s}(t_s)\sin(t_f),$$

which matches the form of Equation 16. This expression is now substituted into the right-hand side of the ε^1 equation, Equation 19, to find the forcing applied to x_1. This substitution may generate $A(t_s)\sin(t_f)$ or $B(t_s)\cos(t_f)$

type terms on the right-hand side, where typically $A(t_s)$ and $B(t_s)$ are differential expressions in terms of X_{0c} and X_{0s}. To ensure that the multiple scales method does not result in the secular terms that are the limitation of the regular perturbation method, $A(t_s)$ and $B(t_s)$ must be set to zero. This results in conditions on X_{0c} and X_{0s} in the form of (typically) first-order differential equations. Satisfying these conditions such that $A(t_s) = B(t_s) = 0$ has the effect of ensuring that the response at the resonant frequency, (or the forcing frequency for the forced vibration case), is captured by x_0. Solving these conditions gives the x_0 response. The resulting ε^1 equation can then be solved to find x_1, which gives information about harmonics of the resonant (or forcing) frequency.

Reconsider the damped, unforced oscillator that caused difficulties for the regular perturbation technique

$$\ddot{x} + 2\varepsilon\hat{\zeta}\omega_n\dot{x} + \omega_n^2 x = 0,$$

which is initially at rest with displacement $x(0)$. For this method, the response is in the form $x(t) = x_0(t_s, t_f) + \varepsilon x_1(t_s, t_f)$, where x_0 and x_1 can be found using Equations 18 and 19, which in this case simplify to

$$\varepsilon^0 : \qquad \omega_n^2 \frac{\partial^2 x_0}{\partial t_f^2} + \omega_n^2 x_0 = 0,$$

$$\varepsilon^1 : \qquad \omega_n^2 \frac{\partial^2 x_1}{\partial t_f^2} + \omega_n^2 x_1 = -2\omega_n \frac{\partial^2 x_0}{\partial t_f \partial t_s} - 2\hat{\zeta}\omega_n^2 \frac{\partial x_0}{\partial t_f}.$$

The first equation is satisfied by the expression

$$x_0 = X_{0c}(t_s)\cos(t_f) + X_{0s}(t_s)\sin(t_f). \tag{20}$$

Substituting this equation into the ε^1 expression gives

$$\begin{aligned} \omega_n^2 \frac{\partial^2 x_1}{\partial t_f^2} + \omega_n^2 x_1 &= 2\omega_n \left(\frac{\mathrm{d}X_{0c}}{\mathrm{d}t_s} + \omega_n \hat{\zeta} X_{0c} \right) \sin(t_f) \\ &\quad -2\omega_n \left(\frac{\mathrm{d}X_{0s}}{\mathrm{d}t_s} + \omega_n \hat{\zeta} X_{0s} \right) \cos(t_f). \end{aligned} \tag{21}$$

To avoid secular terms in the x_1 response, the $\cos(t_f)$ and $\sin(t_f)$ terms on the right-hand side must be removed by setting

$$\begin{aligned} \frac{\mathrm{d}X_{0c}}{\mathrm{d}t_s} + \omega_n \hat{\zeta} X_{0c}(t_s) &= 0, \\ \frac{\mathrm{d}X_{0s}}{\mathrm{d}t_s} + \omega_n \hat{\zeta} X_{0s}(t_s) &= 0. \end{aligned}$$

These can be solved to give

$$X_{0c} = X_{0c}(0)\mathrm{e}^{-\omega_n \hat{\zeta} t_s}, \qquad X_{0s} = X_{0s}(0)\mathrm{e}^{-\omega_n \hat{\zeta} t_s}.$$

Hence the expression for x_0, Equation 20, becomes

$$x_0 = \mathrm{e}^{-\zeta\omega_n t}\left(X_{0c}(0)\cos(\omega_n t) + X_{0s}(0)\sin(\omega_n t)\right),$$

where $\hat{\zeta} t_s = \zeta t$ (since $\hat{\zeta} = \zeta/\varepsilon$ and $t_s = \varepsilon t$) and $t_f = \omega_n t$ have been used. The initial conditions that the system is at rest with displacement $x(0)$ results in

$$x_0 = x(0)\mathrm{e}^{-\zeta\omega_n t}\cos(\omega_n t),$$

where the initial conditions are satisfied by x_0 and hence the x_1 initial conditions are zero. Revisiting the ε^1 equation, Equation 21, the $\sin(t_f)$ and $\cos(t_f)$ on the right-hand side are zero. Since the initial conditions are also zero, this gives $x_1 = 0$. The multiple scales technique prediction of the response is therefore

$$x = x(0)\mathrm{e}^{-\zeta\omega_n t}\cos(\omega_n t).$$

It can be seen that, in contrast to the general perturbation technique, the exponential decay term has now been captured accurately. There is, however, still a slight error in the resonant frequency. This could be improved by including very slow timescale terms, see for example Strogatz (2000) or Glendinning (1994). Further discussion on multiple scales is also provided in Verhulst (1989)

Forced Vibration For the case where forcing is present, the equation of motion can be written as

$$\ddot{x} + 2\varepsilon\hat{\zeta}\omega_n\dot{x} + \omega_n^2 x + \varepsilon\hat{\gamma}_x(x,\dot{x}) = \varepsilon\hat{P}\cos(\Omega t). \tag{22}$$

Here, the forcing has been assumed to be small in comparison to the response, such that $P = \varepsilon\hat{P}$. This is consistent with forcing close to resonance, the region in the frequency domain of most interest.

Before the power series and slow and fast timescale expansions for x, $\dot{x}$ and $\ddot{x}$, Equation 17, can be substituted into the equation of motion, a detuning parameter must be introduced. Without the introduction of such a parameter, if Equation 17 is substituted into Equation 22 and the ε^0 and ε^1 terms are balanced, the resulting ε^0 equation is

$$\omega_r^2\frac{\partial^2 x_0}{\partial t_f^2} + \omega_n^2 x_0 = 0. \tag{23}$$

The solution to this is $x_0 = X_{0c}(t_s)\cos(\omega_n t) + X_{0s}(t_s)\sin(\omega_n t)$ (where $t_f = \omega_r t$ has been used). However, for forced vibration at a frequency close to the natural frequency, the dominant response should be at the excitation frequency, Ω, rather than at the natural frequency ω_n. A detuning parameter ensures that the x_0 response is at the excitation frequency.

The forcing frequency Ω can be written as $\Omega = \omega_n(1+\mu)$, where μ is the frequency detuning parameter. This detuning parameter is small since the forcing is near resonant and therefore μ can be written as $\mu = \varepsilon\hat{\mu}$. Setting the response frequency to equal the excitation frequency, $\omega_r = \Omega = \omega_n(1 + \varepsilon\hat{\mu})$, and substituting this into Equation 17 gives the following expansions for the derivatives of x

$$\begin{aligned} x &= x_0 + \varepsilon x_1 + \mathcal{O}(\varepsilon^2), \\ \frac{\mathrm{d}x}{\mathrm{d}t} &= \omega_n \frac{\partial x_0}{\partial t_f} + \varepsilon \left(\hat{\mu}\omega_n \frac{\partial x_0}{\partial t_f} + \omega_n \frac{\partial x_1}{\partial t_f} + \frac{\partial x_0}{\partial t_s} \right) + \mathcal{O}(\varepsilon^2), \\ \frac{\mathrm{d}^2x}{\mathrm{d}t^2} &= \omega_n^2 \frac{\partial^2 x_0}{\partial t_f^2} + \varepsilon \left(2\hat{\mu}\omega_n^2 \frac{\partial^2 x_0}{\partial t_f^2} + \omega_n^2 \frac{\partial^2 x_1}{\partial t_f^2} + 2\omega_n \frac{\partial^2 x_0}{\partial t_f \partial t_s} \right) + \mathcal{O}(\varepsilon^2). \end{aligned} \tag{24}$$

Substituting these expressions into the equation of motion for the forced system, Equation 22, yields

$$\left[\omega_n^2 \frac{\partial^2 x_0}{\partial t_f^2} + \varepsilon \left(2\hat{\mu}\omega_n^2 \frac{\partial^2 x_0}{\partial t_f^2} + \omega_n^2 \frac{\partial^2 x_1}{\partial t_f^2} + 2\omega_n \frac{\partial^2 x_0}{\partial t_f \partial t_s} \right) \right] +$$
$$2\varepsilon\hat{\zeta}\omega_n \left[\omega_n \frac{\partial x_0}{\partial t_f} \right] + \omega_n^2 \left[x_0 + \varepsilon x_1 \right] + \varepsilon\hat{\gamma}_x \left(x_0, \omega_n \frac{\partial x_0}{\partial t_f} \right) + \mathcal{O}(\varepsilon^2) = \varepsilon\hat{P}\cos(t_f).$$

As with the unforced case, a Taylor series expansion has been applied to the nonlinear term $\varepsilon\hat{\gamma}_x$.

Balancing the ε^0 and ε^1 terms produces

$$\omega_n^2 \frac{\partial^2 x_0}{\partial t_f^2} + \omega_n^2 x_0 = 0 \tag{25}$$

and

$$\begin{aligned} \omega_n^2 \frac{\partial^2 x_1}{\partial t_f^2} + \omega_n^2 x_1 &= -2\hat{\mu}\omega_n^2 \frac{\partial^2 x_0}{\partial t_f^2} - 2\omega_n \frac{\partial^2 x_0}{\partial t_f \partial t_s} \\ &\quad -2\hat{\zeta}\omega_n^2 \frac{\partial x_0}{\partial t_f} - \hat{\gamma}_x(x_0, \omega_n \frac{\partial x_0}{\partial t_f}) + \hat{P}\cos(t_f), \end{aligned} \tag{26}$$

respectively.

The ε^0 equation can be solved to give

$$x_0 = X_{0c}(t_s)\cos(t_f) + X_{0s}(t_s)\sin(t_f).$$

Note that using the detuning parameter has resulted in the oscillatory component of the x_0 response being at the desired frequency, namely the excitation frequency, with $t_f = \omega_r t = \Omega t$. This expression can then substituted into the ε^1 equation. As with the unforced case, the resulting $\sin(t_f)$ and $\cos(t_f)$ terms on the right-hand side of the equation are then set to zero, to ensure that the solution for x_1 is not secular and that the response at frequency Ω is contained completely in x_0. Once this has been done, the remaining equation can be used to find the harmonic content of the response.

An example of the use of this technique is given in Section 5.

4 Normal Forms

The method of normal forms can be used to transform the equations of motion of weakly nonlinear systems into a form which is easier to solve. Specifically, the aim is to apply a nonlinear transformation that removes, for each mode, all terms in the equation of motion that result in harmonics of the natural frequency (in the case of the unforced system) or the dominant response frequency (in the case of forced systems). Transforming these terms out of the equations of motion, for, say, the nth mode, allows the use of a trial solution of the form $U_n\cos(\omega_{rn}t - \phi_n)$ to solve the equation exactly, thereby removing the need for a harmonic balance type approximation.

Until recently, the method of normal forms was only formulated for systems of first order differential equations. Thus the first step was to transform the equations of motion into their state-space representation. However, recent work by Neild and Wagg (2011) has demonstrated that the normal form technique can be carried out on second-order nonlinear oscillators directly. This has several potential advantages, especially for problems relating to mechanical or structural vibration where linear analytical techniques are traditionally carried out in a second-order formulation. The discussion here will be limited to second-order normal form. For information on the first-order normal form see, for example, Nayfeh (1993), Jezequel and Lemarque (1991) and Wagg and Neild (2010).

In the following discussion, the normal form technique is applied to an N degree-of-freedom system expressed in the matrix form

$$M\ddot{\mathbf{x}} + C\dot{\mathbf{x}} + K\mathbf{x} + \mathbf{\Gamma}_x(\mathbf{x}, \dot{\mathbf{x}}, \mathbf{r}) = P_x\mathbf{r},$$

where $\mathbf{x}$ is the $N \times 1$ displacement vector. The $N \times N$ mass, damping and stiffness matrices are represented as M, C and K respectively and Γ_x is a $N \times 1$ vector of nonlinear terms, which are assumed to be small. If the forcing is assumed to be sinusoidal, it can be represented as $P_x\mathbf{r}$, where P_x is a $N \times 2$ forcing amplitude matrix and $\mathbf{r} = \{r_p \;\; r_m\}^T$ is a 2×1 forcing vector with $r_p = \mathrm{e}^{\mathrm{i}\Omega t}$ and $r_m = \mathrm{e}^{-\mathrm{i}\Omega t}$. The subscripts p and m indicate the sign of the complex exponential term, plus and minus respectively. In contrast to the multiple scales analysis, here forcing at any frequency is considered, not just near-resonance forcing.

In the following analysis, while it is not essential, it is convenient to assume that the damping terms are small. Grouping the damping matrix term with the nonlinear term Γ_x gives

$$M\ddot{\mathbf{x}} + K\mathbf{x} + \mathbf{N}_x(\mathbf{x}, \dot{\mathbf{x}}, \mathbf{r}) = P_x\mathbf{r}, \tag{27}$$

where $\mathbf{N}_x = \mathbf{\Gamma}_x(\mathbf{x}, \dot{\mathbf{x}}, \mathbf{r}) + C\dot{\mathbf{x}}$. Note that $\mathbf{N}_x$ is small, and will later be written as order ε^1.

In the normal forms technique we use a series of transformations:

- Modal transformation using modes for the unforced linear equivalent system: $\mathbf{x} \to \mathbf{q}$.
- Force transformation: $\mathbf{q} \to \mathbf{v}$.
- Nonlinear near-identity transformation: $\mathbf{v} \to \mathbf{u}$.

The first of these transforms results in the linear modal representation of the equations of motion. In the modal form, the linear terms in these equations contain no cross-coupling. However, typically, cross-coupling does exist between the nonlinear terms. The second transform is also linear and, for each mode, removes any forcing terms in the modal equation of motion that are away from the modal resonance and places them in the $\mathbf{q} \to \mathbf{v}$ transform. The third transformation moves the non-resonant terms that are present in the nonlinear terms and places them in a nonlinear transformation.

4.1 Linear modal transformation: $\mathbf{x} \to \mathbf{q}$

First the undamped linear terms are decoupled using a linear modal transform. The unforced linear form of the equation of motion is

$$\ddot{\mathbf{x}} + M^{-1}K\mathbf{x} = 0.$$

This can be expressed as an eigenvalue/vector equation for matrix $M^{-1}K$,

$$\omega_{nn}^2\mathbf{X}_n = M^{-1}K\mathbf{X}_n,$$

where the eigenvalues are the squares of the undamped natural frequencies and the eigenvectors their corresponding mode shapes $\mathbf{X}$. Taking the mode shape matrix Φ, in which the nth column is $\mathbf{X}_n$, this equation can be written as

$$\Phi\Lambda = M^{-1}K\Phi, \tag{28}$$

where Λ is a diagonal eigenvalue matrix, in which the nth diagonal element is ω_{nn}^2. By applying the transform $\mathbf{x} = \Phi\mathbf{q}$, where $\mathbf{q}$ are the modal coordinates, and pre-multiplying by Φ^T, Equation 27 may be written as

$$(\Phi^T M\Phi)\ddot{\mathbf{q}} + (\Phi^T C\Phi)\dot{\mathbf{q}} + (\Phi^T K\Phi)\mathbf{q} + \Phi^T\mathbf{N}_x(\Phi\mathbf{q}, \Phi\dot{\mathbf{q}}, \mathbf{r}) = \Phi^T P_x\mathbf{r}.$$

Rearranging, and noting that by definition $M^{-1}K\Phi = \Phi\Lambda$, gives

$$\ddot{\mathbf{q}} + \Lambda\mathbf{q} + \mathbf{N}_q(\mathbf{q}, \dot{\mathbf{q}}, \mathbf{r}) = P_q\mathbf{r}, \tag{29}$$

where

$$N_q(\mathbf{q}, \dot{\mathbf{q}}, \mathbf{r}) = (\Phi^T M\Phi)^{-1}\Phi^T N_x(\Phi\mathbf{q}, \Phi\dot{\mathbf{q}}, \mathbf{r})$$
$$P_q = (\Phi^T M\Phi)^{-1}\Phi^T P_x.$$

Here the original equations of motion were for discrete locations, as are typical when using FE or spring-mass models. If, instead, the Galerkin technique is applied to partial differential equations (as was the case for the cable example), the resulting equations of motion are in the form of Equation 29, so this first transformation is unnecessary.

4.2 Force transformation: $\mathbf{q} \rightarrow \mathbf{v}$

Now a transform is applied to remove non-resonant forcing terms, i.e. for the nth mode remove forcing terms that are at frequencies well away from ω_{nn}. The transform takes the form

$$\mathbf{q} = \mathbf{v} + [e]\mathbf{r},$$

where $[e]$ has size $N \times 2$. Substituting this into the modal equation of motion, Equation 29, gives

$$\ddot{\mathbf{v}} + [e]WW\mathbf{r} + \Lambda\mathbf{v} + \Lambda[e]\mathbf{r} + \mathbf{N}_q(\mathbf{v} + [e]\mathbf{r}, \dot{\mathbf{v}} + [e]W\mathbf{r}, \mathbf{r}) = P_q\mathbf{r}.$$

Here W is a 2×2 diagonal matrix with the first and second diagonal values being $\mathrm{i}\Omega$ and $-\mathrm{i}\Omega$ respectively (recall that $\mathbf{r} = [\mathrm{e}^{\mathrm{i}\Omega t} \;\; \mathrm{e}^{-\mathrm{i}\Omega t}]^T$). This gives the dynamic equation in terms of $\mathbf{v}$

$$\ddot{\mathbf{v}} + \Lambda\mathbf{v} + \mathbf{N}_v(\mathbf{v}, \dot{\mathbf{v}}, \mathbf{r}) = P_v\mathbf{r},$$

where the nonlinear term has been transformed using

$$N_v(\mathbf{v}, \dot{\mathbf{v}}, \mathbf{r}) = N_q(\mathbf{v} + [e]\mathbf{r}, \dot{\mathbf{v}} + [e]W\mathbf{r}, \mathbf{r}).$$

The relationship between the forcing and the transformation matrix is

$$[e]WW + \Lambda[e] + P_v = P_q,$$

which can be rewritten as

$$[\tilde{e}] + P_v = P_q. \tag{30}$$

Recalling that Λ is a diagonal matrix with the nth diagonal element taking the value ω_{nn}^2, the nth row ($n = 1, 2 \ldots N$) and k^{th} column ($k = 1, 2$) of $[\tilde{e}]$ may be written in terms of the corresponding element in $[e]$ using

$$e_{n,k} = \tilde{e}_{n,k}/(\omega_{nn}^2 - \Omega^2). \tag{31}$$

One of two options are now chosen to satisfy Equation 30 element by element. For the $(n, k)^{\text{th}}$ element, if the forcing is close to the natural frequency (i.e. $\Omega \approx \omega_{nn}$), option 1 is used to keep the forcing in the equation of motion. This gives

$$\text{Option 1:} \quad e_{n,k} = 0, \quad P_{v,n,k} = P_{q,n,k}, \tag{32}$$

where the n, k subscripts indicates the $(n, k)^{\text{th}}$ element. If, however, the $(n, k)^{\text{th}}$ element corresponds to a forcing term that is away from resonance then option 2 is used to remove the forcing term from the equation of motion by writing

$$\text{Option 2:} \quad e_{n,k} = P_{q,n,k}/(\omega_{nn}^2 - \Omega^2), \quad P_{v,n,k} = 0, \tag{33}$$

which satisfies Equations 30 and 31.

4.3 Nonlinear near-identity transformation: $\mathbf{v} \to \mathbf{u}$

To reflect the assumption that the nonlinear and damping terms are small, $\mathbf{N}_v$ can be expressed as a power series of ε starting with an ε^1 term, giving

$$\begin{aligned} &\ddot{\mathbf{v}} + \Lambda\mathbf{v} + \mathbf{N}_v(\mathbf{v}, \dot{\mathbf{v}}, \mathbf{r}) = P_v\mathbf{r}, \\ \text{with:} \quad &\mathbf{N}_v(\mathbf{v}, \dot{\mathbf{v}}, \mathbf{r}) = \varepsilon\mathbf{n}_{v1}(\mathbf{v}, \dot{\mathbf{v}}, \mathbf{r}) + \varepsilon^2\mathbf{n}_{v2}(\mathbf{v}, \dot{\mathbf{v}}, \mathbf{r}) + \ldots \end{aligned} \tag{34}$$

Now the equation of motion is in a form in which the near-identity nonlinear transform can be applied. This transform takes the form $\mathbf{v} = \mathbf{u} + \mathbf{h}(\mathbf{u}, \dot{\mathbf{u}}, \mathbf{r})$.

To reflect the fact that this transform is near-identity, $\mathbf{h}$ is written as a power series of ε starting with an ε^1 term, yielding

$$\begin{aligned} &\mathbf{v} = \mathbf{u} + \mathbf{h}(\mathbf{u}, \dot{\mathbf{u}}, \mathbf{r}), \\ \text{with:} \quad &\mathbf{h} = \varepsilon \mathbf{h_1}(\mathbf{u}, \dot{\mathbf{u}}, \mathbf{r}) + \varepsilon^2 \mathbf{h_2}(\mathbf{u}, \dot{\mathbf{u}}, \mathbf{r}) + \dots . \end{aligned} \tag{35}$$

The purpose of applying this transform is to simplify the dynamic equation, Equation 30, into a form that can be solved using a single frequency trial solution for each mode, thereby eliminating the need for a harmonic balance type approximation. The transformed dynamic equation is expressed as

$$\begin{aligned} &\ddot{\mathbf{u}} + \Lambda \mathbf{u} + \mathbf{N}_u(\mathbf{u}, \dot{\mathbf{u}}, \mathbf{r}) = P_u \mathbf{r}, \\ \text{with:} \quad &\mathbf{N}_u(\mathbf{u}, \dot{\mathbf{u}}, \mathbf{r}) = \varepsilon \mathbf{n}_{u1}(\mathbf{u}, \dot{\mathbf{u}}, \mathbf{r}) + \varepsilon^2 \mathbf{n}_{u2}(\mathbf{u}, \dot{\mathbf{u}}, \mathbf{r}) + \dots . \end{aligned} \tag{36}$$

Again the nonlinear and damping terms have been expressed as a power series of ε starting with ε^1 to reflect the assumption that they are small. Ideally $\mathbf{N}_u = 0$, such that the equation is linear, but usually this can not be achieved without invalidating the assumption that the transform is near-identity.

Consider these three equations: the dynamic equation in $\mathbf{v}$, Equation 34; the transform equation, Equation 35; and the resulting dynamic equation, Equation 36. State vector $\mathbf{v}$ can be eliminated from Equation 34 using Equation 35 and then $\ddot{\mathbf{u}}$ can be eliminated using Equation 36, to produce

$$\begin{aligned} P_u \mathbf{r} - \varepsilon \mathbf{n}_{u1}(\mathbf{u}, \dot{\mathbf{u}}, \mathbf{r}) + \varepsilon \frac{\mathrm{d}^2}{\mathrm{d}t^2}(\mathbf{h}_1(\mathbf{u}, \dot{\mathbf{u}}, \mathbf{r})) + \varepsilon \Lambda \mathbf{h}_1(\mathbf{u}, \dot{\mathbf{u}}, \mathbf{r})) + \\ \varepsilon \mathbf{n}_{v1}(\mathbf{u}, \dot{\mathbf{u}}, \mathbf{r}) = P_v \mathbf{r} + \mathcal{O}(\varepsilon^2). \end{aligned}$$

Note that in deriving this equation the Taylor series expansion

$$\mathbf{n}_{v1}(\mathbf{u} + \varepsilon \mathbf{h}_1 + \dots, \dot{\mathbf{u}} + \varepsilon \frac{\mathrm{d}}{\mathrm{d}t}(\mathbf{h}_1) + \dots, \mathbf{r}) = \mathbf{n}_{v1}(\mathbf{u}, \dot{\mathbf{u}}, \mathbf{r}) + \mathcal{O}(\varepsilon^1)$$

has been used. Equating the zero and first-order powers of ε produces

$$\begin{aligned} \varepsilon^0 : \quad & P_u \mathbf{r} = P_v \mathbf{r}, \\ \varepsilon^1 : \quad & \mathbf{n}_{u1}(\mathbf{u}, \dot{\mathbf{u}}, \mathbf{r}) - \frac{\mathrm{d}^2}{\mathrm{d}t^2}(\mathbf{h}_1(\mathbf{u}, \dot{\mathbf{u}}, \mathbf{r})) = \Lambda \mathbf{h}_1(\mathbf{u}, \dot{\mathbf{u}}, \mathbf{r}) + \mathbf{n}_{v1}(\mathbf{u}, \dot{\mathbf{u}}, \mathbf{r}). \end{aligned} \tag{37}$$

The ε^0 equation is satisfied by setting $P_u = P_v$. To satisfy the ε^1 equation, the form of the response of the states $u_1, u_2 \dots u_N$ needs to be considered. The aim of the $\mathbf{v} \to \mathbf{u}$ transform is to remove non-resonant nonlinear terms from the equation of motion (their effects are represented in the transform), such that the response for each state $u_1, u_2 \dots u_N$ is at a single response frequency, $\omega_{r1}, \omega_{r2}, \dots \omega_{rN}$. Since the differential equation in $\mathbf{u}$ is second-order, the trial solutions for the states must consist of both positive and negative complex exponential terms. The state vector $\mathbf{u}$ is therefore split into components $\mathbf{u} = \mathbf{u}_p + \mathbf{u}_m$ allowing the trial solutions for the nth state to be

$$u_n = u_{np} + u_{nm} : \quad u_{np} = (U_n \mathrm{e}^{-\mathrm{i}\phi_n}/2)\mathrm{e}^{\mathrm{i}\omega_{rn}t}, \quad u_{nm} = (U_n \mathrm{e}^{\mathrm{i}\phi_n}/2)\mathrm{e}^{-\mathrm{i}\omega_{rn}t}, \tag{38}$$

for $1 \leq n \leq N$ where the ω_{rn} terms are positive. This results in the form of solution $u_n = U_n \cos(\omega_{rn}t - \phi_n)$ and therefore U_n is taken to be real to ensure a real response to the real excitation. The time derivatives of $\mathbf{u}$ may now be written as $\dot{\mathbf{u}} = \Upsilon(\mathbf{u}_p - \mathbf{u}_n)$ and $\ddot{\mathbf{u}} = \Upsilon^2(\mathbf{u}_p + \mathbf{u}_n)$, where Υ is a diagonal matrix with the nth diagonal element being $\mathrm{i}\omega_{rn}$.

Since, after applying the forcing transformation $\mathbf{q} \rightarrow \mathbf{v}$, only the resonant forcing terms are present in the dynamic equation, it is expected that the response of u_n will be close to the nth natural frequency (i.e. it will be resonant). Therefore, we can say $\omega_{rn} \approx \omega_n$. A frequency detuning parameter can be introduced with the result that, to order ε, $\Lambda = -\Upsilon^2$ can be written (see Neild and Wagg, 2011, for more details). This allows the order ε^1 equation to be expressed as

$$\mathbf{n}_{u1}(\mathbf{u}, \dot{\mathbf{u}}, \mathbf{r}) - \frac{\mathrm{d}^2}{\mathrm{d}t^2}(\mathbf{h}_1(\mathbf{u}, \dot{\mathbf{u}}, \mathbf{r})) = -\Upsilon^2 \mathbf{h}_1(\mathbf{u}, \dot{\mathbf{u}}, \mathbf{r}) + \mathbf{n}_{v1}(\mathbf{u}, \dot{\mathbf{u}}, \mathbf{r}). \tag{39}$$

This detuning approximation does not effect the form of the dynamic equation in $\mathbf{u}$, but it does improve the prediction of the response amplitudes at harmonics of the response frequency – this is discussed in Xie *et al.* (2011).

To proceed, a vector $\mathbf{u}^*$ (of length L) is specified. It contains all the combinations of u_{np}, u_{nm} $(1 \leq n \leq N)$, r_p and r_m terms that are present in $\mathbf{n}_{v1}(\mathbf{u}, \dot{\mathbf{u}}, \mathbf{r})$. This allows the following matrix expressions to be defined

$$\begin{aligned} \mathbf{n}_{v1}(\mathbf{u}, \dot{\mathbf{u}}, \mathbf{r}) &= [n_v]\mathbf{u}^*(\mathbf{u_p}, \mathbf{u_m}, \mathbf{r}), \\ \mathbf{n}_{u1}(\mathbf{u}, \dot{\mathbf{u}}, \mathbf{r}) &= [n_u]\mathbf{u}^*(\mathbf{u_p}, \mathbf{u_m}, \mathbf{r}), \\ \mathbf{h}_1(\mathbf{u}, \dot{\mathbf{u}}, \mathbf{r}) &= [h]\mathbf{u}^*(\mathbf{u_p}, \mathbf{u_m}, \mathbf{r}), \end{aligned} \tag{40}$$

where $[n_v]$, $[n_u]$ and $[h]$ are of size $N \times L$ and $[n_u]$ and $[h]$ are currently unknown. This results in the ε^1 equation (from Equation 39)

$$[n_u]\mathbf{u}^* - [h]\frac{\mathrm{d}^2\mathbf{u}^*}{\mathrm{d}t^2} = -\Upsilon^2[h]\mathbf{u}^* + [n_v]\mathbf{u}^*. \tag{41}$$

To simplify the double derivative term, the form of $\mathbf{u}^*$ is considered. The general form of the ℓth element in vector $\mathbf{u}^*$ may be written as

$$u_\ell^* = r_p^{m_{\ell p}} r_m^{m_{\ell m}} \prod_{n=1}^{N} \left\{ u_{np}^{s_{\ell np}} u_{nm}^{s_{\ell nm}} \right\}.$$

The derivative of u_ℓ^* is therefore

$$\begin{aligned} \frac{\mathrm{d}u_\ell^*}{\mathrm{d}t} &= i\Omega \left(\frac{\partial u_\ell^*}{\partial r_p} r_p - \frac{\partial u_\ell^*}{\partial r_m} r_m \right) + \sum_{n=1}^{N} \mathrm{i}\omega_{rn} \left(\frac{\partial u_\ell^*}{\partial u_{np}} u_{np} - \frac{\partial u_\ell^*}{\partial u_{nm}} u_{nm} \right) \\ &= \mathrm{i} \left[(m_{\ell p} - m_{\ell m})\Omega + \sum_{n=1}^{N} \{ (s_{\ell np} - s_{\ell nm})\omega_{rn} \} \right] u_\ell^* = \tilde{\Upsilon}_\ell u_\ell^*. \qquad (42) \end{aligned}$$

The key feature of this equation is that the derivative of u_ℓ^* is linearly related to u_ℓ^*. Hence the double time derivative of $\mathbf{u}^*$ can be written as

$$\frac{\mathrm{d}^2\mathbf{u}^*}{\mathrm{d}t^2} = \tilde{\Upsilon}^2 \mathbf{u}^*$$

where $\tilde{\Upsilon}$ is a diagonal matrix of size $L \times L$ in which the ℓth diagonal element is given by $\tilde{\Upsilon}_\ell$ (see Equation 42).

Using this information, Equation 41 may now be written as

$$[n_u] = [n_v] - [\tilde{h}] \quad \text{where} \quad [\tilde{h}] = \Upsilon^2[h] - [h]\tilde{\Upsilon}^2. \qquad (43)$$

In this equation, as with $[h]$, $[\tilde{h}]$ is size $N \times L$. The element in the nth row and ℓ^{th} column of $[\tilde{h}]$ may be written as

$$\tilde{h}_{n,\ell} = \left(\left[(m_{\ell p} - m_{\ell m})\Omega + \sum_{n=1}^{N} \{ (s_{\ell np} - s_{\ell nm})\omega_{rn} \} \right]^2 - \omega_{rn}^2 \right) h_{n,\ell} = \beta_{n,\ell} h_{n,\ell}, \qquad (44)$$

using Equation 42 and recalling Υ is a diagonal matrix with the nth diagonal element being $\mathrm{i}\omega_{rn}$. In this equation $h_{n,\ell}$ and $\beta_{n,\ell}$ are the elements in the nth row and ℓ^{th} column of matrices $[h]$ and $[\beta]$ respectively. Note $[\tilde{h}] \neq [\beta][h]$, instead $\tilde{h}_{n,\ell} = \beta_{n,\ell} h_{n,\ell}$.

Now $[n_u]$ and $[h]$ can be selected by considering the size of the $\beta_{n,\ell}$ terms. It is desirable for $[n_u]$ to contain as many zeros as possible so that the dynamic equation in $\mathbf{u}$ is as simple as possible. The restriction on this is that the nonlinear terms in the near-identity transform must be small, of order ε^1 or higher.

There are, therefore, two options to satisfy Equation 43. Where possible each term in $[n_u]$ is set to zero:

$$\text{Option 1 (preferred):} \quad n_{u,n,\ell} = 0, \quad h_{n,\ell} = n_{v,n,\ell}/\beta_{n,\ell}, \tag{45}$$

in which, for example, $n_{u,n,\ell}$ is the (n,ℓ) element in $[n_u]$. However, in the cases where the term in $\mathbf{u}$ is *near-resonant*, $\beta_{n,\ell}$ is small and hence $h_{n,\ell}$ would be large if $n_{u,n,\ell} = 0$ is set. To avoid breaking the near-identity constraint, these terms are kept in the equation of motion by setting

$$\text{Option 2 (near-resonant terms):} \quad n_{u,n,\ell} = n_{v,n,\ell}, \quad h_{n,\ell} = 0, \tag{46}$$

i.e. these terms are not affected by the transform.

By adopting this method, the equation of motion for $\mathbf{u}$ can be solved exactly using the trial solution in the form $u_n = U_n \cos(\omega_{rn} t - \phi_n)$ for the nth mode. Information regarding the response of each mode at other frequencies is contained within the transform equation $\mathbf{v} = \mathbf{u} + \mathbf{h}(u, \dot{u}, r)$.

In the next section the normal form technique will be applied to the Duffing oscillator and then in Section 7 to a two-degree-of-freedom system.

5 Duffing Oscillator

The Duffing oscillator dynamics, which contain a cubic stiffness nonlinearity,

$$\ddot{x} + 2\zeta\omega_n\dot{x} + \omega_n^2 x + \alpha x^3 = P\cos(\Omega t + \phi), \tag{47}$$

appear in many different nonlinear systems. Two closely related example applications, in which this cubic stiffness nonlinearity is deliberately introduced into a system to improve performance, are in energy harvesting and passive structural control.

Energy harvesting is where energy is extracted from the environment and used to power devices. The energy source might be for example heat, light or structural vibration. In the case of structural vibration, one method of extracting energy from a vibrating structure is to attach a tuned mass oscillator to the structure. Being tuned, the motion of the mass will be large and, if the mass is attached to a set of magnets, this motion can be used to induce voltage in a nearby coil, thereby extracting energy. A schematic of such a device is shown in Figure 9, in which the magnets are arranged such that there is a full reversal of magnetic flux once every oscillation of the cantilever (see Barton *et al.*, 2010). By using an iron-core stator the magnet interacts with the stator with the effect of reducing the effective stiffness at low amplitudes. This gives an equation of motion in the form

$$m\ddot{y} + c\dot{y} + ky + \gamma(y, \dot{y}) = -m\ddot{\delta}, \tag{48}$$

where y is the displacement of the mass relative to the structure, δ is the displacement of the structure and c and k contain both cantilever and linear magnetic effects (including the coupling with the coil which has the effect of introducing additional damping). The term $\gamma(y, \dot{y})$ describes the nonlinear magnetic interaction with the coil and may be approximated to $\gamma(y, \dot{y}) = \alpha y^3$. Hence the system is a forced Duffing Oscillator with a support excitation (see Barton *et al.*, 2010). It is desirable to have large relative motion y as the current induced in the coil is proportional to $\dot{y}$. Note this assumes that the motion of the structure is unaffected by the presence of the energy harvesting device, a reasonable assumption for a large structure containing an energy harvester designed to extract energy for, say, powering a structural monitoring transducer. The nonlinear term, which can be altered via redesign of the stator, is potentially beneficial to the performance of the harvesting device as the bandwidth of the device can be increased in its presence.

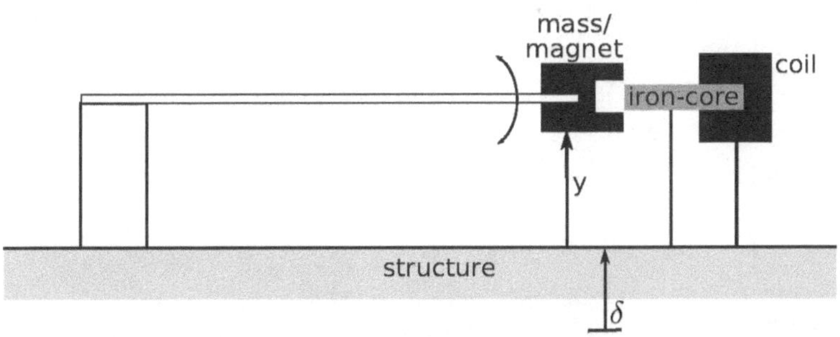

Figure 9. Schematic of a cantilevered vibrating mass energy harvester attached to a structure.

A second application, in which the presence of a cubic stiffness nonlinearity can potentially be beneficial to the performance of a system, is in the field of vibration isolation. It is often desirable for a mount between an oscillating structure and a device to be designed such that, despite the motion of the structure, the motion of the device is minimised. Typically, this isolation mount will have stiffness and damping properties which, coupled with the device's mass, will result in resonant behaviour. In steady-state operation, the mount will be designed such that it does not resonate at the oscillating frequency of the structure. However if, for example, the source of the structural vibration is a machine and the mount is designed to have low natural frequency for good high frequency performance, then at start-up or shut-down of the machine the excitation of the mount will pass through its resonance. Close to resonance it is desirable for y to be small such that the overall motion of the device is small. A cubic nonlinearity can potentially reduce the effects of passing through this resonance, as discussed, for example, in Kovacic *et al.* (2008).

To show how the presence of nonlinearity can be beneficial in both these applications, consider the standard Duffing equation, Equation 47. Figure 10 shows the response of the system when linear and when nonlinear. For illustrative purposes, arbitrarily defining the bandwidth as being limited to the region where the response exceeds 0.5, it can be seen that the presence of nonlinearity affects the bandwidth. If the solution remains on the upper curve then the bandwidth is increased. This is useful for the energy harvesting application, where it is desirable to have a large response over a reasonable bandwidth so that, in the event of detuning of the excitation frequency, the device will still operate well. In contrast, for vibration isolation it is desirable to be on the lower curve, because it results in only a small relative motion across the resonance peak.

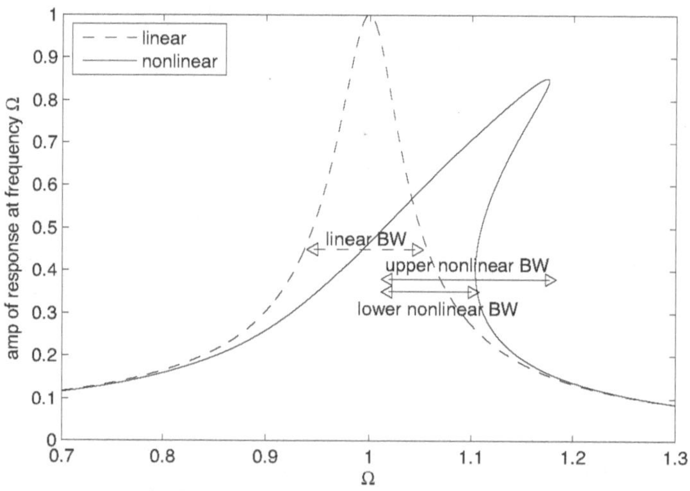

Figure 10. Response of the Duffing oscillator showing the potential bandwidth for the two cases of $\alpha = 0$ and $\alpha = 0.7$, with $\omega_n = 1$, $\zeta = 0.03$ and a forcing amplitude of 0.06.

Here, aspects of the forced Duffing oscillator are studied as an example of a typical nonlinearity that can be introduced into structures to improve their performance. Both the multiple scales and the normal forms techniques are considered.

5.1 Application of Multiple Scales

The equation of motion for a forced Duffing oscillator is given by

$$\ddot{x} + 2\varepsilon\hat{\zeta}\omega_n\dot{x} + \omega_n^2 x + \varepsilon\hat{\gamma}_x(x,\dot{x}) = \varepsilon\hat{P}\cos(\Omega t), \qquad \hat{\gamma}_x = \hat{\alpha}x^3,$$

where we assume that the forcing is close to resonance such that $P = \varepsilon\hat{P}$ and a small detuning term can be introduced: $\Omega = \omega_n(1+\mu)$ with $\mu = \varepsilon\hat{\mu}$. In addition, since the forcing is near resonance, the response frequency, ω_r, is set to equal the forcing frequency, Ω.

The time derivatives of x that have been generated by introducing (1) the power series expansion of x, $x = x_0(t) + \varepsilon x_1(t) + \ldots$, (2) the fast and slow timescales, $x_0(t) = x_0(t_f, t_s)$ etc with $t_f = \Omega t$ and $t_s = \varepsilon t$ respectively, and (3) the detuning parameter, may now be substituted into the equation of motion. These time derivatives, Equation 24, can be substituted into the equation of motion to yield

$$\begin{aligned}\left[\omega_n^2\frac{\partial^2 x_0}{\partial t_f^2} + \varepsilon\left(2\mu\omega_n^2\frac{\partial^2 x_0}{\partial t_f^2} + \omega_n^2\frac{\partial^2 x_1}{\partial t_f^2} + 2\omega_n\frac{\partial^2 x_0}{\partial t_f\partial t_s}\right)\right] &+ \\ 2\varepsilon\hat{\zeta}\omega_n\left[\omega_n\frac{\partial x_0}{\partial t_f}\right] + \omega_n^2\left[x_0 + \varepsilon x_1\right] + \varepsilon\hat{\alpha}\left[x_0^3\right] + \mathcal{O}(\varepsilon^2) &= \varepsilon\hat{P}\cos(t_f)\end{aligned}$$

Balancing the ε^0 and ε^1 terms and ignoring higher order terms, produces

$$\begin{aligned}\varepsilon^0: \quad \omega_n^2\frac{\partial^2 x_0}{\partial t_f^2} + \omega_n^2 x_0 &= 0,\\ \varepsilon^1: \quad \omega_n^2\frac{\partial^2 x_1}{\partial t_f^2} + \omega_n^2 x_1 &= -\omega_n^2 2\hat{\mu}\frac{\partial^2 x_0}{\partial t_f^2} - 2\omega_n\frac{\partial^2 x_0}{\partial t_f\partial t_s}\\ &\quad -2\hat{\zeta}\omega_n^2\frac{\partial x_0}{\partial t_f} - \hat{\alpha}x_0^3 + \hat{P}\cos(t_f),\end{aligned}$$

The solution to the ε^0 equation is

$$x_0 = X_{0c}(t_s)\cos(t_f) + X_{0s}(t_s)\sin(t_f).$$

Substituting this solution for x_0 into the ε^1 equation, gives

$$\begin{aligned}\omega_n^2 \frac{\partial^2 x_1}{\partial t_f^2} + \omega_n^2 x_1 = & \left(2\omega_n^2 \hat{\mu} X_{0s} + 2\omega_n \frac{\mathrm{d}X_{0c}}{\mathrm{d}t_s} + 2\omega_n^2 \hat{\zeta} X_{0c} - \frac{3}{4}\hat{\alpha} X_0^2 X_{0s}\right)\sin(t_f) \\ & + \left(2\omega_n^2 \hat{\mu} X_{0c} - 2\omega_n \frac{\mathrm{d}X_{0s}}{\mathrm{d}t_s} - 2\omega_n^2 \hat{\zeta} X_{0s} - \frac{3}{4}\hat{\alpha} X_0^2 X_{0c} + \hat{P}\right)\cos(t_f) \\ & + \frac{\hat{\alpha}}{4}(3X_{0s}^2 - X_{0c}^2) X_{0c}\cos(3t_f) + \frac{\hat{\alpha}}{4}(X_{0s}^2 - 3X_{0c}^2) X_{0s}\sin(3t_f),\end{aligned} \tag{49}$$

where $X_0 = \sqrt{X_{0c}^2 + X_{0s}^2}$ is the amplitude of the x_0 response.

To ensure the response at the forcing frequency Ω is contained entirely in x_0, the amplitudes of the $\cos(t_f)$ and $\sin(t_s)$ terms on the right-hand side of Equation 49 must be set to zero. This results in the following conditions on X_{0c} and X_{0s}

$$\begin{aligned}\frac{\mathrm{d}X_{0c}}{\mathrm{d}t_s} &= -\frac{1}{\omega_n}\left(\omega_n^2 \hat{\mu} X_{0s} + \omega_n^2 \hat{\zeta} X_{0c} - \frac{3}{8}\hat{\alpha} X_0^2 X_{0s}\right), \\ \frac{\mathrm{d}X_{0s}}{\mathrm{d}t_s} &= \frac{1}{\omega_n}\left(\omega_n^2 \hat{\mu} X_{0c} - \omega_n^2 \hat{\zeta} X_{0s} - \frac{3}{8}\hat{\alpha} X_0^2 X_{0c} + \frac{1}{2}\hat{P}\right).\end{aligned}$$

To find the steady-state solution, the left-hand side of each of these equations is set to zero. This produces

$$\begin{aligned}\left(\omega_n^2 \hat{\mu} - \frac{3}{8}\hat{\alpha} X_0^2\right) X_{0s} + \left(\omega_n^2 \hat{\zeta}\right) X_{0c} &= 0, \\ \left(\omega_n^2 \hat{\mu} - \frac{3}{8}\hat{\alpha} X_0^2\right) X_{0c} - \left(\omega_n^2 \hat{\zeta}\right) X_{0s} &= -\frac{1}{2}\hat{P}.\end{aligned}$$

Note that all the terms contain one constant with a circumflex. These can all be replaced with the original constants by multiplying both equations by ε. Squaring and adding these two equations gives an equation relating the amplitude of response to the forcing amplitude and frequency:

$$X_0^2 \left[\left(\omega_n^2 \mu - \frac{3}{8}\alpha X_0^2\right)^2 + \left(\omega_n^2 \zeta\right)^2\right] = \frac{1}{4}P^2. \tag{50}$$

This can be solved by considering a range of X_0 values and finding the corresponding P for a given μ (if the response to excitation amplitude relationship is wanted), or solving the quadratic in μ, which has either two or no real solutions, for a given P (to give a resonance curve for a given excitation amplitude).

Since the multiple scales technique results in first-order differential equations for the amplitudes of the response X_{0c} and X_{0s}, these equations can also be used directly to study the stability of the solutions. Expressing the differential equations in matrix form gives

$$\frac{\mathrm{d}}{\mathrm{d}t_s}\left\{\begin{array}{c} X_{0c} \\ X_{0s} \end{array}\right\} = \frac{1}{\omega_n}\left\{\begin{array}{c} -\omega_n^2\hat{\mu}X_{0s} - \omega_n^2\hat{\zeta}X_{0c} + \frac{3}{8}\hat{\alpha}X_0^2X_{0s}, \\ \omega_n^2\hat{\mu}X_{0c} - \omega_n^2\hat{\zeta}X_{0s} - \frac{3}{8}\hat{\alpha}X_0^2X_{0c} + \frac{1}{2}\hat{P} \end{array}\right\}$$

$$\text{or:}\quad \frac{\mathrm{d}\mathbf{X}}{\mathrm{d}t_s} = f(\mathbf{X}, t).$$

To study the local stability of a steady-state solution of a system, the system is perturbed a small amount away from the solution and, if locally stable, the system will be attracted back to the solution. Taking the steady-state solution to be $\mathbf{X} = \bar{\mathbf{X}}$, then consider the system perturbed by a small amount, $\mathbf{X_p}$, away from the solution such that $\mathbf{X} = \bar{\mathbf{X}} + \mathbf{X_p}$. We can then write

$$\frac{\mathrm{d}}{\mathrm{d}t_s}\{\bar{\mathbf{X}} + \mathbf{X_p}\} = f(\bar{\mathbf{X}} + \mathbf{X_p}, t) = f(\bar{\mathbf{X}}, t) + Df_x(\bar{\mathbf{X}}, t)\mathbf{X_p},$$

where a Taylor series expansion has been used and where $Df_x(\mathbf{X}, t)$ is the Jacobian of f evaluated at $\mathbf{X}$. By definition $\mathrm{d}\bar{\mathbf{X}}/\mathrm{d}t_s = f(\bar{\mathbf{X}}, t)$, hence this can be removed from the equation to give

$$\frac{\mathrm{d}\mathbf{X_p}}{\mathrm{d}t_s} = Df_x(\bar{\mathbf{X}}, t)\mathbf{X_p}.$$

Therefore the eigenvalues of the Jacobian, $Df_x(\bar{\mathbf{X}}, t)$, determine the stability of the steady-state solution $\bar{\mathbf{X}}$. For a stable solution, the real part of the eigenvalues must be negative such that the perturbation decays away and the system returns to the solution $\bar{\mathbf{X}}$. The Jacobian is given by

$$Df_x(\mathbf{X}, t) = \frac{1}{\omega_n}\left[\begin{array}{cc} -\omega_n^2\hat{\zeta} + \frac{3}{4}\hat{\alpha}X_{0c}X_{0s} & -\omega_n^2\hat{\mu} + \frac{3}{8}\hat{\alpha}(3X_{0s}^2 + X_{0c}^2) \\ \omega_n^2\hat{\mu} - \frac{3}{8}\hat{\alpha}(3X_{0c}^2 + X_{0s}^2) & -\omega_n^2\hat{\zeta} - \frac{3}{4}\hat{\alpha}X_{0c}X_{0s} \end{array}\right],$$

recalling that $X_0^2 = X_{0c}^2 + X_{0s}^2$. The eigenvalue equation for the Jacobian matrix is

$$\lambda^2 + 2\omega_n\hat{\zeta}\lambda^1 + \left[\omega_n^2(\hat{\zeta}^2 + \hat{\mu}^2) - \frac{3}{2}\hat{\mu}\hat{\alpha}X_0^2 + \frac{27}{64\omega_n^2}\hat{\alpha}^2X_0^4\right]\lambda^0 = 0,$$

where λ is an eigenvalue. Since the λ^2 and λ^1 coefficents are positive, the boundary of stability is when the λ^0 coefficent is zero giving

$$(\zeta^2 + \mu^2)\omega_n^4 - \frac{3}{2}\mu\omega_n^2\alpha X_0^2 + \frac{27}{64}\alpha^2X_0^4 = 0,$$

Consider the case where the frequency is kept constant and the excitation is increased. If the excitation frequency is slightly above the natural frequency, multiple solutions can exist and the shape of the response is similar to that for the cable example shown in Figure 4. Using Equation 50, the fold points can be identified (points A and C in Figure 4) by finding $\partial P^2/\partial X_0^2$ and setting it to zero. This yields

$$\frac{\partial P^2}{\partial X_0^2} = (\zeta^2 + \mu^2)\omega_n^4 - \frac{3}{2}\mu\omega_n^2\alpha X_0^2 + \frac{27}{64}\alpha^2 X_0^4 = 0.$$

This is identical to the equation for the points where the local stability of the solution is lost. Hence the fold points correspond to the points where solution stability is reversing. This analysis has used the condition that if $\partial P^2/\partial X_0^2 = 0$ then $\partial P/\partial X_0 = 0$.

In addition, multiple scales gives information regarding the response at other frequencies. Since the $\cos(t_f)$ and $\sin(t_f)$ terms on the right-hand side of Equation 49 have been set to zero, the equation simplifies to

$$\omega_n^2 \frac{\partial^2 x_1}{\partial t_f^2} + \omega_n^2 x_1 = \frac{\hat{\alpha}}{4}\left((3X_{0s}^2 - X_{0c}^2)X_{0c}\cos(3t_f) + (X_{0s}^2 - 3X_{0c}^2)X_{0s}\sin(3t_f)\right),$$

which is a linear differential equation in x_1 that can be solved in the usual way.

5.2 Application of Normal Forms

Reconsider the system

$$\ddot{x} + 2\varepsilon\hat{\zeta}\omega_n\dot{x} + \omega_n^2 x + \varepsilon\hat{\alpha}x^3 = P\cos(\Omega t) = P_x\mathbf{r},$$

where both the damping and the nonlinearity αx^3 are small and $P_x = [P/2 \quad P/2]$. Since the system has a single degree-of-freedom, the linear modal transform $\mathbf{x} \to \mathbf{q}$ is a unity transform giving

$$\ddot{q} + 2\varepsilon\hat{\zeta}\omega_n\dot{q} + \omega_n^2 q + \varepsilon\hat{\alpha}q^3 = P_q\mathbf{r},$$

where $\mathbf{q} = q$ and $P_q = P_x$.

Now the forcing transform $\mathbf{q} = \mathbf{v} + [e]\mathbf{r}$ is applied, where for this example $[e]$ is a 1×2 matrix. Using Equation 31, it can be seen that if the forcing is near resonance then $[e] = [0 \quad 0]$ and $P_v = P_q$ (i.e. option 1, Equation 32). Otherwise, with forcing away from resonance, $[e] = P_q/(\omega_n^2 - \Omega^2)$ and $P_v = 0$ (i.e. option 2, Equation 33). So, considering near-resonant forcing produces

$$\ddot{v} + 2\varepsilon\hat{\zeta}\omega_n\dot{v} + \omega_n^2 v + \varepsilon\hat{\alpha}v^3 = P_v\mathbf{r},$$

with $\mathbf{q} = \mathbf{v}$ where $\mathbf{v} = v$ and $P_v = P_q$.

The near-identity transformation, $\mathbf{u} = \mathbf{v} + \mathbf{h}(\mathbf{u}, \dot{\mathbf{u}}, \mathbf{r})$, is now applied. The nonlinear and damping terms are given by

$$\mathbf{n}_{v1}(\mathbf{v}, \dot{\mathbf{v}}, \mathbf{r}) = 2\hat{\zeta}\omega_n \dot{v} + \hat{\alpha} v^3. \tag{51}$$

To select a suitable transformation, $\mathbf{n}_{v1}(\mathbf{u}, \dot{\mathbf{u}}, \mathbf{r})$ is considered. From Equation 40, it is rewritten as

$$\mathbf{n}_{v1}(\mathbf{u}, \dot{\mathbf{u}}, \mathbf{r}) = [n_v]\mathbf{u}^*(\mathbf{u_p}, \mathbf{u_m}, \mathbf{r}),$$

where $\mathbf{u} = \mathbf{u_p} + \mathbf{u_m}$. Considering the terms in $\mathbf{n}_{v1}$, $[n_v]$ and $\mathbf{u}^*$ are

$$\begin{aligned} [n_v] &= \left[\begin{array}{cccccc} \mathrm{i}2\hat{\zeta}\omega_n\omega_r & -\mathrm{i}2\hat{\zeta}\omega_n\omega_r & \hat{\alpha} & 3\hat{\alpha} & 3\hat{\alpha} & \hat{\alpha} \end{array}\right], \\ \mathbf{u}^* &= \left\{\begin{array}{cccccc} u_p & u_m & u_p^3 & u_p^2 u_m & u_p u_m^2 & u_m^3 \end{array}\right\}^T. \end{aligned}$$

From Equation 44, matrix $[\beta]$ is

$$[\beta] = \omega_{rn}^2 \left[\begin{array}{cccccc} 0 & 0 & 8 & 0 & 0 & 8 \end{array}\right].$$

This matrix is used to select whether option 1, Equation 45, is used to transform the nonlinear terms out of the equation of motion, or whether option 2, Equation 46, is used to avoid contravening the assumption that the transform is a near-identity one. Option 1 is selected for the 3rd and 6th elements, for the other elements option 2 must be selected as $[\beta]$ is zero for these elements. This gives

$$\begin{aligned} [h] &= \left[\begin{array}{cccccc} 0 & 0 & \frac{\hat{\alpha}}{8\omega_{rn}^2} & 0 & 0 & \frac{\hat{\alpha}}{8\omega_{rn}^2} \end{array}\right] \\ [n_u] &= \left[\begin{array}{cccccc} \mathrm{i}2\hat{\zeta}\omega_n\omega_r & -\mathrm{i}2\hat{\zeta}\omega_n\omega_r & 0 & 3\hat{\alpha} & 3\hat{\alpha} & 0 \end{array}\right] \end{aligned}$$

A new equation for the dynamics can now be written (from Equation 36 and the definition of $[n_u]$, Equation 40) as

$$\ddot{u} + 2\zeta\omega_n\dot{u} + \omega_n^2 u + 3\alpha\left[u_p^2 u_m + u_p u_m^2\right] = P_u \mathbf{r},$$

where $P_u = P_v = P_q = P_x = [P/2 \;\; P/2]$. Combining the transformations, using Equations 35 and 40, yields

$$\mathbf{x} = \mathbf{q} = \mathbf{v} = \mathbf{u} + \frac{\alpha}{8\omega_r^2}\left[u_p^3 + u_m^3\right]$$

Concentrating on the dynamic equation, algebraically it is convenient to introduce a linear time-shift, such that, rather than having a response

$$u = u_p + u_m: \quad u_p = \left(\frac{U}{2}\mathrm{e}^{-\mathrm{i}\phi}\right)\mathrm{e}^{\mathrm{i}\omega_r t}, \quad u_m = \left(\frac{U}{2}\mathrm{e}^{\mathrm{i}\phi_n}\right)\mathrm{e}^{-\mathrm{i}\omega_r t}$$

from Equation 38, the response is a pure cosine. This can be achieved by shifting the time origin using $t \rightarrow t + \phi/\omega_r$, to give

$$u = u_p + u_m : \quad u_p = \frac{U}{2}\mathrm{e}^{\mathrm{i}\omega_r t}, \quad u_m = \frac{U}{2}\mathrm{e}^{-\mathrm{i}\omega_r t},$$

Applying the same time-shift to the forcing results in $p = P\cos(\Omega t) \rightarrow p = P\cos(\Omega t + \phi)$, or

$$P_u = \left[\begin{array}{cc} \frac{P}{2} & \frac{P}{2} \end{array}\right] \quad \rightarrow \quad P_u = \left[\begin{array}{cc} \mathrm{e}^{\mathrm{i}\phi}\frac{P}{2} & \mathrm{e}^{-\mathrm{i}\phi}\frac{P}{2} \end{array}\right].$$

Substituting the time-shifted expressions for the forcing and response into the equation of motion gives

$$\begin{aligned}[\omega_n^2 - \Omega^2]\frac{U}{2}[\mathrm{e}^{\mathrm{i}\Omega t} + \mathrm{e}^{-\mathrm{i}\Omega t}] + \mathrm{i}2\zeta\omega_n\Omega\frac{U}{2}[\mathrm{e}^{\mathrm{i}\Omega t} - \mathrm{e}^{-\mathrm{i}\Omega t}] + \frac{3\alpha U^3}{8}[\mathrm{e}^{\mathrm{i}\Omega t} + \mathrm{e}^{-\mathrm{i}\Omega t}] \\ = \frac{P}{2}[\mathrm{e}^{\mathrm{i}\phi}\mathrm{e}^{\mathrm{i}\Omega t} + \mathrm{e}^{-\mathrm{i}\phi}\mathrm{e}^{-\mathrm{i}\Omega t}].\end{aligned}$$

It can be seen that, owing to the nonlinear transform, there are no higher frequency terms. Hence the equations can be balanced exactly by considering the $\mathrm{e}^{\mathrm{i}\Omega t}$ and $\mathrm{e}^{-\mathrm{i}\Omega t}$ terms:

$$\begin{aligned}[\omega_n^2 - \Omega^2]\frac{U}{2} + \mathrm{i}2\zeta\omega_n\Omega\frac{U}{2} + \frac{3\alpha U^3}{8} &= \frac{P}{2}\mathrm{e}^{\mathrm{i}\phi}, \\ [\omega_n^2 - \Omega^2]\frac{U}{2} - \mathrm{i}2\zeta\omega_n\Omega\frac{U}{2} + \frac{3\alpha U^3}{8} &= \frac{P}{2}\mathrm{e}^{-\mathrm{i}\phi}.\end{aligned}$$

Noting that these equations are a complex conjugate pair, the real and imaginary components can be taken, which yields

$$[\omega_n^2 - \Omega^2]\frac{U}{2} + \frac{3\alpha U^3}{8} = \frac{P}{2}\cos(\phi), \quad 2\zeta\omega_n\Omega\frac{U}{2} = \frac{P}{2}\sin(\phi).$$

Squaring and adding these equations eliminates ϕ to give

$$\left([\omega_n^2 - \Omega^2]\frac{U}{2} + \frac{3\alpha U^3}{8}\right)^2 + (\zeta\omega_n\Omega U)^2 = \left(\frac{P}{2}\right)^2,$$

which can be used to calculate the response amplitude based on the forcing amplitude and frequency.

The transformation contains the additional frequency terms. This can be seen by substituting in for u to give

$$x = U\cos(\Omega t) + \frac{\alpha U^3}{32\omega_r^2}\cos(3\Omega t).$$

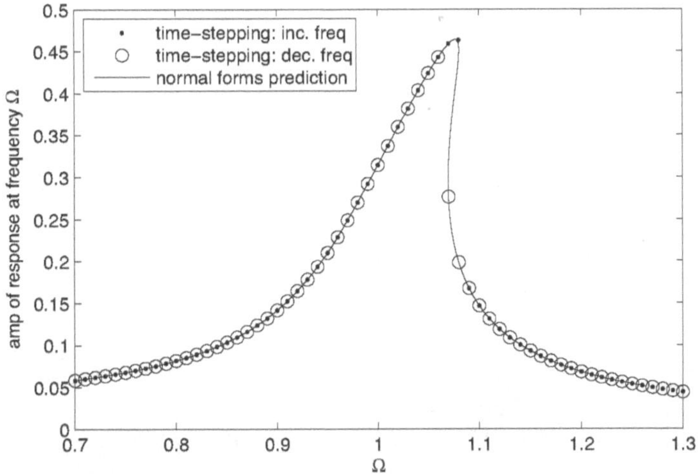

Figure 11. Amplitude of response at the forcing frequency, Ω, as a function of forcing frequency for the nonlinear oscillator – a comparison between the normal form prediction and a time-stepping simulation ($P = 0.03$, $\alpha = 1$, $\omega_n = 1$, $\zeta = 0.03$).

This implies that the response amplitude at the driving frequency is unaffected by the transform – it remains U.

Hence, the normal forms technique provides expressions for the amplitude of response at the driving frequency and at the harmonics, in this example at three times the driving frequency.

The predictions given by these equations can be compared to a time-stepping simulation of the dynamics of the system. Figures 11 and 12 show the amplitude of response at the driving and at three times the driving frequency respectively. It can be seen that the agreement is very good.

6 Acoustics Based Damage Detection

Another example of utilising nonlinear behaviour is the detection of damage in structures. Various techniques exist, or are being developed, to monitor the state of structures either globally using low frequency vibration or more locally using acoustics or ultrasonic excitation.

Research has been conducted on vibration-based nonlinear damage detection where, for example, cracks can open and close during vibration. During bending, when open the cracks reduce the stiffness of the structure

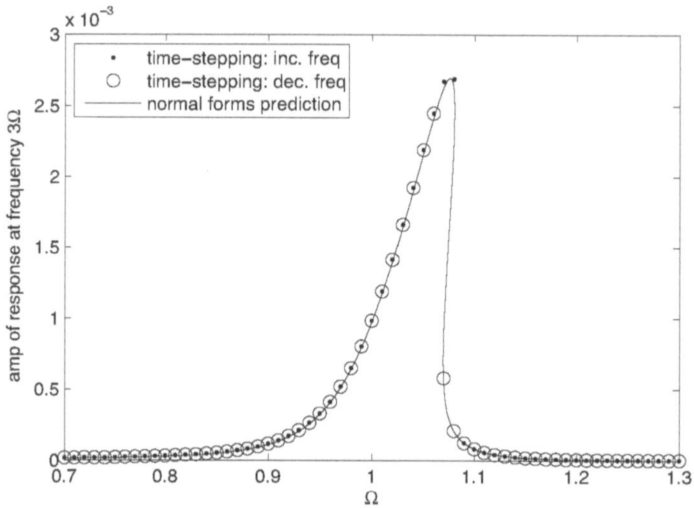

Figure 12. Amplitude of response at 3 times the forcing frequency, 3Ω, as a function of forcing frequency for the nonlinear oscillator – a comparison between the normal form prediction and a time-stepping simulation ($P = 0.03$, $\alpha = 1$, $\omega_n = 1$, $\zeta = 0.03$.)

whereas when closed the cracks are in compression and so do not affect the stiffness. If statically the cracks are partially open (or closed), due to self-weight or loading, then the proportion of an oscillation in the open and closed regions will depend on the amplitude of oscillation: in the extremes at small amplitudes of oscillation the crack will not close whereas at large amplitudes the crack will be closed for nearly half the oscillation (see, for example, Neild *et al.*, 2001). The average stiffness (and hence the natural frequency) is therefore amplitude dependent and so the degree of cracking can be monitored by measuring the amplitude-frequency relationship. Alternatively, harmonics in the response, generated from the stiffness-amplitude relationship, can be monitored. As well as these nonlinear techniques, linear techniques such as monitoring the natural frequency over the lifetime of the structure also exist. However it is often reported that these are very sensitive to environmental conditions such as temperature and humidity.

Linear ultrasonic and acoustic methods can be used to detect cracks or delaminations in structures, as these features cause partial reflection of the incident signals. However, they are relatively insensitive to pre-cracking damage such as early-stage fatigue. Prior to cracking, fatigue often results in materials exhibiting a small amount of strain nonlinearity. One result of the presence of this material bulk nonlinearity is that, when a specimen is excited at one frequency, harmonics of this frequency are present in the response. This nonlinear behaviour is often quantified in terms of the the amplitude of response to the fundamental (or driving) frequency, A_1, and the amplitude of the second harmonic of the driving frequency, A_2, which is generated via the material nonlinearity. It is this example we will consider here. Perturbation analysis is used to gain an understanding of how the nonlinearity will vary as the excitation passes through the sample.

The one-dimensional linear wave equation is formed by combining partial differential equations relating to the dynamics and the load-strain relationship, which are given by

$$\rho\frac{\partial^2 u}{\partial t^2} = \frac{\partial P}{\partial x}, \qquad P = E\frac{\partial u}{\partial x}$$

respectively. Here u denotes the particle displacement, x is particle location, P is the dynamic pressure, ρ is the density and E is the Young's Modulus.

The underlying stress-strain behaviour of most materials will contain a small degree of nonlinearity, such that the stress-strain relationship will often exhibit quadratic behaviour. The effect, along with damping, can be included in the wave equation by writing

$$\rho\frac{\partial^2 u}{\partial t^2} = \frac{\partial P}{\partial x}, \qquad P = E\frac{\partial u}{\partial x} + E^*\left(\frac{\partial u}{\partial x}\right)^2 + B\frac{\partial^2 u}{\partial x \partial t},$$

using the stress-strain relationship, $\sigma = E\epsilon + E^*\epsilon^2$, and introducing the linear viscous damping parameter B. Eliminating P gives

$$\frac{\partial^2 u}{\partial t^2} = c^2\frac{\partial^2 u}{\partial x^2}\left(1 + \frac{2E^*}{E}\frac{\partial u}{\partial x}\right) + \bar{B}\frac{\partial^3 u}{\partial x^2 \partial t} \tag{52}$$

where $c = \sqrt{E/\rho}$ and $\bar{B} = B/\rho$. More generally, for waves travelling across a material c is taken to be the travelling wave velocity. The damping parameter $\bar{B}$ is normally measured experimentally using linear techniques. It is the nonlinear parameter E^*/E that is measured using nonlinear ultrasonic techniques and it is this parameter which is monitored over the life-time of the structure to indicate early-stage fatigue.

To develop an analytical solution to this partial differential equation, the perturbation method is used first. In this example a truncated two-term power-series expansion of the response is examined. The first term to be the linear response, that is, the response if the nonlinearity was set to zero, while the second term is a perturbation away from the linear response, namely:

$$u = u_l + u_{nl} = u_l + \varepsilon \hat{u}_{nl}. \tag{53}$$

The subscripts l and nl indicate the linear and nonlinear components, while $u_{nl} = \varepsilon \hat{u}_{nl}$ has been used to indicate that u_{nl} is small compared with u_l.

Making the assumption that the nonlinear term in Equation 52 is small, order ε, and substituting Equation 53 gives

$$\frac{\partial^2 u_l}{\partial t^2} + \varepsilon \frac{\partial^2 \hat{u}_{nl}}{\partial t^2} = c^2 \left(\frac{\partial^2 u_l}{\partial x^2} + \varepsilon \frac{\partial^2 \hat{u}_{nl}}{\partial x^2}\right) \left(1 + \varepsilon \frac{2\hat{E}^*}{E} \left(\frac{\partial u_l}{\partial x} + \varepsilon \frac{\partial \hat{u}_{nl}}{\partial x}\right)\right)$$
$$+ \bar{B}\left(\frac{\partial^3 u_l}{\partial x^2 \partial t} + \varepsilon \frac{\partial^3 \hat{u}_{nl}}{\partial x^2 \partial t}\right),$$

where $E^* = \varepsilon \hat{E}^*$ reflects the assumption that the nonlinear term is order ε. Note this assumption is that $\frac{2E^*}{E} \frac{\partial u}{\partial x}$ is much smaller than unity, but this does not require that E^* itself is small (typically it will not be). Here, the damping has not been assumed to be small, this ensures that the linear component response captures the decaying nature of the overall response. In reality, the damping is small compared to the undamped linear terms but it is large relative to the nonlinear terms. Balancing the order ε^0 and ε^1 terms gives

$$\varepsilon^0: \quad \frac{\partial^2 u_l}{\partial t^2} = c^2 \frac{\partial^2 u_l}{\partial x^2} + \bar{B} \frac{\partial^3 u_l}{\partial x^2 \partial t}, \tag{54}$$

$$\varepsilon^1: \quad \frac{\partial^2 u_{nl}}{\partial t^2} = c^2 \frac{\partial^2 u_{nl}}{\partial x^2} + 2c^2 \frac{E^*}{E} \frac{\partial^2 u_l}{\partial x^2} \frac{\partial u_l}{\partial x} + \bar{B} \frac{\partial^3 u_{nl}}{\partial x^2 \partial t}. \tag{55}$$

Equation 54 is identical to the case where the system is linear (i.e. setting E^* to zero in Equation 52). The result of applying Equation 53 is to incorporate the assumption that as the nonlinearity is small it does not influence the linear response. Equation 55, which contains the nonlinear response, is a linear equation in terms of u_{nl} with excitation terms provided by the linear response. Physically, the approximation that has been made here is that the amplitude of the nonlinear response is sufficiently small that nonlinear terms containing u_{nl} can be ignored.

6.1 Order ε^0 equation

To solve Equation 54, the linear damped equation, the damping is now assumed to be light and the input signal is a forward travelling cosine wave. At $x = 0$ the excitation may be written as

$$u_l(0, t) = U_0 \cos(-kct)$$

where $k = \omega/c$ is the wave number, c is the travelling wave velocity, U_0 is the input signal amplitude and ω is the excitation frequency. Note that ω is of order 10^6 and c is of order 10^3.

Next consider when $x > 0$. Given the condition that the damping is light, attenuation in the medium would make the wave decay exponentially with distance. Hence a trial solution of the following form is adopted:

$$u_l(x,t) = U_0 \mathrm{e}^{-\alpha x} \cos\left(k(x-ct)\right), \tag{56}$$

where α is the (currently unknown) exponential decay rate with distance travelled. Because α is much smaller then k, slow and fast distance scales are introduced, as used in the multiple scales technique (although the scales are normally in terms of time rather than distance). Writing the slow scale (x_s) and the fast scale (x_f) as

$$x_s = \alpha x, \qquad x_f = kx,$$

we can write Equation 56 as $u(x,t) = u(x_s, x_f, t)$, where, in line with the multiple scales method, we assume that x_s and x_f can be treated as independent variables due to there different scales. Note that typically α and k are of order 10^0 and 10^3 respectively. Hence we can write

$$\frac{\partial u_l}{\partial x} = k\frac{\partial u_l}{\partial x_f} + \alpha\frac{\partial u_l}{\partial x_s}, \tag{57}$$

$$\frac{\partial^2 u_l}{\partial x^2} = k^2\frac{\partial^2 u_l}{\partial x_f^2} + 2k\alpha\frac{\partial^2 u_l}{\partial x_s \partial x_f} + \alpha^2\frac{\partial^2 u_l}{\partial x_s^2}. \tag{58}$$

Using these relationships, Equation 54 becomes

$$\begin{aligned}\frac{\partial^2 u_l}{\partial t^2} &= c^2(k^2\frac{\partial^2 u_l}{\partial x_f^2} + 2k\alpha\frac{\partial^2 u_l}{\partial x_s \partial x_f} + \alpha^2\frac{\partial^2 u_l}{\partial x_s^2}) \\ &\quad + \bar{B}\frac{\partial}{\partial t}(k^2\frac{\partial^2 u_l}{\partial x_f^2} + 2k\alpha\frac{\partial^2 u_l}{\partial x_s \partial x_f} + \alpha^2\frac{\partial^2 u_l}{\partial x_s^2}).\end{aligned} \tag{59}$$

By considering the form of the trial solution, given in Equation 56, it can be seen that the term on the left-hand side of Equation 59 cancels with the first term on the right-hand side. Since $\alpha \ll k$, the third, fifth and sixth terms on the right-hand side are small in comparison to the second, fourth and fifth terms respectively and so are removed resulting in

$$0 = 2c^2 k\alpha\frac{\partial^2 u_l}{\partial x_s \partial x_f} + \bar{B}k^2\frac{\partial^3 u_l}{\partial x_f^2 t}. \tag{60}$$

When the multiple scales is applied to vibration problems, it is usual to ignore only the $\alpha^2 \partial^2 u_l / \partial x_s^2$ term, taking x_s as order ε smaller than x_f and maintaining accuracy to order ε^1. However, in this example, additional simplifications have been made, since the decay rate α is expected to be multiple orders smaller than the wave number k.

Substituting Equation 56 into Equation 60, gives

$$u_l = U_0 \mathrm{e}^{-\alpha x} \cos\left(k(x - ct)\right), \qquad \alpha = \frac{k^2}{2c}\bar{B}. \tag{61}$$

The main physical effect of making the $\alpha \ll k$ approximation is that the wave velocity for the damped response has remained the same as that for the undamped response.

6.2 Order ε^1 equation

Now consider the nonlinear perturbation from the linear response, as described by Equation 55. This equation can be reordered to resemble a linear equation in u_{nl} with a forcing in terms of u_l

$$\frac{\partial^2 u_{nl}}{\partial t^2} - c^2 \frac{\partial^2 u_{nl}}{\partial x^2} - \bar{B} \frac{\partial^3 u_{nl}}{\partial x^2 \partial t} = 2c^2 \frac{E^*}{E} \frac{\partial u_l}{\partial x} \frac{\partial^2 u_l}{\partial x^2}. \tag{62}$$

The right-hand side of Equation 62 can be simplified using Equations 57 and 58 to give

$$2c^2 \frac{E^*}{E} \frac{\partial u_l}{\partial x} \frac{\partial^2 u_l}{\partial x^2} = 2c^2 \frac{E^*}{E} k^3 \frac{\partial u_l}{\partial x_f} \frac{\partial^2 u_l}{\partial x_f^2} = c^2 \frac{E^*}{E} k^3 U_0^2 \mathrm{e}^{-2\alpha x} \sin\left(2k(x - ct)\right), \tag{63}$$

where again $\alpha \ll k$ has been used to eliminate small terms and Equation 61 has been used to eliminate u_l.

The terms on the left-hand side of Equation 62 are identical to those in Equation 54, albeit in terms of u_{nl} rather than u_l. Splitting the response into fast and slow scales and using Equation 63, Equation 62 can be written as

$$\begin{aligned} \frac{\partial^2 u_{nl}}{\partial t^2} &- c^2 \left(k^2 \frac{\partial^2 u_{nl}}{\partial x_f^2} + 2k\alpha \frac{\partial^2 u_{nl}}{\partial x_s \partial x_f} + \alpha^2 \frac{\partial^2 u_{nl}}{\partial x_s^2} \right) \\ &- \bar{B} \frac{\partial}{\partial t} \left(k^2 \frac{\partial^2 u_{nl}}{\partial x_f^2} + 2k\alpha \frac{\partial^2 u_{nl}}{\partial x_s \partial x_f} + \alpha^2 \frac{\partial^2 u_{nl}}{\partial x_s^2} \right) \\ &= c^2 \frac{E^*}{E} k^3 U_0^2 \mathrm{e}^{-2\alpha x} \sin\left(2k(x - ct)\right) \end{aligned} \tag{64}$$

By inspecting the form of the right-hand side, it can be seen that the requirement is for the trial solution to have a fast scale function (itself a function of x_f and t) in the form of a summation of a $\sin(2k(x-ct))$ and a $\cos(2k(x-ct))$ term. In this way the first two terms on the left-hand side of Equation 64 cancel out. Now simplify the equation further using $\alpha \ll k$ to give

$$-2c^2 k\alpha \frac{\partial^2 u_{nl}}{\partial x_s \partial x_f} - \bar{B}k^2 \frac{\partial^3 u_{nl}}{\partial x_f^2 \partial t} = c^2 \frac{E^*}{E} k^3 U_0^2 \mathrm{e}^{-2\alpha x} \sin(2k(x-ct)) . \quad (65)$$

From inspection of this equation the trial solution

$$u_{nl}(x,t) = U_{nl} f(\alpha x) \cos(2k(x-ct)) \quad (66)$$

is selected, where $f(\alpha x)$ is a, as yet unknown, slow scale function. Substituting this solution into Equation 65 results in the amplitude relationship

$$U_{nl} = \frac{E^*}{E} \frac{k^2 U_0^2}{4\alpha}, \quad (67)$$

along with the following differential equation in terms of $x_s = \alpha x$

$$4f(x_s) + \frac{\mathrm{d}f}{\mathrm{d}x_s} = \mathrm{e}^{-2\alpha x}.$$

This can be solved, remembering that the harmonic content at $x = 0$ is zero, to give

$$f(x_s) = \frac{1}{2}\left(\mathrm{e}^{-2\alpha x} - \mathrm{e}^{-4\alpha x}\right) \quad (68)$$

Now the linear solution and the nonlinear perturbation can be combined using Equation 53 together with Equations 61, 66, 67 and 68 to produce

$$u = U_0 \mathrm{e}^{-\alpha x} \cos(k(x-ct)) + \frac{E^*}{E} \frac{k^2 U_0^2}{8\alpha} \left(\mathrm{e}^{-2\alpha x} - \mathrm{e}^{-4\alpha x}\right) \cos(2k(x-ct)) . \quad (69)$$

6.3 Nonlinear Parameter

It is common to use the dimensionless measure of material bulk nonlinearity, β, given as

$$\beta = \frac{8}{k^2 x} \frac{A_2}{A_1^2}, \quad (70)$$

where $k = \omega/c$ is the wave number and dividing by A_1^2 and the propagation distance, x, removes the dependence on these two parameter on the assumption that the material is undamped (see, for example, Melngailis and Maradudin, 1963).

However based on the perturbation analysis, the assumption that the material is undamped can be removed. Firstly E^*/E can be related to β by considering the case were the damping tends to zero in Equation 69. Using a Taylor series expansion for the harmonic term, this gives

$$A_1 = U_0, \quad A_2 = \frac{E^*}{E}\frac{k^2 U_0^2}{4}x$$

Substituting these amplitude relationships into Equation 70 gives $\beta = 2E^*/E$. Taking this to be the definition of β, eliminating E^*/E using the harmonic amplitude A_2 from Equation 69 and then eliminating U_0 using Equation 61 gives an equivalent expression for β for a damped structure, β_d:

$$A_2 = \frac{E^*}{E}\frac{k^2 U_0^2}{8\alpha}\left(\mathrm{e}^{-2\alpha x} - \mathrm{e}^{-4\alpha x}\right), \; A_1 = U_0 \mathrm{e}^{-\alpha x} \rightsquigarrow \; \beta_d = \frac{16\alpha}{k^2\left(1-\mathrm{e}^{-2\alpha x}\right)}\frac{A_2}{A_1^2}.$$

This relationship potentially gives a more accurate way of extracting the true nonlinear stess-strain parameter E^*/E from an experimental measurement of A_1 and A_2. This analysis is extended to include the effects of excitation windows in Liu *et al.* (submitted), where experimental and numerical validation is also given, and related work discussing the effects of transducer dynamics is presented by Liu *et al.* (in press).

7 Parametric excitation

In the presence of a nonlinearity, energy can move between modes of vibration. This has already been demonstrated in Section 2 where a cable excited in the vertical in-plane direction can result in non-zero responses in out-of-plane modes. The final example to be studied in this chapter involves exploiting parametric excitation in order to dissipate energy. If a system is vibrating in one mode, parametric excitation can be used to excite a different mode and hence reduce the amplitude of response in the original mode. An overview of such techniques is given in Ecker (2010) and many theoretical and numerical studies are also available, for example Dohnal (2007, 2009).

To motivate this study the example of a cantilever with a controllable axial tip force will be considered. The first two modes of the system will be derived and the parametric terms, which are introduced by the presence of tip forcing, identified. A normal form analysis will then be performed on a slightly simplified version of the equations using arbitrary parameter values to show the potential of using parametric excitation to increase the effective damping of a mode.

Consider a vertical cantilever which is subject to external forcing $\psi(x)p(t)$, where ψ defines the distribution of the forcing spatially, and a vertical tip forcing $f(t)$, then

$$EI\frac{\partial^4 w}{\partial x^4} + \rho A\frac{\partial^2 w}{\partial t^2} + f\frac{\partial^2 w}{\partial x^2} = \psi(x)p(t). \tag{71}$$

First, the unforced response ($f = p = 0$) of this Euler-Bernoulli beam is investigated. The response can be written as a summation of modal responses:

$$w(x,t) = \sum_k \phi_k(x)q_k(t). \tag{72}$$

The modeshapes $\phi_k(x)$ can be found using the Separation of Variable technique. The orthogonality conditions for these modeshapes can be found by substituting Equation 72 into the unforced version of Equation 71:

$$EI\frac{\partial^4 w}{\partial x^4} + \rho A\frac{\partial^2 w}{\partial t^2} = 0,$$

multiplying the result by the modeshape ϕ_n (where n is arbitrary) and integrating over the length of the beam. Decoupling of the modes (which is achieved in the Separation of Variable technique) requires

$$\int_0^l \phi_n\phi_k \mathrm{d}x = 0 \quad \text{and} \quad \int_0^l \phi_n\phi_k'''' \mathrm{d}x = 0 \quad \text{for } k \neq n, \tag{73}$$

where a dot and a prime indicate differentiation with respect to time and x respectively. This results in the uncoupled modal equation of motion for the nth mode

$$\int_0^l \phi_n^2 \mathrm{d}x\ \ddot{q}_n = c^2 \int_0^l \phi_n\phi_n'' \mathrm{d}x\ q_n.$$

Next, the full equation of motion is considered, to which the Galerkin technique is applied. Firstly, Equation 72 is substituted into Equation 71, the result is multiplied by the modeshape ϕ_n (where n is arbitrary) and integrating over the length of the beam to give the following equation for the nth mode:

$$EI\sum_{k=1}^{\infty}\int_0^l \phi_n(x)\phi_k''''(x)\mathrm{d}x\ q_k(t) + \rho A\sum_{k=1}^{\infty}\int_0^l \phi_n(x)\phi_k(x)\mathrm{d}x\ \ddot{q}_k(t)$$
$$+f\sum_{k=1}^{\infty}\int_0^l \phi_n(x)\phi_k''(x)\mathrm{d}x\ q_k(t) = \sum_{k=1}^{\infty}\int_0^l \phi_n(x)\psi(x)p(t)\mathrm{d}x,$$

To evaluate this equation, modeshapes are needed. As the modeshapes for the nonlinear system are not known, the modeshapes of the unforced linear-equivalent Euler-Bernoulli beam are used. The result of using these mode-shapes is that the modal equations of motion have cross-coupling terms. Using the orthogonality conditions for the unforced linear-equivalent beam, Equation 73, the equation for the nth mode becomes

$$EI\int_0^l \phi_n\phi_n''''\mathrm{d}x\; q_n+\rho A\int_0^l \phi_n^2\mathrm{d}x\; \ddot{q}_n+f\sum_{k=1}^{\infty}\int_0^l \phi_n\phi_k''\mathrm{d}x\; q_k = p\sum_{k=1}^{\infty}\int_0^l \phi_n\psi\mathrm{d}x. \tag{74}$$

Taking the first two modes and making the simplifying assumption that the external forcing p is applied at the node of the second mode gives

$$\begin{aligned}\ddot{q}_1+2\zeta_1\omega_{n1}\dot{q}_1+\omega_{n1}^2q_1+f\alpha_{11}q_1+f\alpha_{12}q_2 &= \gamma_1p(t),\\ \ddot{q}_2+2\zeta_2\omega_{n2}\dot{q}_2+\omega_{n2}^2q_2+f\alpha_{21}q_1+f\alpha_{22}q_2 &= 0,\end{aligned} \tag{75}$$

where modal damping has been added. The parameter γ_1 represents a modal distribution of the forcing for the first mode, the equivalent term for the second mode is zero as the forcing is applied to a node of the second mode. The values of ω_{n1}, ω_{n2}, γ_1 and the α coefficients can be found by comparing these equations with Equation 74.

The case where the frequency of the external forcing is close to the first natural frequency, that is where $\gamma_1p(t) = P\cos(\Omega t)$ with $\Omega \approx \omega_{n1}$, is now considered using the normal form technique. Since the equations are in modal form (using the linear unforced modes), the first transform $\mathbf{x}\rightarrow\mathbf{q}$ is not required. The second transform $\mathbf{q}\rightarrow\mathbf{v}$ modifies the forcing. In this example, this transform is a unity transform as the forcing term is applied only to the first mode and is close to the natural frequency of the first mode:

$$P\cos(\Omega t) = P_v\mathbf{r},\quad \text{where:}\quad P_v = P\begin{bmatrix}1/2 & 1/2\\ 0 & 0\end{bmatrix},\quad \mathbf{r}=\begin{Bmatrix}\mathrm{e}^{\mathrm{i}\Omega t}\\ \mathrm{e}^{-\mathrm{i}\Omega t}\end{Bmatrix}.$$

The transform is given by

$$\mathbf{q}=\mathbf{v}+[e]\mathbf{r},\quad \text{where:}\quad [\tilde{e}]+P_v=P_q,\quad e_{n,k}=\tilde{e}_{n,k}/(\omega_{nn}^2-\Omega^2).$$

For this example, the potential values for $[e]$ are written as

$$\text{potential values:}\quad [e]=\begin{bmatrix}\tilde{e}_{1,1}/(\omega_{n1}^2-\Omega^2) & \tilde{e}_{1,2}/(\omega_{n1}^2-\Omega^2)\\ 0 & 0\end{bmatrix}.$$

As both terms on the top row are divided by $\omega_{n1}^2-\Omega^2$ and $\Omega\approx\omega_{n1}$, Option 1, Equation 32, is selected for these terms. Hence $[e]=0$ and $P_v=P_r$

resulting in $\mathbf{v} = \mathbf{q}$. This gives

$$\ddot{\mathbf{v}}+\Lambda\mathbf{v}+\mathbf{N}_v(\mathbf{v},\dot{\mathbf{v}},f) = P_v\mathbf{r}, \quad \mathbf{N}_v(\mathbf{v},\dot{\mathbf{v}},f) = \left\{ \begin{array}{c} \alpha_{11}fv_1 + \alpha_{12}fv_2 + 2\omega_{n1}\zeta_1\dot{v}_1 \\ \alpha_{21}fv_1 + \alpha_{22}fv_2 + 2\omega_{n2}\zeta_2\dot{v}_2 \end{array} \right\},$$

where $\mathbf{N}_v$ is a function of f as well as the states.

Once the nonlinear near-identity transform has been applied the modal responses will not contain harmonics, these will have been transformed out of the dynamic equations, and may be written as

$$u_1 = U_1\cos(\Omega t - \phi_1), \qquad u_2 = U_2\cos(\omega_{r2}t - \phi_2).$$

The response of the first mode has been taken to be at the forcing frequency, since the forcing frequency is close to the natural frequency of the first mode, and the response on the second mode is at an, as yet, undetermined frequency. The tip, or parametric, forcing is controlled by the user. Consider the case where it is set to $f = F\cos([\omega_{n2} - \Omega]t - \phi)$. These forcing and response terms can be written in exponential form resulting in

$$\begin{aligned} u_1 &= u_{1p} + u_{1m} \quad \text{with} \quad u_{1p} = \frac{U_1}{2}\mathrm{e}^{\mathrm{i}(\Omega t - \phi_1)}, u_{1m} = \frac{U_1}{2}\mathrm{e}^{-\mathrm{i}(\Omega t - \phi_1)} \\ u_2 &= u_{2p} + u_{2m} \quad \text{with} \quad u_{2p} = \frac{U_2}{2}\mathrm{e}^{\mathrm{i}(\omega_{r2}t - \phi_2)}, u_{2m} = \frac{U_2}{2}\mathrm{e}^{-\mathrm{i}(\omega_{r2}t - \phi_2)} \\ f &= f_p + f_m \quad \text{with} \quad f_p = \frac{F}{2}\mathrm{e}^{\mathrm{i}([\omega_{n2}-\Omega]t - \phi)}, f_m = \frac{F}{2}\mathrm{e}^{-\mathrm{i}([\omega_{n2}-\Omega]t - \phi)}. \end{aligned} \tag{76}$$

With reference to Equation 34, the nonlinear term in the equation of motion can be written as $\mathbf{N}_v = \varepsilon\mathbf{n}_{v1}$. Using Equation 40, the function $\mathbf{n}_{v1}$ is now written in vector matrix form, but in terms of $\mathbf{u}$ rather than $\mathbf{v}$. This yields $\mathbf{n}_{v1}(\mathbf{u},\dot{\mathbf{u}},f) = [n_v]\mathbf{u}^*(\mathbf{u_p},\mathbf{u_m},f_p,f_m)$ where

$$[n_v]^T = \begin{bmatrix} \alpha_{11} & \alpha_{21} \\ \alpha_{11} & \alpha_{21} \\ \alpha_{11} & \alpha_{21} \\ \alpha_{11} & \alpha_{21} \\ \alpha_{12} & \alpha_{22} \\ \alpha_{12} & \alpha_{22} \\ \alpha_{12} & \alpha_{22} \\ \alpha_{12} & \alpha_{22} \end{bmatrix}, \quad \mathbf{u}^* = \left\{ \begin{array}{c} f_pu_{1p} \\ f_mu_{1p} \\ f_pu_{1m} \\ f_mu_{1m} \\ f_pu_{2p} \\ f_mu_{2p} \\ f_pu_{2m} \\ f_mu_{2m} \end{array} \right\}.$$

Note that in Section 4 the damping terms were also included in these matrices. In this example, as the damping is modal, the damping terms will all be resonant and so will remain in the equations of motion after the nonlinear transform has been applied. They have therefore been excluded from $[n_v]$ and $\mathbf{u}^*$ for brevity.

Using Equation 44, the matrix $[\beta]$ can now be calculated as

$$[\beta]^T = \begin{bmatrix} \omega_{n2}^2 - \Omega^2 & \omega_{n2}^2 - \omega_{r2}^2 \\ (-\omega_{n2} + 2\Omega)^2 - \Omega^2 & (-\omega_{n2} + 2\Omega)^2 - \omega_{r2}^2 \\ (\omega_{n2} - 2\Omega)^2 - \Omega^2 & (\omega_{n2} - 2\Omega)^2 - \omega_{r2}^2 \\ \omega_{n2}^2 - \Omega^2 & \omega_{n2}^2 - \omega_{r2}^2 \\ (\omega_{n2} - \Omega + \omega_{r2})^2 - \Omega^2 & (\omega_{n2} - \Omega + \omega_{r2})^2 - \omega_{r2}^2 \\ (-\omega_{n2} + \Omega + \omega_{r2})^2 - \Omega^2 & (-\omega_{n2} + \Omega + \omega_{r2})^2 - \omega_{r2}^2 \\ (\omega_{n2} - \Omega - \omega_{r2})^2 - \Omega^2 & (\omega_{n2} - \Omega - \omega_{r2})^2 - \omega_{r2}^2 \\ (-\omega_{n2} + \Omega - \omega_{r2})^2 - \Omega^2 & (-\omega_{n2} + \Omega - \omega_{r2})^2 - \omega_{r2}^2 \end{bmatrix},$$

where $\omega_{r1} = \Omega$ has been used as the first mode is directly forced at frequency Ω which is close to the natural frequency. Given that the response frequencies are close to the natural frequencies, $\omega_{r2} \approx \omega_{n2}$ and $\omega_{r1} = \Omega \approx \omega_{n1}$. Using this, it can be seen that the $[1,6], [1,7], [2,1]$ and $[2,4]$ terms in $[\beta]$ are close to zero. For these terms option 2, Equation 46, is selected and the terms remain in the equations of motion. Other terms can also be close to zero if one of the following relationships between the two natural frequencies are met:

$$\omega_{n2} = a\omega_{n1} \quad \text{where:} \quad a = 0, \frac{1}{3}, 1, 3.$$

Assuming none of these conditions are met, for all the other terms option 1, Equation 45, can be selected. This prefered option results in the terms being removed from the equation of motion.

Applying the relevant option to each term results in the transformed equations of motions:

$$\begin{aligned} \ddot{u}_1 + 2\zeta_1\omega_{n1}\dot{u}_1 + \omega_{n1}^2 u_1 + \alpha_{12}(f_m u_{2p} + f_p u_{2m}) &= P\cos(\Omega t), \\ \ddot{u}_2 + 2\zeta_2\omega_{n2}\dot{u}_2 + \omega_{n2}^2 u_2 + \alpha_{21}(f_p u_{1p} + f_m u_{1m}) &= 0. \end{aligned}$$

Addressing the second equation first, the α_{21} terms may be viewed as a forcing and may be written as

$$\alpha_{21}(f_p u_{1p} + f_m u_{1m}) = \frac{\alpha_{21}}{2} U_1 P \cos(\omega_{n2} t - \phi - \phi_1).$$

With this forcing, inspecting the trial solution for u_2, Equation 76, it can be seen that $\omega_{r2} = \omega_{n2}$. The equation of motion for the second mode can now be solved yielding

$$u_2 = U_2 \cos(\omega_{n2} t - \phi_2), \quad \text{where} \quad \phi_2 = \phi + \phi_1 - \frac{\pi}{2}, \quad U_2 = \frac{\alpha_{21} U_1 F}{4\zeta_2 \omega_{n2}^2}.$$

Now considering the equation of motion for the first mode, the parametric α_{12} terms may be rewritten as

$$\alpha_{12}(f_m u_{2p} + f_p u_{2m}) = -\frac{F^2\alpha_{21}\alpha_{12}}{8\zeta_2\omega_{n2}^2}U_1\sin(\Omega t - \phi_1) = \frac{F^2\alpha_{21}\alpha_{12}}{8\zeta_2\omega_{n2}^2\Omega}\dot{u}_1.$$

Since this term is proportional to $\dot{u}_1$, it can be viewed as an additional damping term. Therefore the parametric forcing has the effect of increasing the damping of the first mode. Interestingly, the analysis indicates that the phase of the tip forcing, ϕ (see Equation 76), does not effect this result.

Figure 13 shows the response of the first mode of a system, with dynamics governed by Equation 75 using arbitrary parameter values, when excited close to resonance. It can be seen that time-stepping results compared very closely to the normal forms prediction. In addition it can be seen that the response peak is lower and hence the damping is higher than the case where the parametric excitation is not present, $F = 0$. Simulations for a range of tip forcing phases, ϕ, were conducted and as indicated by the normal forms analysis no changes to the degree of additional damping added was observed.

While this result looks promising, for a practical system, it does not tell the full story. Often it is the displacment response of the system, rather than the modal response, that is important. One example of where this may affect the result is that it appears desirable for the modal damping of the second mode, ζ_2, to be as small as possible such that the added damping to the first mode is maximised. However, as the second mode is responding at resonance (recall that $\omega_{r2} = \omega_{n2}$) a reasonable level of modal damping, ζ_2, is needed to ensure the response amplitude, U_2, is not excessive. The displacement results may be further affected by the presence of harmonic responses in both modes due to the nonlinear behaviour. These can be calculated using the normal forms technique by considering the nonlinear $\mathbf{v} \to \mathbf{u}$ transform.

Promising experimental results from the cantilever beam system, described at the start of this section and used to motivate the normal forms analysis, have been performed by Pumhossel and Ecker (2007) and Ecker and Pumhossel (2009). These confirm that axial tip forcing can result in higher modal damping for a targeted mode through parametric excitation of a second mode.

8 Conclusions

In this chapter two approximate methods for studying the response of weakly nonlinear systems have been discussed. In addition, three exam-

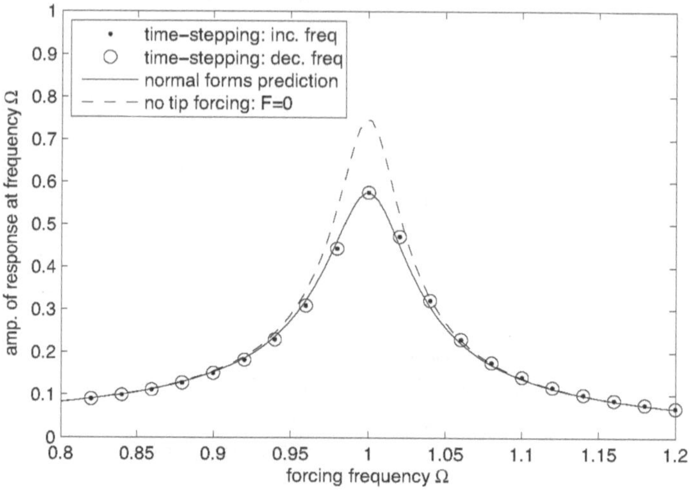

Figure 13. Response of the first mode of a two-mode system with parametric nonlinearity for the case where the excitation is close to the first natural frequency. The system parameters used are $\omega_{n1} = 1$, $\omega_{n2} = 4.3$, $\zeta_1 = 0.02$, $\zeta_2 = 0.05$, $\alpha_{11} = \alpha_{12} = \alpha_{21} = \alpha_{22} = 10$ and the external and parametric forcing amplitudes, P and F respectively, are $P = F = 0.03$.

ples of systems, in which nonlinear dynamic behaviour can be exploited, have been discussed and analysed using the approximate methods.

Considering perturbation methods, it was shown that using regular perturbation theory can result in a poor prediction of the vibration response in certain circumstances such as when different timescales are present. The multiple-scales technique overcomes this limitation and was shown to predict the response of the Duffing oscillator well. In addition to predicting the steady-state response, the stability of the various response solutions were examined using the technique. Then both regular perturbation theory and multiple scales was used to analyse how harmonics of acoustic signals are generated as the signals propagate through a material with a nonlinear stress-strain relationship – the application being detection of early fatigue damage.

The normal forms technique, in which a series of transforms are applied to the equations of motion, was also discussed. It was shown that, through the application of these transforms, all the non-resonant terms can be removed from the equations of motion allowing it to be solved using a straightforward trial solution at the response frequency. Information regarding the response at other frequencies can be extracted from the nonlinear transformation. The technique was shown to accurately predict the response of the Duffing oscillator at the forcing frequency and at three times the forcing frequency. In addition, the technique, which can conveniently handle multiple modes, was successfully used to analyse parametric excitation between modes. This analysis demonstrated the potential use of parametric excitation as a means to increase modal damping.

Bibliography

T. Bakri, R. Nabergoj, A. Tondl and F. Verhulst, Parametric excitation in nonlinear dynamics. *International Journal of Non-Linear Mechanics*, 2004, 39, 311–329.

D.A.W. Barton, S.G. Burrow and L.R. Clare, Energy harvesting from vibrations with a nonlinear oscillator. *Journal of Vibration and Acoustics*, 2010, 132, paper 021009.

M. Cartmell, *Introduction to Linear, Parametric and Nonlinear Vibrations*, Chapman and Hall, London, 1990.

F. Dohnal, Suppressing self-excited vibrations by synchronous and time-periodic stiffness and damping variation. *Journal of Sound and Vibration*, 2007, 307, 137–152.

F. Dohnal, Optimal dynamic stabilisation of a linear system by periodic stiffness excitation. *Journal of Sound and Vibration*, 2009, 320, 777–792.

H. Ecker, Exploring the use of parametric excitation, *Tenth International conference on Recent Advances in Structural Dynamics (RASD)*, 2010, Southampton, paper: Keynote 1.

H. Ecker and T. Pumhossel, Experimental results on parametric excition damping of an axially loaded cantilever beam. *Proceedings of the ASME 2009 International Design Engineering Technical Conferences & Computers and Information in Engineering Conference (IDETC/CIE)*, 2009, San Diego, California, USA.

P. Glendinning, *Stability, instability and chaos*, Cambridge University Press, 1994.

A. Gonzalez-Buelga, S.A. Neild, D.J. Wagg and J.H.G. Macdonald, Modal stability of inclined cables subjected to vertical support excitation. *Journal of Sound and Vibration*, 2008, 318, 565–579.

L. Jezequel and C.H. Lemarque, Analysis of nonlinear dynamic systems by the normal form theory. *Journal of Sound and Vibration*, 1991, 149(3), 429–459.

I. Kovacic, M.J. Brennan and T.P. Waters, A study of a nonlinear vibration isolator with a quasi-zero stiffness characteristic. *Journal of Sound and Vibration*, 2008, 315, 700–711.

B. Krauskopf, H.M. Osinga and J. Galan-Vioque (Editors), *Numerical Continuation Methods for Dynamical Systems: Path following and boundary value problems*, Springer, 2007.

S. Liu, S.A. Neild, A.J. Croxford and Z. Zhou, Effects of damping on harmonic generation due to bulk material nonlinearity. Submitted to *NDT& E international.*

S. Liu, A.J. Croxford, S.A. Neild and Z. Zhou, Effects of Experimental Variables on the Nonlinear Harmonic Generation Technique. *IEEE Transactions on Ultrasonics, Ferroelectrics and Frequency Control*, in press.

J.H.G. Macdonald, M.S. Dietz, S.A. Neild, A. Gonzalez-Buelga, A.J. Crewe and D.J. Wagg, Generalised modal stability of inclined cables subjected to support excitations. *Journal of Sound and Vibration*, 2010a, 329(21), 4515–4533.

J.H.G. Macdonald, M.S. Dietz and S.A. Neild, Dynamic excitation of cables by deck and/or tower motion. *Proceedings of the ICE - Bridge Engineering*, 2010b, 163(BE2), 101–111.

M.R. Marsico, V. Tzanov, D.J. Wagg, S.A. Neild and B. Krauskopf, Bifurcation analysis of parametrically excited inclined cable close to two-to-one internal resonance. *Journal of Sound and Vibration*, 2011, 330(24), 6023–6035.

J. Melngailis and A. Maradudin, Diffration of light by ultrasound in anharmonic crystals. *Physical Review*, 1963, 131(5), 1972–1975.

A.H. Nayfeh, *Method of Normal Forms*, Wiley, 1993.

S.A. Neild, P.D. McFadden and M.S. Williams, A discrete model of a vibrating beam using a time-stepping approach. *Journal of Sound and Vibration*, 2001, 239, 99–121.

S.A. Neild and D.J. Wagg, Applying the method of normal forms to second order nonlinear vibration problems. *Proceedings of the Royal Society, Part A*, 2011, 467, 1141–1163.

T. Pumhossel and H. Ecker, Active damping of vibrations of a cantilever beam by axial force control. *Proceedings of the ASME International Design Engineering Technical Conferences & Computers and Information in Engineering Conference (IDETC/CIE)*, 2007, paper DETC2007-34638.

D. Reed, J. Yu, H. Yeh and S. Gardarsson, Investigation of tuned liquid dampers under large amplitude excitation. *ASCE Journal of Engineering Mechanics*, 1998, 124, 405–413.

T.T. Soong and G.F. Dargush, *Passive energy dissipation systems in structural engineering*, Wiley, 1997.

S.H. Strogatz, *Nonlinear Dynamics and Chaos*, Westview, 2000.

F. Verhulst, *Nonlinear Differential Equations and Dynamical Systems*, Springer, 1989.

D.J. Wagg and S.A. Neild, *Nonlinear Vibration with Control – for flexible and adaptive structures*, Springer, 2010.

P. Warnitchai, Y. Fujino and T. Susumpow, A nonlinear dynamic model for cables and its application to a cable-structure system. *Journal of Sound and Vibration*, 1995, 187(4), 695–712.

Z. Xie, S.A. Neild and D.J. Wagg, The selection of the linearized natural frequency for the second-order normal form method. *Proceedings of IDETC/CIE ASME International Design Engineering Technical Conferences*, Washington, DC. USA, 28-31 August 2011.

Vibration Isolation

Lawrie Virgin
Department of Mechanical Engineering, Duke University,
Durham, NC 27708, USA.

Abstract The ability to *isolate* a structure or machine from the undesirable effects of applied motion (especially vibration) has wide application. Suspension systems are incorporated into large buildings to protect them from earthquake excitation, mountain bikes and vehicles in general are designed to minimize the transfer of unwanted accelerations from the terrain to the occupants, and sensitive equipment often needs to be isolated from ambient vibrations in the surrounding environment. This chapter explores the ways in which specifically nonlinear components can be utilized to advantage in vibration isolation in the context of steady excitation (Platus (1991); Rivin (2006); Virgin and Davis (2003); Virgin et al. (2008)). There have been other attempts to take advantage of nonlinearity in a vibration isolation context. A zero-spring-rate suspension system (Woodard and Housner (1991)) was developed in which a clever arrangement of (linear) springs acted together such that, under preload, they behaved in a nonlinear geometric sense. The system is essentially the same as a negative-stiffness mechanism described by Platus (1991) and developed commercially. In most design situations there is a trade-off between constraints, and in the case of vibration isolation, the springs need to sufficiently soft for dynamic transmissibility requirements but sufficiently stiff that they can provide static support. Other approaches have been studied by Virgin and Davis (2003); Winterflood et al. (2002); Zhang et al. (2004); and Carrella et al. (2006).

1 Introduction

Conventional vibration isolation systems are based on the classical (linear) harmonic oscillator and the avoidance of resonance. Consider the simple schematic shown in Figure 1(a). Here, we assume the 'supporting surface' moves, and we wish to prevent as much of this motion as possible from being transmitted to the 'object'. If we assume that the supporting surface moves in a purely vertical direction, it is harmonic, and that the 'vibration isolator'

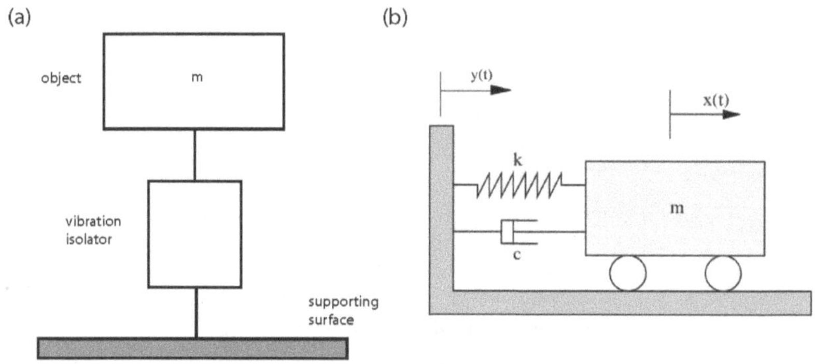

Figure 1. (a) A schematic of a supported mass, (b) the spring-mass-damper model.

consists of components that can be modeled as a spring and damper, then we have the classical system shown in Figure 1(b). Given a harmonic input of the form $y(t) = Y \sin \omega t$, the mass responds according to the well-known transmissibility expression:

$$\frac{X}{Y} = \left[\frac{1 + (2\zeta\Omega)^2}{(1 - \Omega^2)^2 + (2\zeta\Omega)^2} \right]^{1/2} . \tag{1}$$

This is plotted in Figure 2 as a function of the frequency ratio $\Omega = \omega/\omega_n$, where $\omega_n = \sqrt{k/m}$, for a number of representative damping ratios, ζ. In terms of vibration isolation the goal is unambiguous: minimize the ratio $|X/Y|$.

Thus, the transmissibility is small for relatively high frequency ratios, i.e., for $\Omega > \sqrt{2}, X/Y < 1.0$. Given a forcing frequency, ω, a typical design option would be to mount the device on a soft spring to induce a low natural frequency, ω_n, and thus a large Ω. However, the spring should obviously be stiff enough to support the mass statically (since typically $mg = kx$ when the weight of the mass needs to be supported), and this often places practical limits on the spring stiffness. Furthermore, the stiffness associated with a linear spring is fixed, and thus an isolator that is effective for certain forcing frequency ranges may not be effective under changing conditions, and in fact may lead to enhanced transmissibility associated with resonance effects. It is clearly advantageous to be able to alter, or tailor, the stiffness

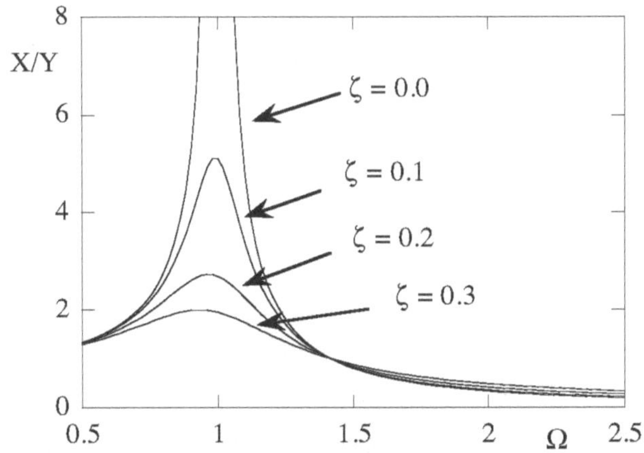

Figure 2. Displacement transmissibility of a spring-mass-damper as a function of frequency ratio.

in order to maintain a high frequency ratio. This is where nonlinear effects can be taken advantage of.

Figure 3 shows how the stiffness of a nonlinear spring (the local slope of the force, F, versus - deflection, δ, relation) can be tuned, even to the extent that a (near) zero stiffness can be achieved, as illustrated in part (c).

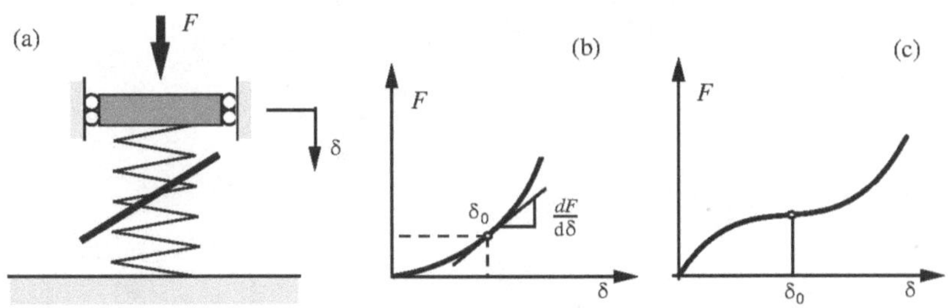

Figure 3. The changing stiffness associated with a pre-loaded, nonlinear spring.

In this chapter we shall present two cases in which this concept is explored. The first involves the use of post-buckled struts. In this the case the axial stiffness of the strut depends on the degree of (post)-buckling. This concept is then extended to very heavily deformed (but still elastic) structural components, whose stiffness can be altered quite easily.

2 Elastically buckled isolator elements

2.1 Axial stiffness of a post-buckled strut

Axially loaded structures typically possess nonlinear characteristics, especially close to, or beyond, initial buckling. Usually, buckling is viewed as an undesirable feature, in particular when this precipitates a total loss of stiffness and collapse (Virgin (2007)). However, certain structural forms, e.g., plates, possess a significant amount of post-buckled strength which can be utilized in design. For example, thin panels comprising the stressed skin components of aircraft fuselage or wings may be allowed to undergo mild elastic buckling and still retain their load-carrying capacity. In this section we develop the notion of using post-buckled struts as the spring components in a vibration isolation system, as indicated schematically in Figure 4. In their post-buckled state they are axially soft and yet able to support at least their buckling load statically. This concept will also be extended to the use

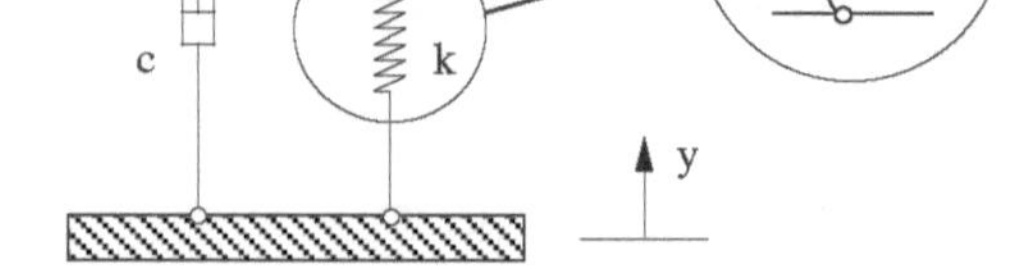

Figure 4. Schematic showing the post-buckled strut as an isolator spring.

of plate-like spring components.

Consider a thin elastic structural element, pinned at both ends and subject to constant axial loading as shown in Figure 5. Any structure exhibiting stable post-buckled behavior can be used in this situation. This is the classic Euler strut (elastica) with flexural rigidity EI, length L, and P is the

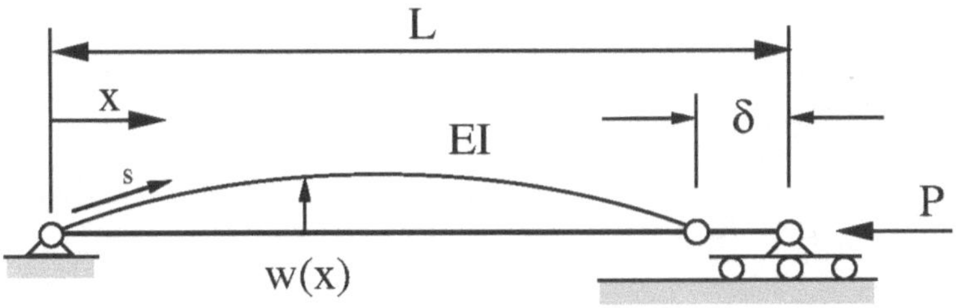

Figure 5. Schematic of a continuous buckled strut.

axial load. For moderately large deflections an approximate energy analysis can be used to obtain the axial force versus end shortening relation. The postbuckled equilibrium configuration is given by

$$\frac{P}{P_e} = 1 + \frac{\pi^2}{8}\left(\frac{Q}{L}\right)^2, \tag{2}$$

where

$$P_e = EI\left(\frac{\pi}{L}\right)^2. \tag{3}$$

P_e is the classical Euler critical load, and provided the axial load remains greater than this value, the strut deflects into one of two symmetrically located equilibria as given by equation 2.

However, we need to consider stiffness in the axial direction since this is the direction in which the force acts. Based on the lowest buckling mode shape $w(x) = Q\sin(\pi x/L)$, the geometric relation between the lateral deflection Q and the end shortening δ can be established:

$$\left(\frac{\delta}{L}\right) = \frac{\pi^2}{4}\left(\frac{Q}{L}\right)^2 + \frac{3\pi^4}{64}\left(\frac{Q}{L}\right)^4, \tag{4}$$

i.e., the end shortening is approximately related to the square of the lateral deflection. Eliminating Q in Equations 2 and 4 leads to

$$\frac{P}{P_e} = \frac{1}{3}\left[2 + \sqrt{1 + 3\left(\frac{\delta}{L}\right)}\right] \approx 1 + \frac{1}{2}\left(\frac{\delta}{L}\right). \tag{5}$$

This is a result based on moderate lateral deflections, i.e., up to about 20% of the length, where the half-sine mode shape is appropriate. The analysis can be extended to much larger deflections, and this will be covered using an elastica formulation later in the chapter. Equations 2 and 5 are shown by the solid curves in Figure 6 (a) and (b) respectively.

The post-buckled stiffness is only mildly influenced by the presence of initial imperfections (unlike in the vicinity of the critical point), and adopting a single-mode initial geometry ($w_0(x) = Q_0 \sin(\pi x/L)$) leads to the dashed lines in Figure 6(a) and (b). If we load the strut axially to just above its elastic critical load, for example, $P/P_e = 1.05$ then (for $Q_0 = 0$) we have $Q/L \approx 0.2$ and $\delta/L \approx 0.1$. If we use this point as our basic equilibrium datum, we observe a (low slope) linear relation for small oscillations about this point based on new local coordinates:

$$\bar{p} = \frac{P}{P_e} - 1.05, \bar{\delta} = \frac{\delta}{L} - 0.1, \tag{6}$$

and indicated in Figure 6(b). This strut then, is able to support a relatively high axial load (sufficient to cause buckling) but exhibits the desirable soft spring characteristic. It is also noted that adjusting the pre-load can be used to tune the stiffness depending on achieving a desirable frequency ratio.

2.2 A simple experiment

An experimental verification using a variable speed vertical shaker, with a mass (dead weight) and two thin steel struts confirms the analysis. The measured post-buckled response is shown in Figure 6(c) and (d) in terms of lateral and axial deflection. The two struts were 268 mm long, 19 mm wide, 0.66 mm thick (and thus $I = 4.55 \text{ x } 10^{-13}m^4$), and taking a typical value for Young's modulus of 200 GPa, an Euler critical elastic buckling load of $25N \equiv 2.55kg$ is computed (based on $P_{cr} = EI\pi^2/L^2$). The experimental result indicates a critical load in this vicinity, and a Southwell plot (Virgin (2007)) can be used to recast the data from Figure 6(c) to estimate a critical load of approximately 23 N. Due to initial geometric imperfections, the "critical load" is manifest as a relatively rapid increase in the deflection due to additional load, i.e., the stiffness reduces but then increases as the load passes through the range of the underlying critical load of the geometrically perfect system. The quadratic relation between lateral deflection and end shortening established by equation 4 is thus confirmed, as well as the locally linear relation between axial load and end shortening.

Suppose we choose a point on this (imperfect) curve (Figure 6(d)) as the base (buckled) equilibrium position: $P = 23.5N \rightarrow (P/P_e \approx 1.0)$ and $\delta = 15.2mm \rightarrow (\delta/L = 0.057)$, This is the (operating) point about which transmissibility is measured. Locally the stiffness is approximately 195 N/m (i.e., the slope of P/P_e v δ/L about the chosen operating point) and since

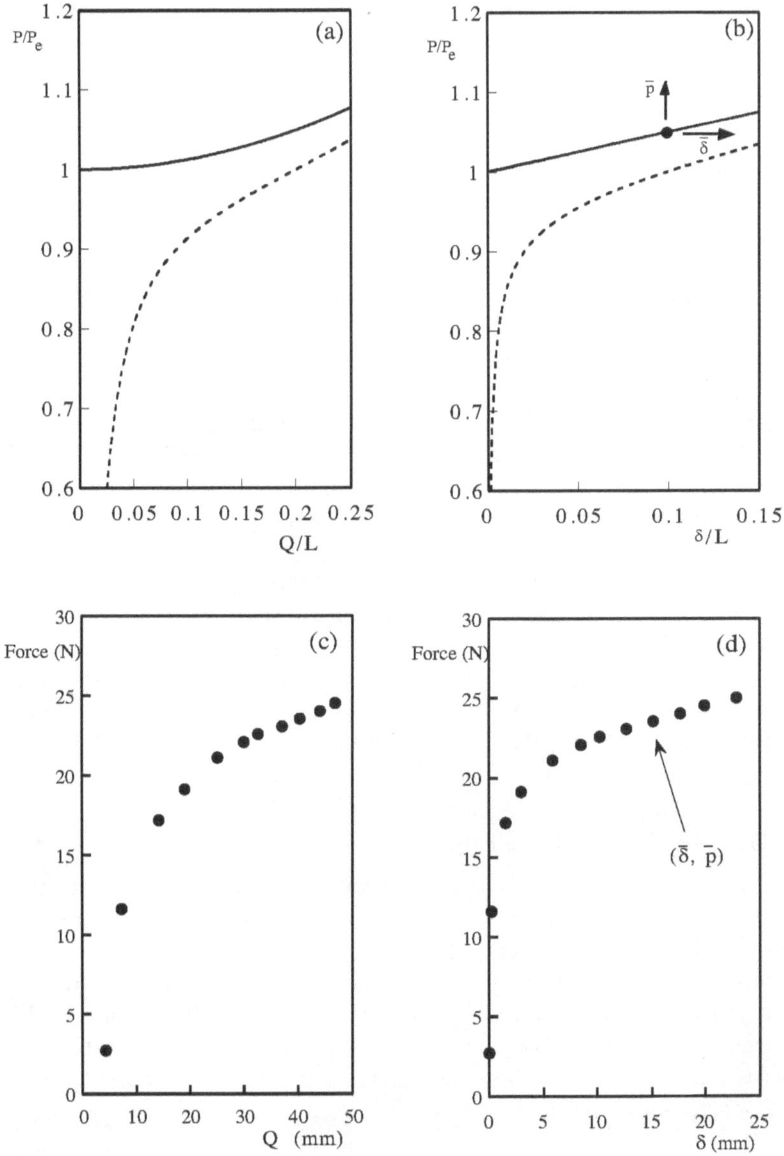

Figure 6. Deflection of the strut as a function of axial load. Analysis: (a) midpoint lateral deflection, (b) end shortening. Experimental data: (c) midpoint lateral deflection, (d) end shortening.

the mass is 2.4 Kg we would thus expect a natural frequency of free vibration close to 9 rad/s $\equiv$ 1.43 Hz. A free decay of this system gave a natural period of approximately 0.68 seconds (and hence $\omega_n = 1.47Hz$), which corresponds quite closely to the estimated value, although we would expect damping to reduce the measured frequency slightly. However, we anticipate effective isolation for forcing frequencies $\omega/\omega_n = \Omega > \sqrt{2} \approx 2.2$ Hz. It is worth noting that replacing the buckled struts with conventional linear springs of the same stiffness would result in a static deflection of approximately 60 mm, i.e., four times the deflection of the struts. Furthermore, any number of struts could be used if a different supported mass was needed, and these results would generalize.

When the base excitation is applied to the strut-supported mass, we desire the transmissibility ratio (X/Y, given by equation 1) to be small. This should be the case for $\Omega > \sqrt{2}$. Experimental results, confirming low transmissibility, are shown in Figure 7 for a range of frequency ratios. Three responses are shown as insets for different frequency ratios. The time series occurring when the forcing frequency is exactly twice the natural frequency (indicated by the star) shows an interesting subharmonic of order two (a parametric response). In general we see a highly attenuated response for higher frequencies, i.e., in this frequency range the mass is effectively isolated from the motion of the base.

An example of a modification to using thin steel struts is the use of 4 symmetrically placed polycarbonate panels as shown in Figure 8. This configuration has the capability of being adjusted in order to provide isolation from rotational sources of excitation.

2.3 Beyond Initial Buckling

The schematic shown in Figure 8 displays a deflected configuration for which the initial post-buckling analysis (e.g., equation 2) has ceased to be applicable. In this case, we can obtain equilibrium configurations for highly deflected shapes using an elastica analysis. This is based on an arc-length description of the deflected shape as shown in right hand part of Figure 8. We can also account for the effects of higher modes and self-weight of the struts, in addition to the single mode analysis, with pinned boundary conditions, given in the previous section.

The Elastica The equations governing the elastic behavior of slender systems are cast in terms of arc-length coordinates (S, θ), with appropriate boundary conditions. Referring to the configuration shown in Figure 9 the

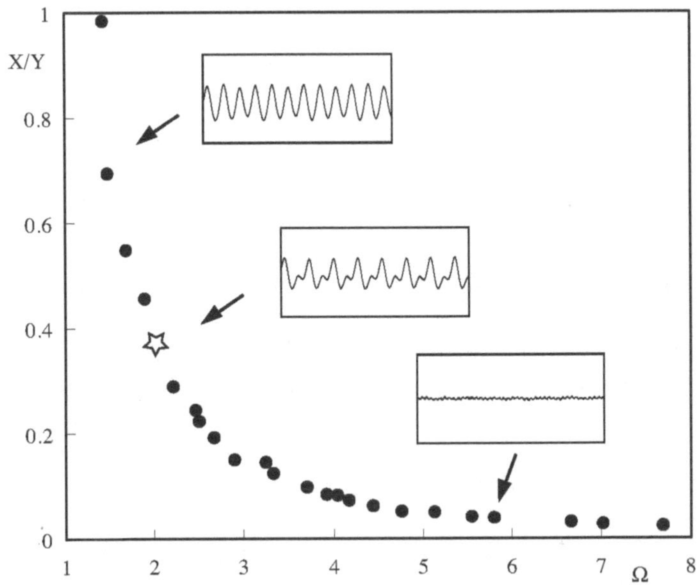

Figure 7. Transmissibility for the displacement of the strut-supported mass.

governing equations can be written as

$$\begin{array}{ll} \partial X/\partial S = \cos\theta, & \partial Y/\partial S = \sin\theta, \\ \partial \theta/\partial S = M/EI, & \partial M/\partial S = Q\cos\theta - P\sin\theta, \\ \partial P/\partial S = -W\sin\beta, & \partial Q/\partial S = -W\cos\beta, \end{array} \tag{7}$$

where the orientation is accounted for by β such that $\beta = \pi/2$ corresponds to a vertical configuration, and the component weight per unit length due to gravity is W. The last two equations can then be augmented for dynamics by including inertia and damping terms.

It is convenient to nondimensionalize the governing equations, and in order to do this we introduce the following variables

$$\begin{array}{c} w = WL^3/EI, \quad x = X/L, \quad y = Y/L, \quad s = S/L, \\ p = PL^2/EI, \quad q = QL^2/EI, \quad m = ML/EI, \end{array} \tag{8}$$

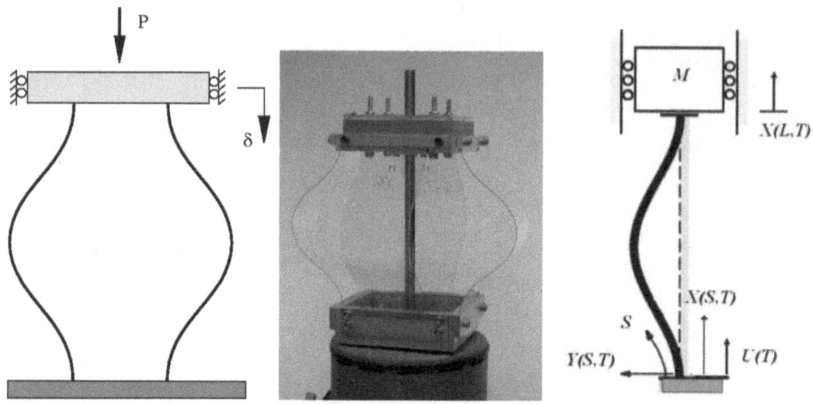

Figure 8. Using four symmetrically placed polycarbonate panels as stiffness elements.

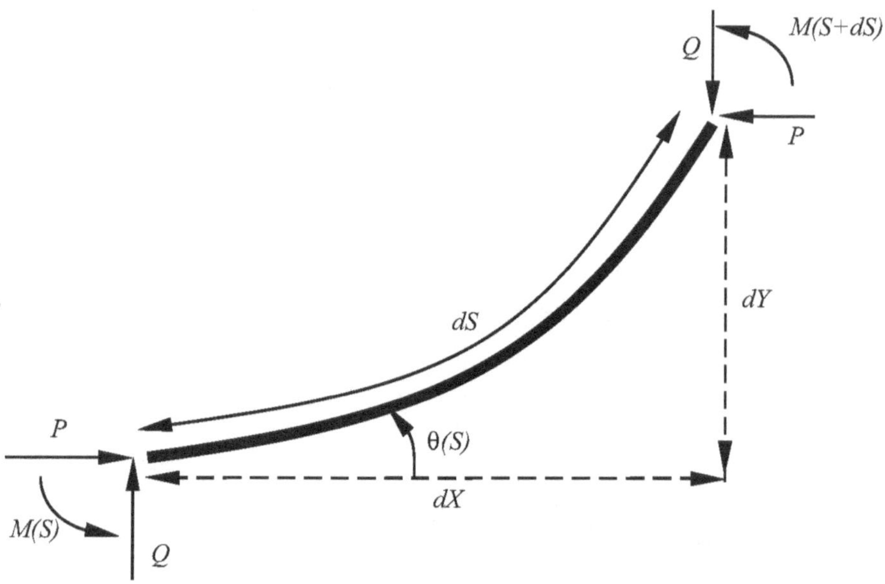

Figure 9. The geometry of an elastica.

and thus the equations become

$$\begin{aligned}
\partial x/\partial s &= \cos\theta, \\
\partial y/\partial s &= \sin\theta, \\
\partial\theta/\partial s &= m, \\
\partial m/\partial s &= q\cos\theta - p\sin\theta, \\
\partial p/\partial s &= -w\sin\beta - \partial^2 x/\partial t^2 - c\partial x/\partial t, \\
\partial q/\partial s &= -w\cos\beta - \partial^2 y/\partial t^2 - c\partial y/\partial t.
\end{aligned} \tag{9}$$

where the additional nondimensional parameters have been introduced:

$$\begin{gathered}
c = CL^2\sqrt{g/WEI}, \\
t = (T/L^2)\sqrt{EIg/W}, \quad \Omega = \omega L^2\sqrt{W/EIg}.
\end{gathered} \tag{10}$$

These equations can be solved by assuming small amplitude (harmonic) vibration about an equilibrium configuration. A convenient approach is based on the shooting method (Stoer and Bulirsch (1980)) incorporating the appropriate boundary conditions.

For use as spring elements in a vertical vibration isolator context it is also convenient to introduce a non-dimensional end-shortening parameter, $\delta = \Delta/L$, resulting from the static deflection due to the mass load, which itself can also be non-dimensionalized using $f = MgL^2/EI$. For the clamped-clamped column configuration we have the boundary conditions

$$\begin{gathered}
x(0,t) = u(t), \quad y(0,t) = y(1,t) = \theta(0,t) = \theta(1,t) = 0, \\
p(1,t) = f + f\frac{\partial^2 x(1,t)}{\partial t^2}.
\end{gathered} \tag{11}$$

Some typical equilibrium configurations are shown in Figure 10 in terms of a load-end shortening relation in part (a) for values of the 'weight' parameter ranging from $w = 0$ for the uppermost curve to $w = 50$ for the lowest curve. We see the familiar critical load for a clamped beam at $f_{cr} = 4\pi^2$, i.e., four times greater than the pinned-pinned case. Part (b) shows some post-buckled equilibrium configurations for $f = 40.5$ to $f = 62.5$ when $w = 0$.

Figure 11 shows both theoretical and experimental results for a specific column. Here, the parameters of the problem result in a value of $w = 1.23$ for which it can be shown that $p_c = 38.86$. The effect of a small initial geometric imperfection can be seen in the vicinity of the critical load. The two different symbols for the experimental data represent deflections in both directions (the system is nominally symmetric). Some sample free vibration mode shapes are also shown in Figure 11 for the specific case $f = 40$ and $w = 0$, again slightly buckled, but assuming negligible self-weight. At this value of the force-deflection relation (the operating point) these mode shapes

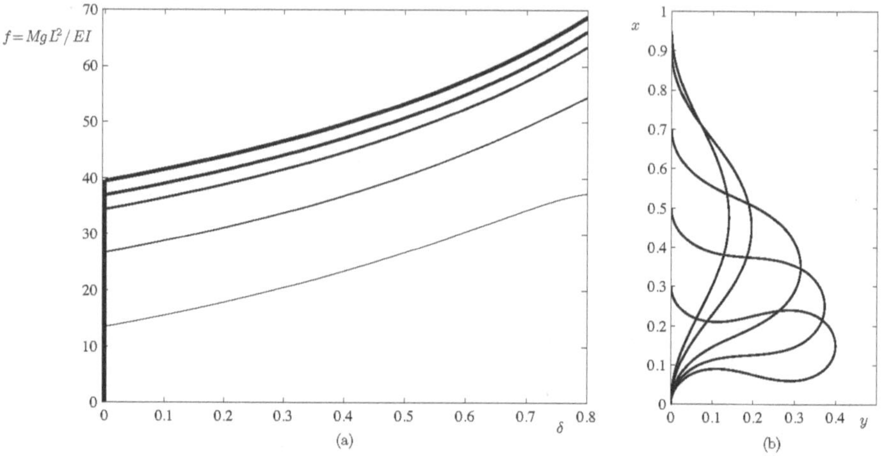

Figure 10. The static behavior of a strut based on an elastica analysis, (a) force-deflection, (b) deflected shapes.

correspond to the natural frequencies $\Omega_1 = 0.698, \Omega_2 = 44.73$ and $\Omega_3 = 75.26$.

If the system is then subjected to base excitation of the form $u(t) = u_0 \sin \Omega t$, in which $u_0 = U_0/L$, and a little linear viscous damping is incorporated, the transmissibility can be computed (again using the shooting method), and Figure 12 shows a typical frequency response for the case $w = 0, f = 40$. The resonant peaks at the natural frequencies can be clearly seen, together with the anti-resonances at $\Omega = 4.13$ and $\Omega = 103.81$.

An equivalent experimental result can be found by placing a mass of $M = 2.537$ Kg on the end of the strut (in this case 4 identical struts for practical reasons) such that $f = 50.25$. For an underlying natural frequency of 0.85 Hz (again corresponding to $w = 1.23$) we then sweep through a range of forcing frequencies to obtain the results displayed in Figure 13. Three numerical curves are shown (for $c = 5, 15$ and 25). The agreement is not especially good. This is probably due to the inherent sensitivity of the strut geometry as well as issues concerning the differences between the struts used in the experimental configuration. However, the basic concept of using a nonlinear stiffness to provide a tunable vibration isolation system is not limited to buckled structures, and the next section opens the scope to consider the use of highly deflected structural configurations for potential

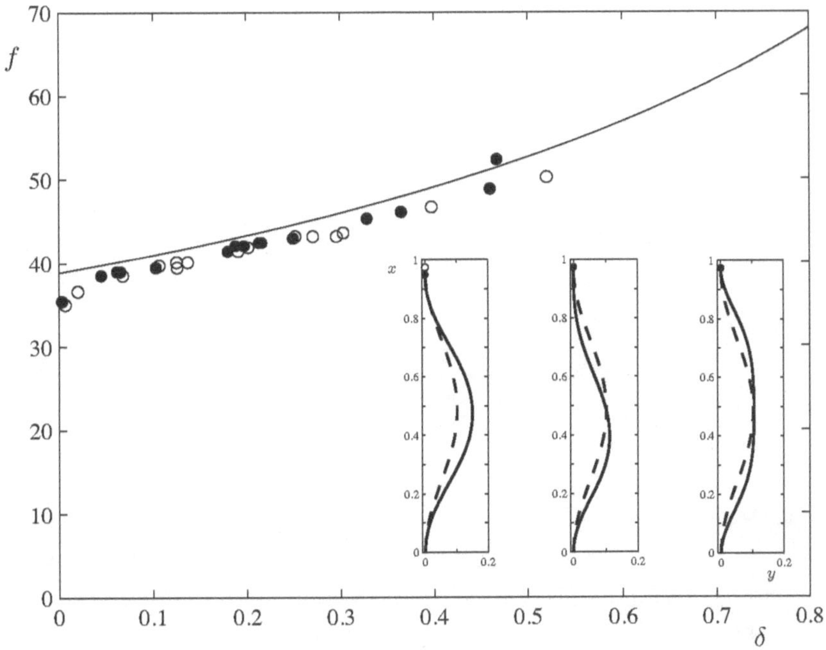

Figure 11. The force-deflection of a strut with experimental data superimposed. The inset shows equilibrium (continuous lines) and vibration mode shapes (dashed lines).

use as suspension elements.

3 Highly deformed elastic elements

The buckled elements described in the previous section have the limitation that the axial load must remain greater than the buckling load, otherwise 'stroke-out' will occur. This is true, of course, for conventional helical springs used in suspension systems. It was also shown in the previous section that the stiffness of an axially loaded element may change (too) rapidly in the vicinity of critical loading. However, we can extend the concept of utilizing highly nonlinear equilibrium configurations of elastic structures

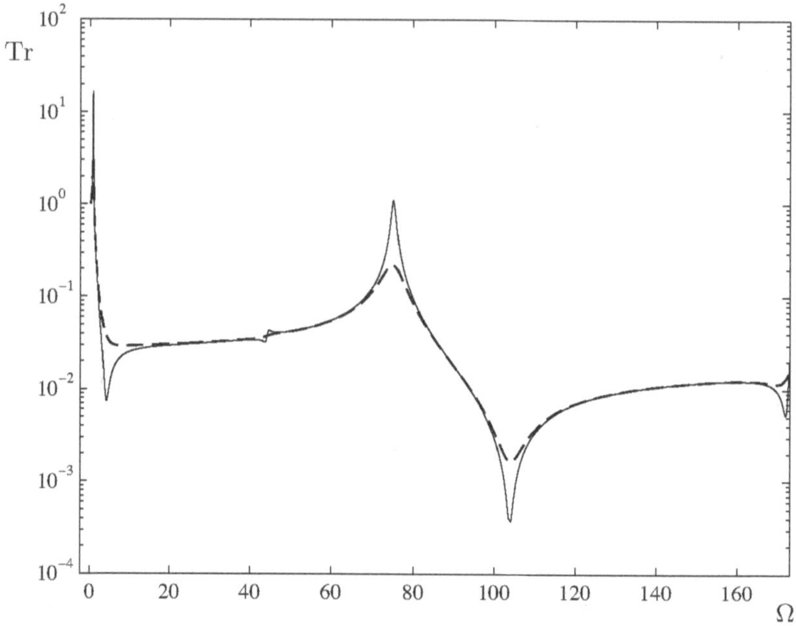

Figure 12. Transmissibility for the clamped-clamped strut, continuous line $c = 1$, dashed line $c = 5$.

not necessarily associated with post-buckling. In this case it is strictly necessary to use an elastica analysis so that no restrictions are placed on the deformation (assuming elastic material behavior), and configurations can be explored that exploit vibration isolation features. Again the local stiffness (defined by a given pre-load and geometry) can be adjusted rather easily.

There are a number of equilibrium configurations in which an elastic strip can be deformed in order to provide a spring support for an isolation system. Some examples are shown in Figure 14. In each of these cases the nonlinear P v δ relation allows some freedom in the choice of stiffness, such that a low stiffness is obtained even with a relatively large pre-load. From this point we shall focus on the specific shape of a 'pinched loop' as indicated by the lower right picture in Figure 14. In this case, the isolator is a thin strip that is bent so that the two ends are clamped together, forming a loop. It will be shown that the stiffness of the system can be conveniently altered. The clamped ends are attached to an excitation source and the

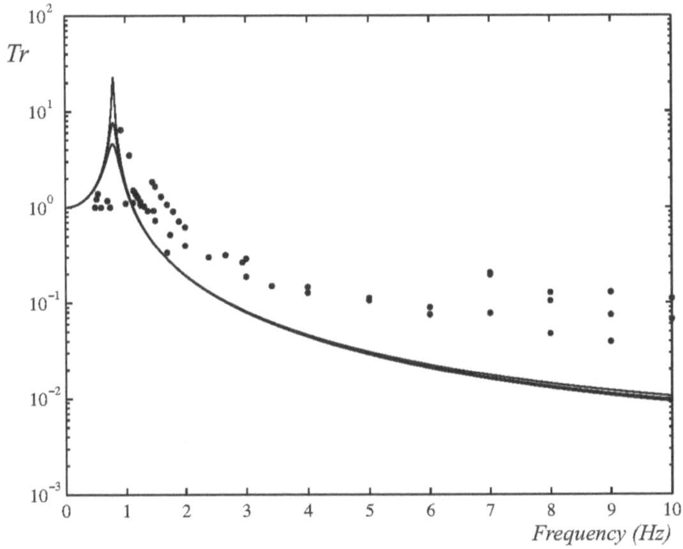

Figure 13. Transmissibility for the clamped-clamped strut with experimental data superimposed.

supported system is attached at the loop midpoint directly above the base.

The strip is modeled as an elastica, and the resulting nonlinear boundary value problem is again solved numerically using a shooting method (Santillan et al. (2005)). First the equilibrium shapes of the loop with varying static loads and lengths are obtained. The analysis reveals a large range of stiffness tunability; the stiffness is dependent on the geometric configuration, which itself is determined by the supported mass, loop length, and loop self-weight. Free vibration frequencies and mode shapes are also found. Finally, the case of forced vibration is studied, and the displacement transmissibility over a large range of forcing frequencies is determined for varying parameter values. Experiments using polycarbonate strips are conducted to verify equilibrium and dynamic behavior.

The basic equations of the loop are the same as in equation 9. However, the boundary conditions are different, and since the length is used as a control parameter it is convenient to introduce a new nondimensional parameter $a = (EI/W)^{1/3}$, and thus $'L'$ is effectively replaced by $'a'$ in

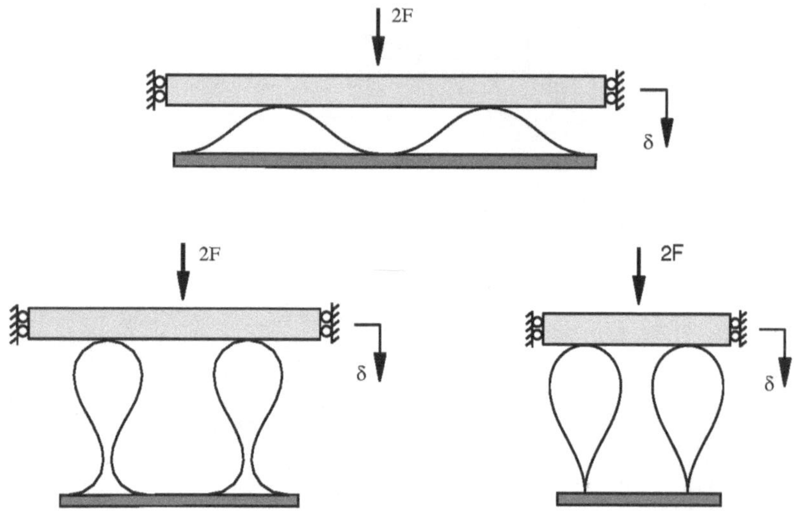

Figure 14. Some potential supporting springs based on highly deformed elastica.

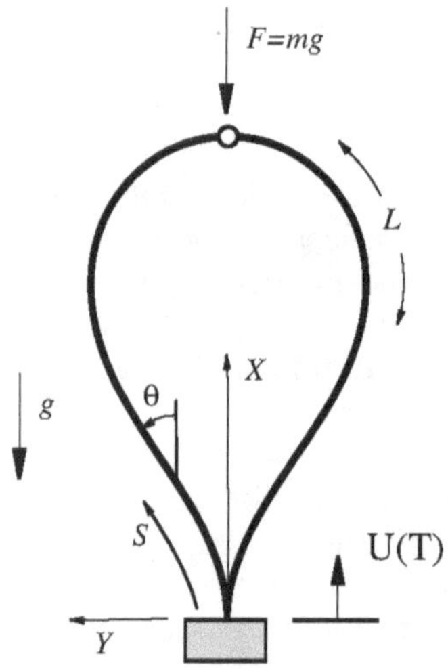

Figure 15. Schematic of the pinched loop spring.

equations 8, and this also simplifies the time scaling to:

$$t = T\sqrt{g/a}, \quad \Omega = \omega\sqrt{a/g}, \tag{12}$$

where ω is a dimensional vibration frequency. To this is added a nondimensional mass load (applied to the top of the loop)

$$f = Fa^2/EI, \tag{13}$$

where $F = Mg$. Due to symmetry only half the loop is analyzed, ($0 \leq s \leq l/2$, where the point load at $s = l/2$ is $f/2$). The governing elastica equations using the defined nondimensional parameters are similar to those given in equations 9. The boundary conditions are given by

$$\begin{gathered} x(0,t) = u(t), \quad y(0,t) = \theta(0,t) = 0, \\ y(l/2,t) = 0, \quad \theta(l/2,t) = -\pi/2, \\ p(l/2,t) = f/2 + (f/2)\partial^2 x(l/2,t)/\partial t^2. \end{gathered} \tag{14}$$

3.1 Stiffness in the vertical direction

For the equilibrium solution, the nondimensional length, l, and supported system mass parameter, f, are specified, and numerical integration is performed together with the shooting method. The resulting vertical mass deflection (or vertical deflection at the top of the loop at $s = l/2$) is called δ, where

$$\delta = x_e(l/2)|_{f=0} - x_e(l/2). \tag{15}$$

Solving the boundary value problem with $f = 0$ (for the case $l = 2$) results in the equilibrium configuration given by the uppermost curve in Figure 16. The other curves in Figure 16 show static shapes for increasing load increments of $f = 10, 20, 30, ..., 90, 98.6$. The final curve corresponding to $f = 98.6$ gives the static configuration just before self-contact. Intuitively we expect the stiffness to be relatively low near this operating point. The second curve from the top in Figure 16 corresponds to this case ($l = 2$) in a plot of f vs. δ. The other curves correspond to loops of varying lengths. Each curve is continued until self-contact occurs. As can be seen, the local vertical stiffness (the slope of the curve about any point in Figure 16(b)) varies from a high value, near the origin, to a low value for a relatively long loop with a pre-load.

Load-deflection results for $l = 1.7416$ are shown in Figure 17, where the data points represent experimental data. Note that upward (stretching) loads (i.e., $f < 0$) are included. Fixing the pre-load f then determines the static deflection δ and thus the loop spring constant is defined as the local slope about this (operating) point.

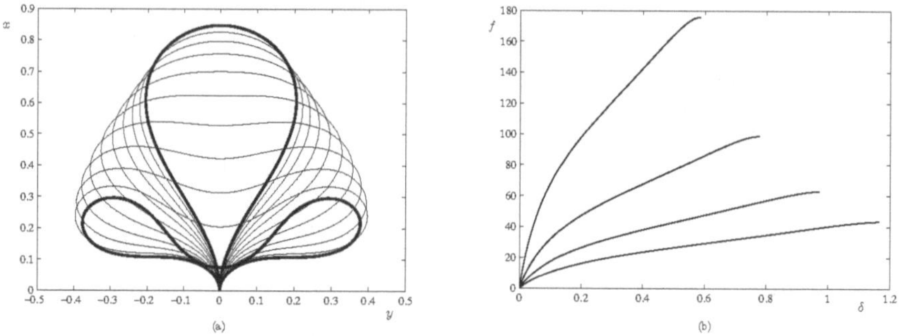

Figure 16. Numerical equilibrium results. (a) Static configurations for $l = 2$ and $f = 0$ (uppermost curve), 10, 20, ..., 90, 98.6. (b) Force-deflection curves for lengths $l = 1.5$ (uppermost curve), 2, 2.5, 3.

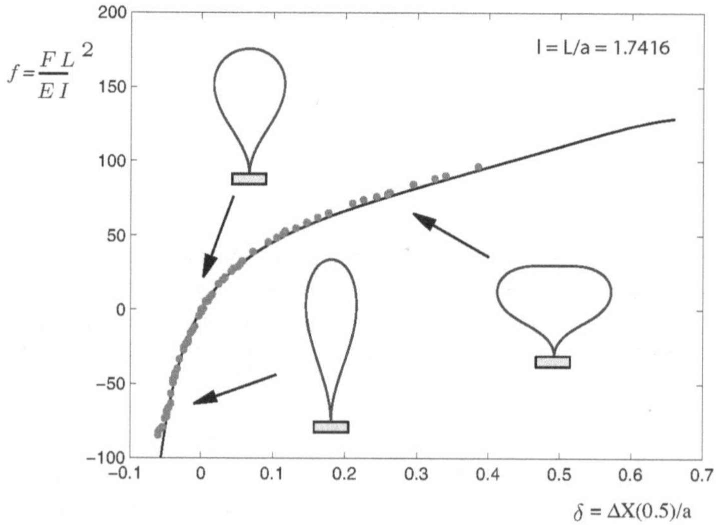

Figure 17. Force-deflection curve for $l = 1.7416$. Solid curve: numerical results; data points: experimental results.

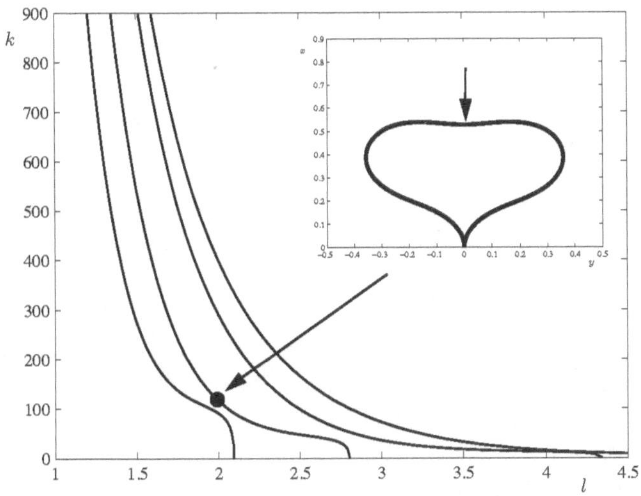

Figure 18. Vertical stiffness ($k = df/d\delta$) as a function of length, l, for (from left to right) $f = 90$, 50, 20, 10.

Figure 18 shows the computed initial spring constant, k, where now the vertical stiffness $k = df/d\delta$ is variable, determined by the length of the loop. Results are plotted for four specific pre-load values f. Not surprisingly, the longer the loop, the less stiff it is (in the vertical direction) for a given point force. Thus, the pre-load (which would typically correspond to additional dead-weight of the system to be isolated) can be used to tailor the stiffness to a value suitable for isolation. Alternatively, and more realistically, the stiffness may be tailored by changing the length of the loop. Also shown in this figure (in the inset) is the specific equilibrium shape when $f = 50, l = 2$.

It should also be mentioned here that the nonlinear nature of the stiffness characteristics can also be problematic if taken too far. For example, if we continue plotting the force-deflection behavior for even larger deflection we get the result shown in Figure 19. However, the final part of this curve is not physical, since self-contact would have occurred by then. It is also likely that the material would suffer plasticity effects if deformed to this extent.

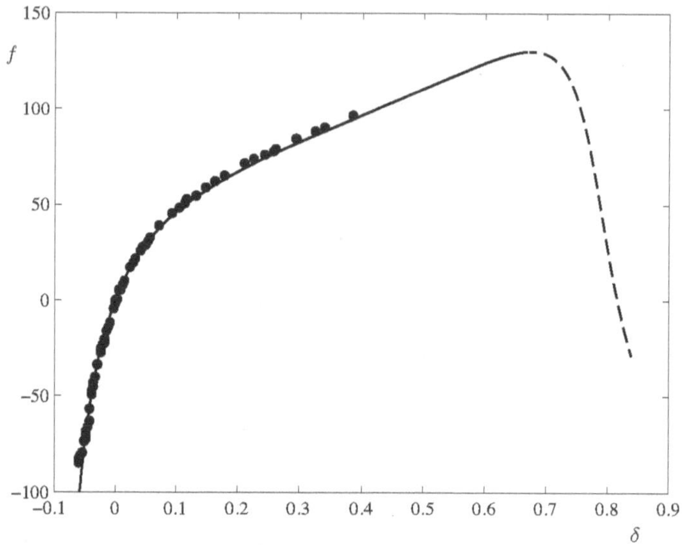

Figure 19. Loss of stiffness for grossly deformed geometry.

3.2 Free Vibration

Extending the analysis for dynamics, free vibration frequencies and mode shapes for the system are found using the equilibrium solution (for given values of l and f and $c = 0$) and adding frequency as an additional unknown parameter in the shooting procedure. The lowest (symmetrical) frequency is plotted in Figure 20 for the four nondimensional lengths $l = 3$, 2.5, 2, and 1.5. For each length, the fundamental frequency is found up to self-contact, which, for $l = 3$ and 2.5, occurs before the system becomes unstable. When this occurs, the frequency drops to zero, and the structure is unable to statically support the mass. Without the mass load ($f = 0$) and a nondimensional length of $l = 2$ the fundamental natural frequency is $\Omega = 21.1$, and it can be shown that the higher (symmetric) frequencies can be found at $\Omega = 59.6$ and 120.6, and they generally reduce with increasing f. Within this figure (in the inset) is an example of the corresponding mode shape for the specific case of $l = 2.5, f = 10, \Omega_1 = 6.1$. The symmetric constraint prevents the appearance of the even number modes. We also see from this figure that, unlike the buckled springs from earlier, this system

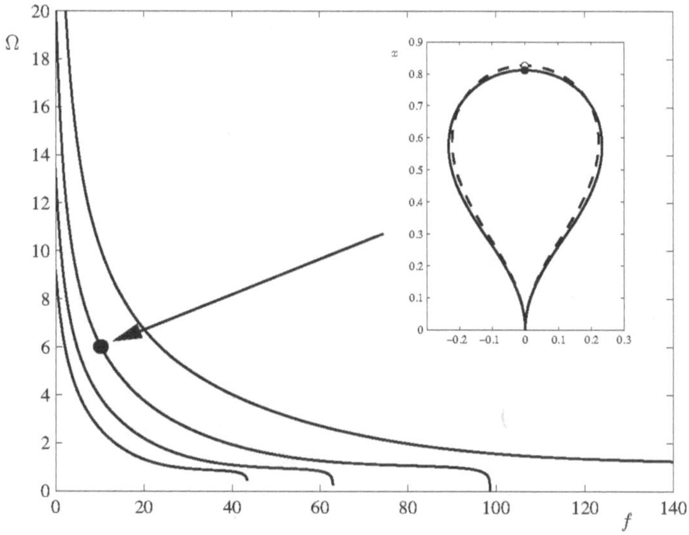

Figure 20. Lowest frequency as a function of f for (from left to right) $l = 3, 2.5, 2, 1.5$.

does not need a minimum load (mass) to be effective.

4 Transmissibility

Forced vibration is again considered by assuming a vertical base excitation, $u(t)$, that is sinusoidal, i.e.,

$$u(t) = u_o \sin \Omega t. \tag{16}$$

The nondimensional forcing amplitude, u_o, is defined as $u_o = U_o/a$, where U_o is the dimensional excitation amplitude. Again the forced elastica equations are solved numerically using the shooting method. The resulting displacement transmissibility is then defined as

$$Tr = \frac{|x_d(l/2)|}{u_o}. \tag{17}$$

Numerical and experimental transmissibility results in the vicinity of the first resonant frequency are shown in Figure 21, where the smooth curve represents numerical results. The dashed horizontal line represents a transmissibility value of unity, i.e, where the amplitude of the input and output are equal. In Figure 21 the results correspond to a nondimensional length

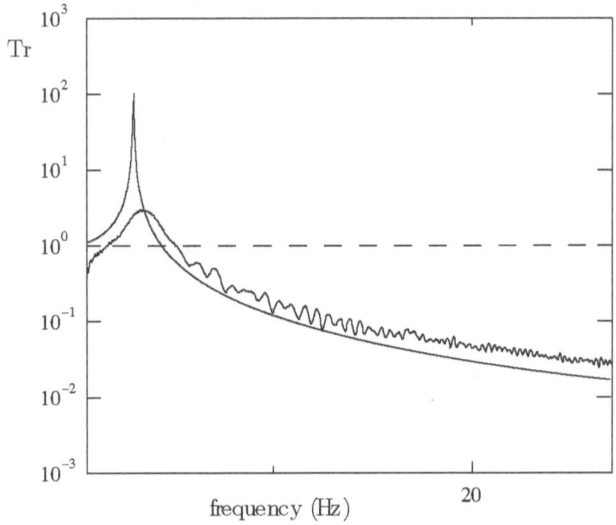

Figure 21. Experimental and numerical transmissibility result for $l = 1.987, f = 29.29$.

$l = 1.987$, where $f = 29.3$. The natural frequency for this case is found numerically using the free vibration equations, giving 3.33 Hz. The experimental transmissibility peaks occur at 3.68 Hz. Clearly there is some discrepancy in the damping model.

There are of course higher frequencies and mode shapes present in the response. If we extend the frequency range for both theory and experiment we get the results shown in Figure 22. The anti-resonances are difficult to detect in the experiments due to the effects of noise.

In both the numerical and experimental results, there is a degree of attenuation of the transmissibility for forcing frequencies above the first resonance, and this resonant frequency can be tuned by either altering the loop length or static mass of the system.

While in this case the mass motion is constrained to be purely vertical, the pinched loop isolator system is one that may also be studied as a horizontal motion isolator or even a system that isolates motion in both di-

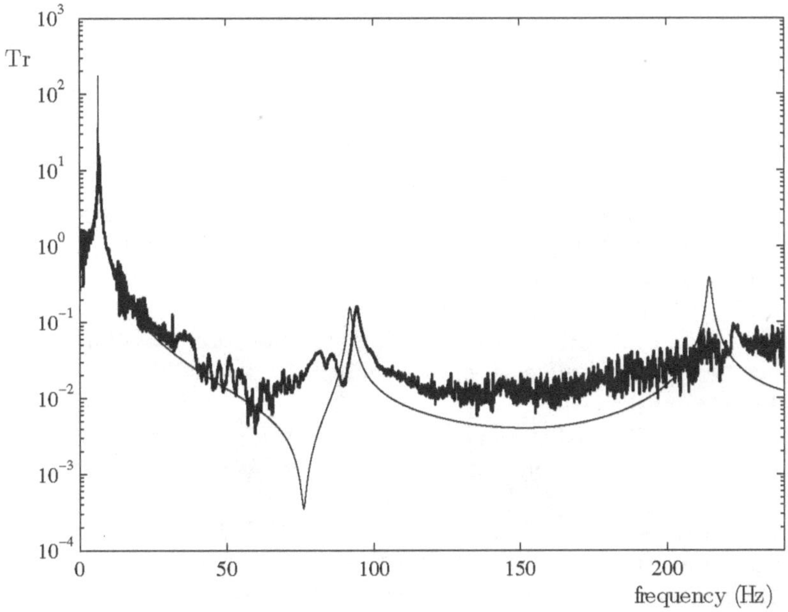

Figure 22. Experimental and numerical transmissibility result for $l = 1.479, f = 29.29$, including the higher frequency range.

rections simultaneously. This system characteristic makes the pinched loop isolator particularly advantageous over a helical spring system, where the stiffness is fixed, and the motion must be limited to be exclusively in one direction.

Acknowledgements

The material on which this chapter is based was developed with the assistance of Sophia Santillan (USNA) and Ray Plaut (Virginia Tech). Their contributions are gratefully acknowledged.

Bibliography

A. Carrella, T.P. Waters, and M.J. Brennan. Optimisation of a passive vibration isolator with quasi-zero-stiffness characteristic. ISVR technical memorandum, No. 960, University of Southampton, 2006.

D.L. Platus. Negative-stiffness-mechanism vibration isolation systems. *Vibration Control in Microelectronics, Optics, and Metrology*, 44, 1991.

E.I. Rivin. Vibration isolation of precision objects. *Sound and Vibration*, 40:12–20, 2006.

S.T. Santillan, L.N. Virgin, and R.H. Plaut. Equilibria and vibration of a heavy pinched loop. *Journal of Sound and Vibration*, 288:81–90, 2005.

J. Stoer and R. Bulirsch. *Introduction to Numerical Analysis*. Springer-Verlag, 1980.

L.N. Virgin. *Vibration of Axially Loaded Structures*. Cambridge University Press, Cambridge, UK, 2007.

L.N. Virgin and R.B. Davis. Vibration isolation using buckled struts. *Journal of Sound and Vibration*, 260:965–973, 2003.

L.N. Virgin, S.T. Santillan, and R.H. Plaut. Vibration isolation using extreme geometric nonlinearity. *Journal of Sound and Vibration*, 315:721–731, 2008.

J. Winterflood, T. Barber, and D.G. Blair. High performance vibration isolation using spring in Euler column buckling mode. *Physics Letters A*, 19:1639–1645, 2002.

S.E. Woodard and J.M. Housner. Nonlinear behavior of a passive zero-spring-rate suspension system. *Journal of Guidance*, 14:84–89, 1991.

J.Z. Zhang, D. Li, M.J. Chen, and S. Dong. An ultra-low frequency parallel connection nonlinear isolator for precision instruments. *Key Engineering Materials*, pages 257–258, 231–236, 2004.

Designing Nonlinear Torsional Vibration Absorbers

Steven W. Shaw
Department of Mechanical Engineering
Michigan State University
East Lansing, MI, 48824, USA

Abstract We consider issues related to the design of mechanical systems that feature nonlinear vibratory behavior. Designs that account for, and also exploit, nonlinear dynamics are considered. The chapter provides a brief overview of general considerations related to these issues, and then considers in detail the case of vibration absorbers that are used to reduce torsional vibrations in rotating systems. These absorbers, which are being developed for use in automotive engines, consist of centrifugally driven masses that ride on user-specified paths relative to a rotor that is being driven by a fluctuating torque. The absorbers are used to reduce the torsional vibrations of the rotor by providing a torque on the rotor that counteracts, at least partially, the applied torque. The mathematical model of this system represents, under assumptions on the system parameters consist with practical applications, a system of N identical nonlinear oscillators with weak damping, weak global coupling, and weak near-resonant excitation. The nonlinearity in the oscillators depends on the path, and the goal of this nonlinear dynamic design problem is to select a path that provides good vibration reduction over a specified range of torque amplitudes. The desired response has all N absorbers behaving identically, that is, in a synchronous manner. However, the structure of this system leads to two distinct types of dynamic instabilities, and both must be avoided for a given design to be feasible. In this work we examine the synchronous absorber response in terms of vibration reduction, torque range, and stability. It is shown that a particular epicycloidal path provides good performance, and can be made stable. These results are a summary of previous work described in (30).

1 Accounting for Nonlinear Behavior in the Design of Vibratory Systems

The engineering design of mechanical systems can be based on static principles when inertial effects are not important, but in many applications one must account for dynamic effects. In some of these applications, linear models provide sufficient information about important dynamic behavior, for example, for determining resonant frequencies and modes of vibrations in structures (11). In such cases, two very convenient properties hold: (i) superposition, which allows one to view each vibrational mode as independent, and thus one can predict behavior near resonances using linear single degree of freedom models, and (ii) scaling of response, so that the response amplitude is directly proportional to the level of excitation driving the system. These principles allow for the use of powerful analytical and computational tools, but they also limit the types of response behaviors that can be predicted by models. In fact, in many cases, linear models break down in terms of predictive capabilities, and nonlinear effects must be taken into account. In these cases, one must turn to more sophisticated analysis tools, since superposition and scaling no longer hold. The ultimate tool for evaluating any engineering system is an experimental study, but these are typically costly, time consuming, not very useful for developing fundamental understanding, and usually not viable for extensive parameter studies. Simulation studies share many of these shortcomings, and while less expensive to run, require the development of a reliable mathematical model. When available, analytical tools, applied to idealized models, are very useful for developing physical insight into general features of system response, and how it depends on system and input parameters. The insight gained about nonlinear system responses from such studies is analogous to that developed by analyzing simple linear systems, and it forms the basis for much of the insight one can use when designing with nonlinear behavior in mind.

A dynamic system responds in a time-varying manner, typically to some type of inputs. These inputs include impulses, such as those that arise in impact problems, harmonic and other time-periodic forces, such as those that arise in rotating systems, and random forces, such as those that arise, for example, from turbulent fluid loading or earthquakes. These inputs cause system responses that must satisfy certain specifications in order to perform as desired. Common types of undesirable responses to dynamic loading include: excessive amplitudes, fatigue failure, and instabilities leading to behavior such as flutter or chaos. In order to design systems that meet the response specifications, one must understand the system input-output characteristics. For linear systems, the understanding of system response to simple inputs, coupled with the superposition principle, provides a powerful tool for determining these characteristics. Similarly, for nonlinear systems, a lot can be learned about what is possible, in terms of system responses, by considering simple examples. However, superposition does not hold, which leads to more rich, and interesting, possibilities for the response. Still, simple examples demonstrate many of the most important, and general, features of nonlinear system response, including behavior near resonance, energy exchange between modes, bifurcations, chaos, etc.

Here we focus on the nonlinear vibratory response of mechanical systems that can be described by a few degrees of freedom. Of primary interest here are nonlinear resonant behaviors in these systems when they are subjected to periodic excitation. In order to design such systems to meet response specifications, one must be able account for nonlinear resonant behavior in some predictive manner. Following a general discussion along these lines, we provide detailed results for a practical example that allows for a thorough investigation of its nonlinear response, and tailoring that response by selection of system parameters, to achieve a certain goal. The systems of interest are centrifugal pendulum vibration absorbers, or CPVAs. The results for CPVAs presented here include modeling, analysis, and simulations; the reader is referred to (32; 23; 24; 29) for experimental results on these systems.

1.1 Nonlinear Resonances

For our general discussion, we focus on the following features of nonlinear resonant behavior: effects of nonlinearities on isolated resonances, nonlinear modal interactions, and transients leading to resonant steady state response. Some fundamental knowledge about nonlinear vibrations, as outlined in Chapter One of this volume, is assumed on the part of the reader.

For an isolated resonance to occur the following conditions must hold: the excitation must have a frequency component near an isolated natural (modal) frequency of the system, the excitation must be able to put energy into that mode, and dissipation must be relatively small. For linear systems operating near such a resonance, dissipation plays the essential role in limiting the response amplitude. However, in lightly damped systems, the effects of frequency pulling that arise from nonlinear stiffness, either hardening or softening, will limit the amplitude above a certain excitation level. In addition to limiting the response near resonance, this frequency pulling, which derives from the fact that nonlinear stiffness generally causes the frequency of free oscillation to depend on amplitude, causes the resonance curve to bend over, thereby widening the range over which large amplitude responses occur. This bending can be large enough to cause bistability in the frequency response, that is, the existence of two (or more) stable steady-state responses coexisting over a certain frequency range, in which case the steady state response ultimately achieved depends on initial conditions. One of these steady states is essentially that of the linearized system, and the other is a large amplitude nonlinear response in which the nonlinear (amplitude dependent) natural frequency roughly matches the driving frequency. These stable operating states are separated, in a particular sense that delineates the initial conditions, by an unstable response. This unstable response is of saddle type, and its stable manifold, also known as its separatrix, or inset, provides the boundary between initial conditions that lead to the two possible stable steady state responses. The details of this frequency response can be found in any text on nonlinear vibrations, e.g., (22; 9)

Of interest here are the features of such a response that affect the design of systems operating near such a nonlinear resonance. One such feature is that in systems with noise or other disturbances, the bistability can be problematic, since a well-behaved system with a linear-looking response away from the linear resonance, can be bumped by the disturbance, via a transient, into a large amplitude, sustained response. In fact, this nonlinear resonant response will persist if one changes the excitation amplitude and/or frequency. The only way to get the system back to the linear response is to provide another disturbance, in which case it may or may not go to the desired response, or to vary the excitation parameters, until the system reaches the point where the nonlinear response branches meet in a saddle-node bifurcation (9). In some systems, with very light damping, this may be at a value that is twice the resonant frequency, in fact, in a frequency range where other modes come into play.

In a linear system model, individual modes are uncoupled. In discrete systems with distinct natural frequencies (eigenvalues) and a special form of damping (4), the modeshapes (eigenfunctions) are independent of the damping and represent standing waves of vibration of the structure, that is, the nodal points are fixed. For more general forms of damping, the modes are traveling waves. For all such cases, each system mode is described by a damped single degree of freedom system. When driven by external excitation, if the frequency (or frequency content) of the excitation is near a natural frequency, the response is dominated by that mode, but will contain small amounts of other system modes. When a system has repeated frequencies, the response near that resonance is typically composed of a linear combination of the associated natural modes, with each mode participating substantially. If the system is nonlinear, such repeated frequencies can result in responses not predicted by linear theory; this occurs in the pendulum absorber example considered below. Also, under some conditions, described in more detail below, system modes with well separated frequencies can interact through nonlinear effects, and the resulting response may involve more than one mode in a substantial way (21).

To make these notions more specific, consider a system with two modes with repeated natural frequencies. Classic examples of this occur in the vibrations of taut strings, square and circular plates, and beams with square or circular cross-section. Let the system modal amplitudes and frequencies be denoted by x_i $i = 1, 2$ and $\omega_1 \approx \omega_2$, respectively, and suppose that the response of each mode is harmonic at a frequency $\omega \approx \omega_1$. If nonlinear coupling between modes exists such that mode j has terms of the form $x_i x_j^2$, then mode j will see a resonant forcing from mode i through this term. Generally, these terms arise from nonlinear coupling, arising from a potential, that allows cross-talk between the modes, which can result in responses not predicted by linear models (21). Similarly, if there exist modes with a 3:1 frequency ratio, the same nonlinear terms, when considered in terms of their Fourier components, generate harmonics that promote energy exchange. Another common internal resonance encountered in mechanical systems is that of a 2:1 frequency ratio, in which case quadratic modal coupling terms lead to the conditions needed for internal resonance. Indeed, there are a multitude of such possibilities for internal resonances, involving multiple modes for which the ratios of natural frequencies are close to rational numbers, p/q, where p and q are positive integers. Generally, resonances with higher values of $p + q$ have smaller basins of attraction, and are easily suppressed by dissipation. Thus, from a mechanical design point of view, the resonances discussed above are the most relevant, and even these occur only in lightly damped systems. A common class of systems where they are encountered are those with specially symmetries, such as the absorber problem considered in detail below.

We now turn to a practical example in which one can design the system nonlinearity to avoid both bistabiliy and modal interactions, either of which leads to undesirable system behavior.

2 Centrifugal Pendulum Vibration Absorbers

Many rotating systems experience engine-order excitation, which are loads that are synced with the rate of rotation in such a manner that the frequency of excitation is proportional to the rotor spin rate, denoted Ω. The excitation order n is the constant of proportionality, so that the frequency of excitation is $n\Omega$. Rotor systems are designed to minimize such loads, for example, by dynamic balancing, but in many cases such loads are unavoidable. For example, internal combustion engines experience engine-order torsional loading on the crankshaft, arising primarily from cylinder gas pressure fluctuations that act on the crankshaft through the piston/connecting rod linkage. In some applications it is essential to reduce these torsional vibrations. For example, in light aircraft engines, these fluctuating loads cause wear and fatigue on the gearing between the engine and propellor. Until quite recently, these torsional vibrations have not been an issue for automotive engines, where torque converters, multi-cylinder engines, flywheels, etc., help soften their effects. However, pressures for improved fuel economy, without sacrificing performance or comfort, have pushed engine designers into more aggressive operating regimes, specifically, those with higher in-cylinder gas pressures and lower engine speeds, both of which improve engine efficiency, but also result in more severe vibrations, caused primarily by powertrain torsionals.

Several methods exist for mitigating the effects of these torsional loads, including simple flywheels, so-called harmonic balancers, and dual-mass flywheels (DMFs). The first of these simply adds DC inertia to the rotor, thereby lowering the levels of vibration, but at the highly undesirable price of more sluggish system responsiveness. The harmonic balancer and DMF are frequency tuned devices that address torsional vibrations only at a specific engine speed corresponding to a troublesome resonance, or, in the case of very soft spring DMFs, they isolate powertrain components downstream from the engine torsional vibrations at low engine speeds. These have proved satisfactory for several decades, but it is becoming increasing clear that more sophisticated methods will be needed to push engine operating envelopes. Passive vibration suppression is the only practical solution for these applications, due to power and reliability requirements.

CPVAs have been around for about 100 years and are widely used for reducing engine-order torsional vibrations in the aerospace industry, specifically light aircraft engines and helicopter rotors (13; 34; 15). Over the past ten years these devices have started to receive attention for use in automotive engines (25; 33). A primary target application are engines that utilize cylinder deactivation, wherein cylinders are shut down when the engine can provide the needed power with fewer cylinders. While the amount of fuel per cylinder must be increased to meet torque demand, efficiency is improved by reductions in engine pumping losses. In a preliminary study on a V8 engine that could cut off four cylinders, the addition of CPVAs allowed the engine to operate all the way down to fully loaded idle conditions, a significant improvement (25). It is expected that CPVAs will find many automotive applications in the next few years.Another application is for boosted engines, which have relatively high cylinder compression and operate with higher efficiency, but also generate strong torsional vibrations. Fuel saving enhancement is also aided by widening the operating range for torque converter lockup; a torque converter is a rotational coupling device between the engine and transmission that is generally quite dissipative. It becomes much more efficient if locked into a rigid connection by a clutch, and the conditions for lockup include low levels of engine torsional vibrations. So, by expanding this range improves fuel economy. While many of these implementation have yet to be realized, the LuK Corporation has developed a combined DMF/CPVA system that smoothes out vibrations, and is now available in some production vehicles (12; 33).

2.1 Basic Operation of Pendulum Absorbers

In its essence, a CPVA is a passive device that, when properly tuned, functions like a torsional tuned mass damper (TMD). Specifically, it is a secondary inertia, in this case a centrifugally driven pendulum, attached to a primary device, in this case a rotor, designed so that when the primary device is subjected to fluctuating loads, in this case engine-order torques, the pendulum responds in such a manner that it cancels, at least partially, the resulting fluctuations of the rotor speed. A key difference between CPVAs and the usual TMD is that the pendulum restoring force is provided by centrifugal effects, not an elastic spring. This restoring force scales like Ω^2, where Ω is the mean rotor spin rate, and thus the natural frequency of the pendulum is proportional to Ω. The constant of proportionality, denoted here by $\tilde{n}$, sets the linear tuning for the pendulum absorber, that is, its small oscillation tuning. For excitation of order n, tuning $\tilde{n} \approx n$ allows the absorber address this excitation at all rotor speeds. When considering engine order excitation, these order-tuned absorbers have a distinct advantage over frequency tuned devices. Typical TMDs, of which the DMF and harmonic balancer are examples, are effective only at one specific engine speed, typically corresponding to a powertrain torsional mode natural frequency.

When implemented as a simple pendulum, a CPVA has limited packaging space, and the center of mass (COM) necessarily follows a circular path. This path leads to a softening nonlinear response that is highly unfavorable in several aspects, as is well known (10; 26; 2), and described in detail below. An appealing aspect of these absorbers is that their nonlinear response can be altered, even tailored, by changing the path followed by the absorber COM. This is achieved by utilizing a bifilar (two point) suspension, like that shown for a helicopter rotor absorber in Figure 1, in which the absorber mass rides on a pair of rollers such that the absorber mass does not rotate relative to the rotor (13; 15; 8; 20). The surfaces along which the rollers ride, on both the absorber mass and the flange attached to the rotor, can be machined to realize a wide variety of paths for the absorber COM. In addition, this arrangement is very convenient for packaging absorbers, since the desired linear tuning is fixed solely by the diameter of the rollers and the curvature of the path at zero amplitude (8; 20).

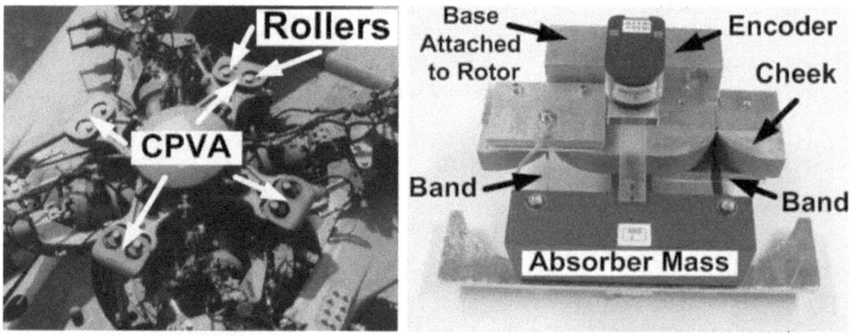

(a) (b)

Figure 1. (a) Top view of a helicopter rotor with four CPVAs, consisting of masses suspended by pairs of rollers. (b) A CPVA with bifilar suspension in which thin steel bands wrap on cheeks such that the absorber mass follows a user-defined path, in this case a particular epicycloid; the encoder is used for measuring absorber position.

The absorber works by reacting to speed fluctuations of the rotor, which cause the absorber to oscillate along its path, and this motion generates a torque on the rotor that cancels, at least partially, when properly designed, the applied fluctuating torque. The torque from the absorber acting back on the rotor scales like T_a $mcA\Omega^2$ where m is the absorber mass, $c = R_o$ is the maximum distance from the absorber COM to the rotor center, A is the peak absorber COM amplitude along its path, and Ω is the rotor speed. In practical designs the product mc is limited by hardware constraints, and thus one would like to push A to large amplitudes. However, one typically wants to achieve a maximum T_a with a minimum inertia mc, hence A will be large. This necessitates designing for nonlinear responses.

Another feature that makes these systems additionally rich is that, in order to realize substantial total absorber mass, and to dynamically balance these masses, several identical absorbers are placed around and/or along the rotor, as seen in the helicopter rotor of Figure 1. If the rotor is considered rigid, which is a valid assumption for most of these systems, this leads to a system with a high level of symmetry and repeated natural frequencies. When analyzing the system response, one must account for this internally resonant structure, in addition to the nonlinear behavior of individual absorbers.

2.2 Mathematical Model

In order to develop useful design guidelines, specifically, to select absorber inertia and path parameters to meet target response specifications, it is useful to have a predictive mathematical model for the response of a rotor fitted with multiple absorbers, expressed in terms of general system parameters. The essential model for evaluating absorber performance consists of equations of motion for a rotor and N identical absorbers, taken to be point masses riding on paths relative to the rotor. This model neglects the rollers used in many bifilar configurations, but these can be accounted for (8; 20). The system equations of motion, details of which can be found in other works, e.g. (8; 30; 19), are developed using the idealized model shown in Figure 2a. These are nondimensionalized to reduce the problem to a set of dimensionless variables and parameters. The model consists of: (i) a rigid rotor spinning at a normalized mean rate $< \nu >= 1$ where $\nu = \dot{\theta}/\Omega$ is the nondimensional rotor speed, and θ is the rotor angle; (ii) the rotor is subjected to a torque with constant and fluctuating components $\varepsilon(\Gamma_o + \Gamma(\theta)) = (T_o + T(\theta))/(J\Omega^2)$, where T's are the the physical torques, J is the rotor inertia, and ε is a scaling parameter defined below; (iii) a set of N identical absorbers each of mass m, modeled as point masses riding on identical paths relative to the rotor, where S_i is the arc-length displacement of absorber i along its path and $s_i = S_i/c$ where $c = R(0)$ is the distance from the rotor center to the furthest point on the absorber path; and (iv) dissipation effects for the absorber and rotor are modeled as equivalent viscous damping. The path for the absorbers is specified by a function $r(s) = R(cs)/c$, which is the non-dimensional radial distance from the rotor center to the absorber when it is at a displacement $S = cs$. Common paths for absorbers include circles (26), which are easy to manufacture, cycloids, proposed and used for helicopter rotors (15), and the *tautochronic epicycloid*, a path that makes the absorber linear in terms of nonlinear detuning (8). More details about absorber paths follows subsequently.

The equations of motion are derived with a Lagrangian approach using generalized coordinates θ and s_i, for $i = 1, \ldots, N$. The effects of damping on the rotor and absorber, and the torques acting on the rotor, are modeled using generalized forces. The applied torque is modeled as a DC component plus a single harmonic of order n, expressed as $\varepsilon(\Gamma = \Gamma_o + \Gamma\cos(n\theta))$. This is a valid assumption for many automotive operating conditions, since many higher harmonics cancel each other in multi-cylinder engines. The dominant excitation order for a four-stroke engine with K cylinders is $n = K/2$, that is, there are n forcing cycles per 2π rotation. Note that this torque, since it depends explicitly on the crank angle, rather than time, could be modeled using a potential. However, since the rotor never reverses direction, we use θ as the independent variable, in place of time, rendering this applied torque as a periodic perturbation in the independent variable. This change of variable results in the torque playing the role of a small perturbation, and makes the system amenable to perturbation analyses; see, for example, (5; 2; 30).

Another important feature of the equations of motion in this formulation is that the rotor speed is a dynamic state, whereas the rotor angle is the independent variable. With θ as the independent variable, the system states are the absorber displacement and speed and the rotor speed, that is, (s_i, s_i', ν), where $(\cdot)' = d(\cdot)/d\theta$. Thus, the basic system is a rotor spinning at nearly constant speed, subjected to a torque that is small (when scaled by the rotor kinetic energy), with a relatively small absorber mass riding on a specified path, with small absorber and rotor dissipation. We use a small parameter ε to conveniently scale these effects, such that when $\varepsilon = 0$ the system collapses to a rotor running freely at constant speed with absorbers of zero mass riding on their paths. For $\varepsilon << 1$ these effects are accounted for by the perturbation analysis described below. For ε we use the ratio of the rotational inertia of an absorber mc^2 to the rotor inertia J, i.e., $\varepsilon = mc^2/J$. For the perturbation scheme to be valid, $N\varepsilon << 1$ must hold, that is, the total absorber inertia must be relatively small. Additional small parameters in the model are $\varepsilon\mu$, the absorber equivalent viscous damping coefficient, $\varepsilon\mu_o$, the rotor equivalent viscous damping, and, as described above the applied torque acting on the rotor. In terms of these states and parameters, the model equations for the absorbers are given by,

$$\nu s_i'' + \big(g(s_i) + s_i'\big)\nu' + \varepsilon\mu s_i' - \nu\frac{1}{2}\frac{dr^2(s_i)}{ds_i} = 0 \quad i = 1, \ldots, N, \tag{1}$$

and that for the rotor is,

$$\begin{aligned}\nu\nu' + \varepsilon\Big\{ \sum_{j=1}^{N} \big[g(s_j)(\nu\nu' s_j' + \nu^2 s_j'') + \frac{dg(s_j)}{ds_j}\nu^2 s_j'^2 + \\ + \frac{dr^2(s_j)}{ds_j}\nu^2 s_j' + \nu\nu' r^2(s_j)\big] + \mu_o\nu - \Gamma_o - \Gamma\cos(n\theta)\Big\} = 0,\end{aligned} \tag{2}$$

where $g(s)$ is a path function, specifically, it is $r(s)$ times the cosine of the angle between the local path tangent at s and a line perpendicular to the radial line passing from the rotor center to s. This function, given by $g(s) = r(s)\sqrt{1-(dr(s)/ds)^2}$, describes the kinematic rotor/absorber coupling, by specifying the angle of the contact force acting between the rotor and the absorber.

Note that for the rotor mean speed to be $\nu = 1$, the bearing resistance and mean torque must balance to leading order, that is, $\mu_o - \Gamma_o = \mathcal{O}(\varepsilon)$, which will be assumed to hold. Then, since the absorber/rotor coupling is small and the rotor speed is nearly constant, as seen since $\nu\nu' = \mathcal{O}(\varepsilon)$, the dynamics can be uncoupled to leading order in ε, as follows. We express the rotor speed as $\nu(\theta) = 1 + \varepsilon w(\theta) + \mathcal{O}(\varepsilon^2)$, substitute this into equations (1) and (2), and expand in ε. The result for the leading order term of the rotor acceleration is,

$$w' = \Gamma\cos(n\theta) - \sum_{j=1}^{N}\left[g(s_j)s_j'' + \frac{dg(s_j)}{ds_j}s_j'^2 + \frac{dr^2(s_j)}{ds_j}s_j'\right], \tag{3}$$

where the first term is direct result of the fluctuating torque, and the remaining terms are the effects arising from the dynamics of the absorbers. Note that the goal of the absorbers is to use their dynamics, that is, the terms in the summation, to cancel the effect of the torque, thus minimizing this quantity. As w' is reduced by some measure, so too are the rotor torsional vibrations.

The expanded form of the rotor speed, $\nu = 1 + \varepsilon w + \mathcal{O}(\varepsilon^2)$, using equation (3), is substituted into the absorber equation and the result expanded, yielding a set of equations governing the response of the absorbers that is uncoupled from the rotor dynamics to leading order in ε (5; 30). These equations, which include the nonlinear effects arising from the specified absorber path $r(s)$, as well as the effects of the rotor acceleration, are given by,

$$s_i'' - \frac{1}{2}\frac{dr^2(s_i)}{ds_i} = \varepsilon\hat{f}_i(\mathcal{S}, \mathcal{S}', \mathcal{S}'', \theta) + \mathcal{O}(\varepsilon^2) \quad i = 1, \ldots, N, \tag{4}$$

where $\mathcal{S}$ is the vector of s_i's and the $\hat{f}_i$'s will be specified below, after providing some details about the absorber path.

These equations represent a set of N identical nonlinear oscillators with weak damping, weak near-resonant forcing, and weak coupling. The coupling between absorbers stems from their common interactions with the rotor, and when described in this form, which hides the rotor dynamics, the absorbers are globally coupled, that is, each absorber is affected by all absorbers, including itself; these effects are described more fully below. This coupling leads to a symmetry in the problem that will be exploited for stability calculations, as in (6; 2; 30). This symmetry also raises the possibility for energy to flow between the absorbers, through nonlinear coupling terms, which cannot be ignored, since the absorbers are lightly damped and driven near resonance. These features set the stage for a variety of interesting dynamic interactions, most of which are undesirable from the point of view of absorber performance.

In terms of determining the system dynamics using this model, we employ a perturbation method to solve for the response of the absorbers using equation (4), and then use equation (3) to determine the corresponding rotor dynamics. Note that rotor torsional vibrations are conveniently described by the magnitude of the angular acceleration, which is given, in terms of the physical and nondimensional variables, by $\ddot{\theta} = \Omega^2 \nu \nu' = \varepsilon \Omega^2 w' + \mathcal{O}(\varepsilon^2)$. Thus, w' captures the rotor vibrations to leading order. It is worth noting that due to nonlinear coupling effects arising from the kinematics of the absorber path, even when the absorber response is nearly harmonic at order n, equation (3) shows that the rotor vibrations can contain significant harmonics at orders n, $2n$, $3n$ (14). Since the goal of the absorber is to reduce the order n vibrations, when we plot the nondimensional rotor acceleration, it will be in terms of the order n harmonic component of w'.

Before proceeding further, we need to describe a parameterization of the absorber path function $r(s)$ that is useful for design.

2.3 Nonlinear Design: The Absorber Path

The path is selected to provide reduction of rotor torsional vibration over a range of torque loads, while avoiding undesirable behaviors. The primary effects of concern are: (i) bistability that can lead to jumps and hysteresis in the response as the torque amplitude or rotor speed are varied; (ii) symmetry-breaking instabilities that lead to nonsynchronous responses in systems of multiple absorbers; and (iii) large amplitude transient responses. All of these can be achieved by simply making mc^2 large, which keeps the absorber in its linear response range, but this leads to undesirable added mass and rotational inertia. Thus, in order to achieve good performance with minimal added inertia, one must consider nonlinear effects when selecting the absorber path.

A convenient form for paths of interest was originally presented by Denman (8) (who credited Borowski), and used in many subsequent studies by the authors, e.g., (30). This formulation is expressed in terms of the local radius of curvature, $\rho(s) = \sqrt{\rho_o^2 - \lambda^2 s^2}$ (all lengths are normalized by c), where the radius of curvature at the path vertex $s = 0$ sets the linear tuning by $\rho_o = 1/(1+\tilde{n}^2)$ (8; 30). The parameter λ affects the path at moderate and large amplitudes, that is, it specifies the absorber nonlinearity arising from its path. For $\lambda = 0$, the path is a circle, which is strongly softening; for $\lambda = 1$, the path is a cycloid, which is mildly hardening; and for $\lambda = \lambda_e = \tilde{n}/\sqrt{1+\tilde{n}^2}$, the path is the tautochrone, which is neither hardening or softening, that is, it is essentially linear, as described in more detail below. Figure 2b depicts sample absorber paths for various $\tilde{n}$ and λ values. While the differences in these paths appear slight, they have a dramatic effect on the nature of the response when the absorber is driven beyond its linear range, as demonstrated below.

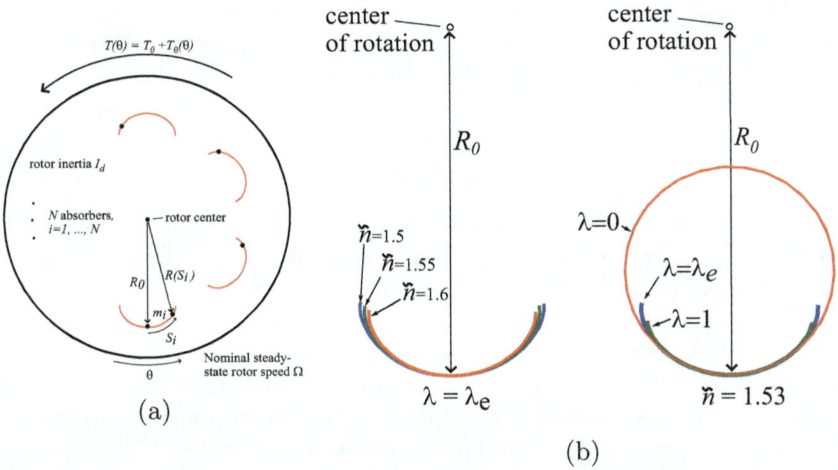

Figure 2. (a) Schematic of a rotor with absorbers riding on user-defined paths, shown in red. (b) Left: three tautochronic paths with $\lambda = \lambda_e \equiv \sqrt{\tilde{n}^2/(1+\tilde{n}^2)}$ for $\tilde{n} = 1.5$, 1.55, and 1.6. Right: three paths with $\tilde{n} = 1.53$ for $\lambda_{circle} = 0, \lambda = \lambda_e = .8371, \lambda_{cycloid} = 1$. Note that $R_o = c$.

The nonlinear path designations are based on a system model in which the rotor runs at constant speed and the absorber has no damping. A convenient way to view this situation is to consider the $\varepsilon = 0$ version of the equations of motion, equations (1-2), which yields $\nu = 1$, a constant rotor speed, and $s'' - \frac{1}{2}\frac{dr^2(s)}{ds} = 0$, a conservative oscillator for the absorber motion, where $-\frac{1}{2}r^2(s)$ plays the role of an effective potential that gives the absorber restoring force. The tautochrone is precisely the path that renders this oscillator as linear for all feasible amplitudes of s. To specify symmetric paths that are near the tautochrone, we take $r^2(s) = 1 - \tilde{n}^2 s^2 + \varepsilon\gamma s^4$. For $\gamma = 0$ the oscillator response is a pure harmonic of frequency $\tilde{n}$, so that the absorber tuning order is $\tilde{n}$.[1] This response is valid out to the absorber amplitude at which the epicycloidal path has a cusp, which is described by the condition $g(s_m) = 0$ that yields the maximum (cusp) amplitude, derived below. Thus, for absorber paths that are near the tautochrone, one does not need to carry out amplitude expansions to achieve a weakly nonlinear system. In this case, the weak nonlinearity from the path is captured in the parameter γ, and additional nonlinearities arise from the rotor absorber coupling, described in detail below.

The influence of the path type on the absorber dynamics are made evident by considering an amplitude expansion of the restoring term and examining the leading order terms. This calculation, using the path parameters $\tilde{n}$ and λ, is given in detail in (30), and it yields,

$$-\frac{1}{2}\frac{dr^2(s)}{ds} = \tilde{n}^2 s - 2\varepsilon\gamma s^3 + (\lambda^2 - \lambda_e^2)\left(\frac{1}{6}(1+\tilde{n}^2)^3 s^3 + \mathcal{O}(s^5)\right), \quad (5)$$

which clearly shows the role of the linear tuning parameter $\tilde{n}$, and the how the leading order nonlinear path term depends on γ and the value of λ relative to λ_e (8; 30). Note that when $\gamma = 0$, circular paths with $\lambda = 0$ are softening, since the leading order nonlinear term from the path has a negative coefficient. In contrast cycloidal paths with $\lambda = 1$ are hardening, since $\lambda_e < 1$, and the leading order nonlinear coefficient is positive. In both cases the level of softening/hardening from the path nonlinearity increases as the linear tuning order $\tilde{n}$ increases, due to the coefficient. The parameter γ provides a convenient way to tune the path away from the tautochrone, where $\gamma > 0$ (< 0) gives a softening (hardening) effect. In this work we will take $\lambda = \lambda_e$ and use $(\tilde{n}, \gamma)$ to tune the path.

This near-tautochronic formulation allows one to retain the effects of the nonlinearities that arise from the rotor/absorber coupling, which are important for absorbers riding out to large amplitudes. If the absorbers ride on non-tautochronic paths with moderate amplitudes, the path nonlinearity will dominate that of the rotor/absorber coupling, which greatly simplifies the model, since the resulting absorber equations are essentially a set of globally coupled Duffing oscillators. That case has been considered in detail in (2; 17).

In summary, the absorber design problem boils down to selecting the maximum absorber mass m and placement c allowed by hardware and re-

[1] Note that in this rescaled version of the system model the mean rotor speed is unity, so that order and frequency are identical.

sponsiveness constraints (i.e., total system rotational inertia), and then selecting the absorber path parameters $(\tilde{n}, \gamma)$. For all absorbers, selection of the linear tuning order $\tilde{n}$ is crucial, since it sets the baseline response, even out to large absorber amplitudes, as shown below. For the nonlinear path characteristic, we use the cubic path coefficient γ to move away from the perfect tautochrone.

For these nearly tautochronic paths, the absorber dynamics are essentially harmonic over their entire feasible range, and thus the perturbation method outlined below can be applied to absorbers moving with large amplitudes.

2.4 Scaling and Perturbation Analysis

The absorber system represents a set of nonlinear oscillators that are weakly coupled through the rotor, each responding to a common applied harmonic force. The scaling and analysis are geared towards bringing out the effects of absorber coupling, damping, excitation, and detuning from resonance to leading order in ε. In order to achieve a form for the absorber equations that is convenient for this perturbation analysis, we employ the ε scaling assumptions outlined above, use the path formulation given in equation (5), and replace the $\mathcal{S}''$ terms in the $\hat{f}_i$ in equation (4) with $-n^2\mathcal{S}$, which is valid to leading order in ε for the (nearly harmonic) responses of interest. These substitutions lead to the desired form of the absorber equations, expressed as follows,

$$s_i'' + \tilde{n}^2 s_i = \varepsilon f_i(\mathcal{S}, \mathcal{S}', \theta) + \mathcal{O}(\varepsilon^2) \qquad i = 1, \ldots, N \tag{6}$$

where

$$\begin{aligned} f_i(\mathcal{S}, \mathcal{S}', \theta) &= 2\gamma s_i^3 + \hat{f}_i(\mathcal{S}, \mathcal{S}', \mathcal{S}'', \theta)|_{\mathcal{S}''=(-n^2\mathcal{S})} \\ &= 2\gamma s_i^3 - \mu s_i' - (s_i' + g_o(s_i))\Bigg(\Gamma\cos(n\theta) \\ &\quad + \frac{1}{N}\sum_{j=1}^{N}\left[2\tilde{n}^2 s_j s_j' + n^2 s_j g_o(s_j) - s_j'^2 \frac{dg_o(s_j)}{ds_j}\right]\Bigg), \end{aligned} \tag{7}$$

where $g_0(s) = \sqrt{1 - \tilde{n}^2(1 + \tilde{n}^2)s^2}$ is the $\varepsilon = 0$ version of $g(s)$.

These equations form the basis of the analysis, whose aim is to predict the response of the absorbers in terms of the system parameters, and use the results to select absorber path parameters. Of interest here are the steady-state amplitudes and phases, and the local stability of these responses. Once the absorber responses are obtained, the rotor acceleration, and thus the level of torsional vibration, can be reconstituted to leading order from w' as given in equation (3), as described above.

It is important to note that the absorbers are tuned such that they addresses the order n excitation, and thus the linear tuning is selected to be $\tilde{n} \approx n$. To make this precise for the perturbation calculations, we introduce the (small) absorber detuning parameter, $\varepsilon\Delta = n^2 - \tilde{n}^2$, which will be used in the following section.

The absorber equations (6) are amenable to the method of averaging. The process starts with the assumption of a nearly harmonic response at the forcing order, motivating the following coordinate transformation to amplitude A_i and phase ψ_i variables,

$$\begin{aligned} s_i(\theta) &= A_i(\theta)\cos(n\theta + \psi_i(\theta)) \\ s_i'(\theta) &= -nA_i(\theta)\sin(n\theta + \psi_i(\theta)). \end{aligned} \tag{8}$$

This transformation is substituted into the absorber equations (6), rendering an equation involving A_i' and ψ_i', that is, the rates of change of the amplitude and phases with respect to θ. Note that the transformation requires, by direct differentiation of s_i and a comparison with s_i' in equation (8), the following condition to hold,

$$A_i'\cos(n\theta + \psi_i) - A_i\psi_i'\sin(n\theta + \psi_i) = 0. \tag{9}$$

This equation and the absorber equation with A_i' and ψ_i' can be solved for (A_i', ψ_i'). All terms in these equations are $\mathcal{O}(\varepsilon)$, except for one term in ψ_i' that is proportional to $n^2 - \tilde{n}^2$, for which we substitute $\varepsilon\Delta$. Thus, for $\varepsilon << 1$, $(A_i', \psi_i') = \mathcal{O}(\varepsilon)$ and (A_i, ψ_i) will be slowly varying. These terms will contain oscillating terms of order (frequency) n in θ, which cause relatively rapid oscillations to be superimposed on the slow drift in (A_i, ψ_i). In order to isolate the slow drift, one averages the (A_i', ψ_i') equations over one forcing period, $2\pi/n$, yielding a set of equations for (a_i, ϕ_i), which are the θ-averages of (A_i, ψ_i) (9). For $\varepsilon << 1$, the expressions for (a_i', ϕ_i') are $\mathcal{O}(\varepsilon)$ and, importantly, do not depend on θ to leading order. Here the averaging process is more cumbersome than is the case for textbook examples, due to the presence of square root terms in g_o, which appear in the f_i's in equation (7). Thus, some terms in the resulting averaged equations must be evaluated numerically, or expressed in terms of amplitude expansions in the a_i's.

The result of the averaging process, to leading order in ε, takes the form,

$$
\begin{aligned}
a_i' &= \frac{\varepsilon}{n}\Big[-\frac{1}{2}\mu n a_i + \Gamma\sin(\phi_i)F_1(a_i) + P_i(Z)\Big] \\
\phi_i' &= \frac{\varepsilon}{n a_i}\Big[\frac{a_i\tilde{n}^2}{8N}\Big(4 - a_i^2\Big(3\tilde{n}^2(\tilde{n}^2+1) + n^2(\tilde{n}^2+3)\Big)\Big) \\
&\quad -\frac{1}{2}a_i\Delta - \frac{3}{4}\gamma a_i^3 + \Gamma\cos(\phi_i)F_2(a_i) + Q_i(Z)\Big]
\end{aligned}
\tag{10}
$$

where Z represents the vector of all (a_j, ϕ_j)'s, and the coupling functions (P_i, Q_i) are given by,

$$
P_i(Z) = \frac{1}{N}\sum_{j=1}^{N}\Big[-\frac{1}{4}\tilde{n}^2 n^2 a_i a_j^2 \sin(2(\phi_j - \phi_i)) + \tilde{n}^2 a_j G_1 + \tilde{n}^2 n^2(\tilde{n}^2+1)a_j^3 H_1\Big],
$$

$$
Q_i(Z) = \frac{1}{N}\sum_{j=1}^{N}\Big[\frac{1}{4}\tilde{n}^2 n^2 a_i a_j^2 \cos(2(\phi_j - \phi_i)) + \tilde{n}^2 a_j G_2 + \tilde{n}^2 n^2(\tilde{n}^2+1)a_j^3 H_2\Big],
$$

which are expressed in terms of functions F_1, F_2, G_1, G_2, H_1, and H_2, which depend on the amplitudes a_i and the phase difference $\alpha_{ji} = \phi_j - \phi_i$, as follows,

$$
\begin{aligned}
F_1(a_i) &= \frac{1}{2\pi}\int_0^{2\pi} \sin^2 x[1 - \tilde{n}^2(1+\tilde{n}^2)a_i^2\cos^2 x]^{\frac{1}{2}}\,dx, \\
F_2(a_i) &= \frac{1}{2\pi}\int_0^{2\pi} \cos^2 x[1 - \tilde{n}^2(1+\tilde{n}^2)a_i^2\cos^2 x]^{\frac{1}{2}}\,dx, \\
G_1(a_i, a_j, \alpha_{ji}) &= \frac{1}{2\pi}\int_0^{2\pi} \cos(x)\sin(x-\alpha_{ji})[1 - \tilde{n}^2(1+\tilde{n}^2)a_j^2\cos^2 x]^{\frac{1}{2}} \\
&\quad \times [1 - \tilde{n}^2(1+\tilde{n}^2)a_i^2\cos^2(x-\alpha_{ji})]^{\frac{1}{2}}\,dx, \\
G_2(a_i, a_j, \alpha_{ji}) &= \frac{1}{2\pi}\int_0^{2\pi} \cos(x)\cos(x-\alpha_{ji})[1 - \tilde{n}^2(1+\tilde{n}^2)a_j^2\cos^2 x]^{\frac{1}{2}} \\
&\quad \times [1 - \tilde{n}^2(1+\tilde{n}^2)a_i^2\cos^2(x-\alpha_{ji})]^{\frac{1}{2}}\,dx,
\end{aligned}
$$

$$H_1(a_i, a_j, \alpha_{ji}) = \frac{1}{2\pi} \int_0^{2\pi} \cos(x)\sin^2(x)\sin(x - \alpha_{ji}) \times [\frac{1 - \tilde{n}^2(1 + \tilde{n}^2)a_i^2\cos^2(x - \alpha_{ji})}{1 - \tilde{n}^2(1 + \tilde{n}^2)a_j^2\cos^2 x}]^{\frac{1}{2}} dx,$$

$$H_2(a_i, a_j, \alpha_{ji}) = \frac{1}{2\pi} \int_0^{2\pi} \cos(x)\sin^2(x)\cos(x - \alpha_{ji}) \times [\frac{1 - \tilde{n}^2(1 + \tilde{n}^2)a_i^2\cos^2(x - \alpha_{ji})}{1 - \tilde{n}^2(1 + \tilde{n}^2)a_j^2\cos^2 x}]^{\frac{1}{2}} dx.$$

This completes the averaging process.

These averaged equations are capable of capturing a wide range of behaviors. They have also proven to do an excellent job at predicting responses in experimental studies (32; 18). Steady state absorber responses correspond to constant amplitudes and phases, which are found by locating the (a_i, ϕ_i) values for which $(a_i', \phi_i') = (0, 0)$. The stability of these fixed points correspond to the stability of the attendant periodic responses in the s_i's (9). These including steady state synchronous responses, in which all absorbers move with identical amplitudes and phases (6; 30), as well as nonsynchronous steady state responses (5; 1). These equations also describe transient responses, which are important for determining amplitudes during system startup (31; 17).

Note that in the averaged equations we do not replace $\tilde{n}$ everywhere by n, as is often done in order to simplify terms in the following expressions, in which case the difference is pushed out to higher orders in ε. In contrast, here we keep n, $\tilde{n}$, and Δ as they appear in the averaging process, even though any two of these three parameters would suffice to specify the absorber and excitation orders. The reason for this approach is that it is generally observed that one obtains more accurate results by keeping these terms in this form, rather than simplifying them, and it does not introduce any inconsistencies (9; 28). In this work we will focus on $\tilde{n}$ and n for parameter studies, since these have clear physical meanings.

The Synchronous Response

The desired response of the absorbers has all absorbers moving in a synchronous manner with constant amplitude and phase, that is, $(a_i, \phi_i) = (a, \phi)$, $\forall i$. This is dynamically equivalent, in terms of vibration reduction, to the case of a single absorber of total mass Nm. However, as shown below, the stability characteristics of this response are affected by the presence of multiple absorbers.

For the synchronous response, equations 10 collapse into a single form, that of a single absorber, as found in (32), and many of the functions simplify. Specifically, for $a_i = a_j = a$ and $\alpha_{ji} = 0$, one can show that $G_1(a,a,0) = 0$, $H_1(a,a,0) = 0$, $G_2(a,a,0) = (4 - 3a^2\tilde{n}^2(1+\tilde{n}^2))/8$, and $H_2(a,a,0) = 1/8$. Functions $F_{1,2}$'s can be determined by numerical integration, or expressed in terms of elliptic functions (17); here we evaluate them numerically.

Evaluating equations (10) for the synchronous response and setting the result equal to zero, one can solve for the following steady-state phase conditions,

$$\begin{aligned} \Gamma \sin(\phi) &= \frac{\mu n a}{2F_1(a)} \\ \Gamma \cos(\phi) &= \frac{aE(a)}{8F_2(a)}. \end{aligned} \tag{11}$$

where

$$E(a) = 4(\Delta - \tilde{n}^2) + a^2\left(6\gamma - n^2\tilde{n}^2(3+\tilde{n}^2) + 3\tilde{n}^4(1+\tilde{n}^2)\right).$$

A convenient expression for the steady state synchronous amplitude, uncoupled from the phase, is found by computing $\Gamma^2\sin^2(\phi) + \Gamma^2\cos^2(\phi) = \Gamma^2$,

$$\Gamma^2 = \frac{a^2}{64}\left\{\left(\frac{4\mu n}{F_1(a)}\right)^2 + \left(\frac{E(a)}{F_2(a)}\right)^2\right\} \tag{12}$$

which relates a to the system and excitation parameters. In order to avoid solving for roots a for given parameters, one can set a, compute $F_{1,2}(a)$, for example, by numerical integration, and then compute the corresponding torque level Γ using equation (12). The corresponding phase can then be determined using equation (11). This allows for relatively easy plotting of (a,ϕ) vs. Γ, which will be presented for examples after we complete a stability analysis.

One can determine the maximum possible absorber amplitude, dictated by the cusp on the epicycloidal path, by solving $g(a_m) = 0$ for a_m. The result for $\gamma = 0$ is simply $a_{mo} = 1/(\tilde{n}\sqrt{1+\tilde{n}^2})$, and an approximation for the more general path can be found by expanding $g(s)$ in $\varepsilon\gamma$, which results in,

$$a_m = a_{mo}\left[1 + \varepsilon\gamma\left(\frac{1+4\tilde{n}^2}{2\tilde{n}^4(1+\tilde{n}^2)^2}\right) + \mathcal{O}(\varepsilon^2)\right], \tag{13}$$

which is quite accurate for a wide range of path parameters. Note that for $\gamma > 0$ (< 0) the maximum absorber amplitude is increased (decreased) from

a_{mo}, consistent with direct examination of the paths shown in Figure 2b. This expression of the limit for the amplitude of the synchronous response is used in the calculations below.

Here we consider the local dynamic stability of the synchronous response. Only stable responses are viable for operation, and, in fact, one should not design a system to operate near an instability, since parameter uncertainties and/or small disturbances can result in the system venturing far from the desired response. Local stability of the synchronous response is determined by linearizing the averaged equations about that response. The synchronous response is represented by a fixed point of the autonomous averaged equations, and thus stability is dictated by the eigenvalues of the corresponding Jacobian.

For convenience we express the averaged equations in the form,

$$z' = \varepsilon H(z) + \mathcal{O}(\varepsilon^2) \qquad z \in \mathcal{R}^N \times \mathcal{S}^N, \tag{14}$$

where elements of z are the averaged amplitudes a_i and phases ϕ_i of the absorbers. A fixed point with $(a_i, \phi_i) = (a, \phi)$ $\forall i$ the synchronous response, denoted as z_s, satisfies $H(z_s) = 0$. Note that other types of steady state responses can occur, most notably, those in which subsets of the N absorbers are mutually synchronous (5; 1). The synchronous response is the most symmetric of these, in that every absorber is interchangeable with every other absorber, leading to a global symmetry of the response. Responses with lower orders of symmetry exist (5; 1), and arise from the synchronous response in a generic way, described by symmetric bifurcation theory (16; 5; 1).

The local stability of the synchronous response is determined by linearizing the averaged equations about z_s, leading to the $2N \times 2N$ Jacobian,

$$\mathcal{J} = \left.\frac{\partial H}{\partial z}\right|_{z=z_s} \tag{15}$$

which depends on (a, ϕ) and the system and excitation parameters. This is a tedious, but worthwhile, calculation, since the symmetry of this response allows one to determine the eigenvalues of $\mathcal{J}$ by evaluating the eigenvalues of a pair of 2×2 matrices, as described below. This Jocabian $\mathcal{J}$, due to symmetries, is a block circulant matrix (7; 6; 27) of the form,

$$\mathcal{J} = \begin{bmatrix} A & B & \dots & B & B \\ B & A & B & \dots & B \\ \vdots & \vdots & \vdots & \vdots & \vdots \\ B & B & \dots & B & A \end{bmatrix} \tag{16}$$

where entries of the 2×2 block matrices A and B can be found in (30). This structure occurs since the (p, q) 2×2 block describes how the synchronous response of absorber p is affected by small deviations from the synchronous response in absorber q. Of course, in the synchronous response, each absorber has the same effect on itself, hence all diagonal entries are identical, and are labeled A. Similarly, each absorber has the same effect on all other absorbers, except itself, so all off-diagonal terms are the same, and labeled as B.

The computation of $\mathcal{J}$ is quite tedious, and involves taking dervatives of the functions $(F_1, F_2, G_1, G_2, H_1, H_2)$ and evaluating them on the synchronous response. For the present results, these are done by using amplitude expansions in a, which provides accurate results for absorber amplitudes nearly up to the cusp. Details of this calculation and further discussion are given in (30).

Given the circulant structure of $\mathcal{J}$, one can prove that the eigenvalues of $\mathcal{J}$ are the two eigenvalues of $[A + (N-1)B]$, plus the two eigenvalues of $[A - B]$ repeated $N-1$ times (6; 2). Thus, the stability of the synchronous response boils down to determining the eigenvalues of two 2×2 matrices. For the absorber system, it can be shown that that the traces of both matrices of interest are equal and negative, specifically,

$$\text{Trace}[A - B] = \text{Trace}[A + (N-1)B] = -\mu, \tag{17}$$

where $\mu > 0$, since the absorbers dissipate energy as they move along their paths (30). A consequence of this fact is that no Hopf bifurcations can occur, which would result in traveling wave types of responses. Therefore, the only types of bifurcations that can occur correspond to an eigenvalue of one of these matrices passing through the origin from negative to positive. These can be found by finding conditions for zero eigenvalues of the determinants.

In this system, the instabilities associated with the eigenvalues of the two 2×2 matrices are qualitatively different. When an eigenvalue of $[A + (N-1)B]$ moves through zero, this corresponds to a single zero eigenvalue, leading to a saddle-node bifurcation of synchronous responses, that is, the merging of two synchronous branches, one of which is stable (since the trace is negative). Note that this bifurcation preserves the symmetry of the synchronous response, and, in fact, this bifurcation can occur in systems with one a single absorber, $N = 1$, in which case $\mathcal{J} = A$ and the eigenvalues of interest are those of A, consistent with $[A + (N-1)B] = A$ for $N = 1$. (Of course, in the case $N = 1$, matrix B is meaningless.) These instabilities will be labeled "J" for the jumps that occur as parameters are varied near these instabilities, as described below.

If an eigenvalue of $[A - B]$ goes through zero, the situation is more complicated, since this implies an instability involving $N-1$ eigenvalues of $\mathcal{J}$ simultaneously moving through zero. This situation is directly related to the symmetry of the synchronous response, and such instabilities correspond to symmetry breaking that give birth to non-synchronous responses, hence we label these as "NS" instabilities. Symmetric bifurcation theory tells us that in these cases one can expect multiple response branches emerging from the synchronous response, and in generic situations these branches will correspond to responses with one order of symmetry lower than the synchronous response (16). For the present system, these responses have two groups of absorbers moving in a mutually synchronous manner (5; 1). For $N = 2$ absorbers, the situation is simple, the two absorbers simply become non-synchronous. For $N = 3$ absorbers, these non-synchronous responses have a group of two absorbers that remain mutually synchronous, and one absorber moving on its own. Of course, the synchronous response still exists, but is dynamically unstable, and none of the absorbers follow that response. Also, note that by permutation considerations, there exist three such responses. For $N = 4$ absorbers, there exist responses with groups of three and one (four of these) as well as groups of two and two (six of these) (1). While many of these branches can be unstable when they emerge from the synchronous response in a subcritical manner, in some cases, for example, with circular path absorbers, many branches are stable over some parameter ranges, in fact, at low torque levels, leading to the possibility of multiple stable steady states of different types coexisting, with the response observed depending on initial conditions (1). For absorbers with nearly tautochronic paths, the most common situation is that one branch emerging from the NS bifurcation is supercritical and stable, and it corresponds to the non-synchronous response that has $N-1$ absorbers mutually synchronous and one absorber moving on it own (5). In fact, if one tunes the system very close to the perfectly tuned tautochrone, $(\tilde{n}, \lambda) = (n, \lambda_e)$, and the system is lightly damped, $\mu << 1$, this instability can occur at very low torques (6; 5). This response is undesirable, since the amplitude of the absorber moving on its own grows quickly as the torque is increased, resulting in a significantly smaller operating range.

Evaluation of the determinants of $[A + (N-1)B]$ and $[A - B]$ involve the system parameters, and the amplitude and phase of the synchronous response, (a, ϕ), which also are determined by the system parameters. The stability of the synchronous response for a given set of parameter values is conveniently found by first setting a value for a and computing the corresponding torque amplitude Γ using equation (12), using equation (11) to compute the phase, and then using those results to evaluate the determinants. This is done by numerical computation for the examples considered below.

It is interesting to note that the values of the determinants is independent of the number of absorbers, for $N > 1$. This follows since the terms in the summation over absorbers are identical, when evaluated on the synchronous response, and since the summations carry a common factor of $1/N$, N cancels out. This implies that the synchronous response and its stability characteristics are independent of the number of absorbers, but depend on the total amount of absorber inertia. However, the absorber damping no doubt depends on how the absorber inertia is divided up, and this will affect the features of the synchronous response. This issue requires further modeling and experimental verification, and is left for future work.

Experimental results obtained using the absorber shown in the right panel of Figure 1 can be found in (32) for steady state dynamics and (18) for transient dynamics. The experimental investigations have proven the accuracy and utility of the mathematical model and analysis described above.

2.5 Examples

The results presented here will consist of response curves and local stability results for synchronous responses. The path parameters γ and $\tilde{n}$ will be varied in order to assess absorber performance in synchronous operation, in terms of rotor torsional vibration, stability, and torque range, which is limited by absorber amplitude. The goal is to find path parameters that lead to stable operation over the largest torque range, while maintaining acceptable rotor vibration. For these purposes, one can evaluate the stability of the synchronous response at some maximum amplitude that will be allowed during operation. To this end we define a maximum allowable absorber amplitude, in terms of the cusp amplitude, as follows,

$$a_o = \alpha \ a_m, \tag{18}$$

where a_m is given in equation (13) and the parameter α, $0 < \alpha \leq 1$, sets the maximum amplitude of operation. The goal is to select path parameters that yield stable response at $a = a_o$ and provide favorable rotor vibration reduction over the range $0 < a < a_o$. For the present system, stability at $a = a_o$ insures that the response will be stable and perform as desired for all $a < a_o$. The actual value selected for α will depend on hardware considerations, since no absorbers can actually attain the cusp on the epicycloid (32); here we use $\alpha = 0.9$.

In order to assess the effectiveness of the absorbers we consider the rotor angular acceleration as described by w', recalling that $\ddot{\theta} = \Omega^2(\varepsilon w' + \mathcal{O}(\varepsilon^2))$. When the absorbers are working, this will be small, but since it is a function of θ, it is convenient to have a measure of its magnitude. Since the absorbers are designed to reduce order n rotor vibrations, we take as our measure of rotor vibration amplitude the order n Fourier component of w', computed as

$$|w'| = \frac{n}{\pi}\text{Abs}[\int_0^{\frac{2\pi}{n}} e^{-in\theta}w'(\theta)d\theta],$$

where Abs[y] is the magnitude of the complex number y. For the analytical predictions we use equation (3) for w', with the absorber synchronous response given by $s = a\cos(n\theta + \phi)$. It is worth noting that when the absorbers are working well, the order n component of w' is reduced to a magnitude that is comparable with those of harmonics at $2n$ and $3n$. This rich harmonic content will be demonstrated in the examples to follow.

To quantify the effectiveness of the absorbers on the rotor system, it is useful to compare the order n rotor angular acceleration with the absorbers locked at their respective vertices, denoted $|w'|_o$, with the w' resulting when the absorbers are free to move, computed as outlined above. This normalizes w' to account for the flywheel effect of the absorbers, that is, the reduction in torsional vibration due to the added inertia of the absorbers. A simple estimate of $|w'|_o$ can be found by assuming an oscillating torque $T_n\cos(n\Omega t)$ applied to a rotor with inertia J, resulting in $|w'|_o = \Gamma$, which is accurate to leading order in ε. Thus, this reference response is the diagonal line when plotted versus the torque level Γ, and this line separates rotor vibration reduction (below the reference) from amplification (above the reference). The latter is, of course, highly undesirable, but can occur for softening paths, when the response jumps to an in-phase response after a type J instability (26; 2; 23; 24).

We now turn to results for systems with excitation orders $n = 3/2$ and $n = 2$. These are chosen since they represent the torques experienced in four-stroke engines with three and four cylinders, respectively. Due to harmonic cancellation in four-stroke engines with M cylinders, the oscillating component of the applied torque is dominated by the order $n = M/2$ harmonic. In these situations, the arrangement of cylinders fixes the excitation order, n, and the designer selects the absorber inertia ε, the path parameters $\tilde{n}$ and γ, and the number of absorbers N. The absorber damping μ is difficult to model, and, in fact, will vary depending on many factors that are beyond the control of the designer, for example, the viscosity of lubricants, which depend on operating temperature. So, in order to assess the performance of a system with a given absorber path, one sets these parameters and considers the system response over a range of torque amplitudes Γ.

Two examples are considered, both with $N = 2$ absorbers, since this is sufficient to capture the NS instability. Detailed results about the synchronous response, including its stability characteristics, how these depend on the path parameters, and assessment of absorber performance are shown for $n = 3/2$. In addition, the analytical predictions are compared against numerical simulations of the equations of motion. Finally, we show stability results in the $(\tilde{n}, \gamma)$ path parameter space for synchronous absorber amplitude a_m, which clearly shows how the two types of instability are influenced by the path. This plot provides important information for the selection of path parameters, and is shown for both $n = 3/2$ and $n = 2$.

We first consider the synchronous response over a range of torque amplitudes, in the form of response curves depicting the absorber amplitude a and the order n harmonic of the rotor angular acceleration, $|w'|$, versus the torque amplitude Γ. The phase ϕ is close to π in all cases, since the absorber, when working properly, is out of phase with respect to the fluctuating torque, and is therefore not shown. Here stable synchronous responses are indicated by solid lines and unstable synchronous responses are shown as dashed lines. Instabilities associated with jump J, or saddle-node bifurcations, are evident since they occur at points of vertical tangency in the response curve, that is, where two response branches, one stable and the other unstable, merge and annihilate one another. Symmetry-breaking NS bifurcations are observed on synchronous response branches that simply become unstable as Γ is varied. Note that emerging from these branches at NS instabilities are multiple non-synchronous response branches, as described above. Based on previous work (5), it is expected that these will be supercritical for paths near the tautochrone, but this has not been thoroughly examined. Finally, note that since these examples have only $N = 2$ absorbers, the non-synchronous responses are relatively simple, in that the two absorbers simply separate in terms of amplitude and/or phase.

Figure 3 considers the effects of the nonlinear path tuning parameter γ on the synchronous response, for a system with a linear absorber tuning order $\tilde{n} = 1.48$. Such linear undertuning of absorbers ($\tilde{n} < n$) is not of practical interest, for a variety of reasons, but is used here since this example demonstrates both types of instability. For the most softening of the paths considered here, $\gamma = 3$, the absorber experiences a J instability. For torque levels above this instability point, absorber encounters with amplitude limiting stops will occur, leading to highly undesirable impact dynamics. In fact, if such motions are initiated, there will be hysteresis in the response, and the torque amplitude will need to be lowered to a level well below the J point in order to recover the desired response. For $\gamma = 2$ the response is very close to the softening/hardening transition, as seen by the near-vertical part of the absorber response curve, but no instabilities occur. For $\gamma = 1$

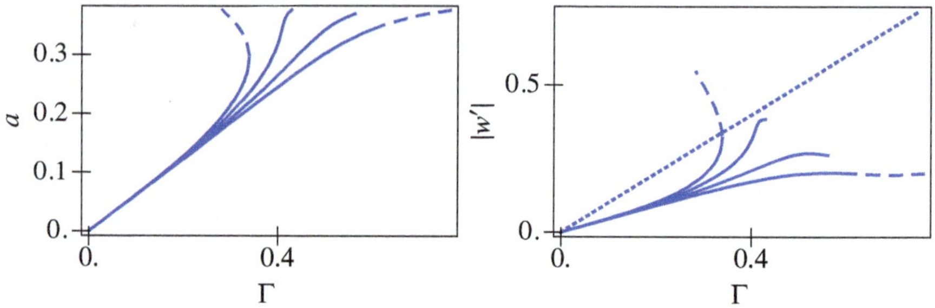

Figure 3. System response versus torque amplitude: absorber amplitude a and order n component of the rotor angular acceleration $|w'|$. The linear tuning is set to $\tilde{n} = 1.48$ and nonlinear tuning takes on values $\gamma = 0, 1, 2, 3$, varying from neutral towards increased softening. In the plot of a the largest amplitude is for the most softening path, $\gamma = 3$, and decreasing γ reduces the amplitude; the other curves can be correlated by noting stability and termination points, which occur when the absorber nears the cusp. The dotted line in the $|w'|$ plot depicts the rotor torsional vibration reference line, $|w'| = \Gamma$, indicating the vibration level if the absorbers are fixed. System parameter values: $\varepsilon = 0.1$, $n = 1.5$, $\mu = 0.3$, $N = 2$, and the $\gamma = 0$ cusp amplitude is given by $a_{cusp,0} \approx 0.38$.

the response is similarly stable, but slightly hardening, while for $\gamma = 0$ the response is more hardening and experiences an NS instability, leading to nonsynchronous responses. It is interesting to note for the lower values, $\gamma = 0$ & 1, the hardening nature of the kinematic coupling between the rotor and absorber dominates the softening of the path nonlinearity. This can be understood by considering that, as the absorber rides along its path, the torque produced by the absorber on the rotor is generated by the normal force between the absorber and rotor, acting along a line of action that is locally perpendicular to the path. The kinematics of this interaction are described in the function $g(s)$ and result in the reduction of the effective torque at large amplitudes. Thus, the absorber has a reduced effect when pushed to large amplitudes by large torques, which is dynamically equivalent to a hardening effect (17). For absorbers with strongly nonlinear paths, e.g., circles, the dominant nonlinear effects arise from the path (2). It is also of interest to note that the synchronous response of circular path absorbers can experience both J and NS instabilities on the same branch (1).

Figure 4 shows results from a case in which we fix $\gamma = 0$ (perfect tautochrones) and vary the linear tuning order $\tilde{n}$, taking values 1.5, 1.51, 1.52,

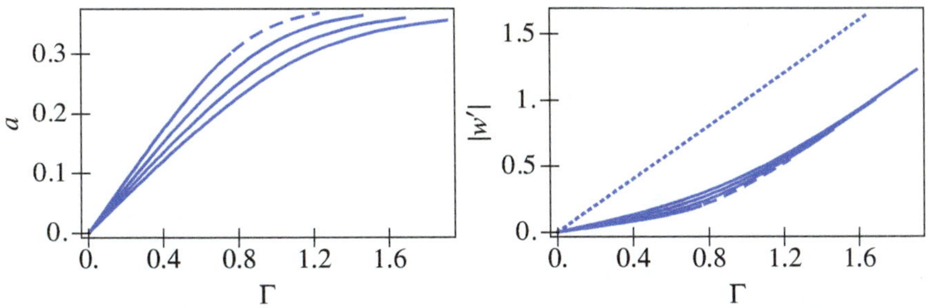

Figure 4. System response versus torque amplitude: absorber amplitude a and order n component of the rotor angular acceleration $|w'|$. The nonlinear tuning is set to $\gamma = 0$ and the linear tuning takes on values $\tilde{n} = 1.50, 1.51, 1.52, 1.53$, varying from perfect tuning towards increased overtuning. The topmost a curve is for perfect tuning, $\tilde{n} = 1.5$ and overtuning, shown in the other curves, extends the response out to larger torques. The linear tuning for the other curves can be correlated by considering the termination points. The dotted line in the $|w'|$ plot is the reference line, $|w'| = \Gamma$. System parameter values: $\varepsilon = 0.1$, $n = 1.5$, $\mu = 0.3$, $N = 2$, and the $\gamma = 0$ cusp amplitude is given by $a_{cusp,0} \approx 0.38$.

and 1.53. This is a practical situation, in which a designer linearly detunes a tautochronic absorber to gain robustness and extend the torque range, as shown below. The torque range is increased as the absorber detuning is moved away from perfect tuning, that is, away from resonance. Intimately related to this observation is the fact that the slope of the absorber as a function of Γ decreases with detuning, allowing it to remain close to linear, and away from a_m, for a larger torque range. While an NS instability occurs for perfect tuning, it quickly disappears and has vanished by $\tilde{n} = 1.51$. This instability disappears in subtle manner near the cusp value as $\tilde{n}$ is varied, but the details of this transition are not important in applications, since the transition happens in a very small parameter window. The most important observation about these results is that the order n rotor vibration response amplitude is quite insensitive to the linear detuning. While there is a slight degradation in performance, in terms of vibration reduction, this effect is insignificant compared to the benefits of the extended torque range.

We now demonstrate the validity of the analysis via simulations of the full equations of motion, equations (1) and (2). In these plots we show absorber and rotor responses versus the rotor angle θ, which plays the role of time. For the rotor we plot w', and for the absorbers we use coordinates

$\frac{1}{2}(s_1 + s_2)$ and $\frac{1}{2}(s_1 - s_2)$, which allow one to clearly distinguish between synchronous and non-synchronous responses (5). The system is started with initial conditions based on the synchronous steady state response predicted from the averaging analysis, with a small difference in initial conditions of the two absorbers, in order to allow the non-synchronous response to grow, should the synchronous response be unstable.

For the parameters used in Figure 5, the synchronous response is stable, as seen by the fact that $\frac{1}{2}(s_1 - s_2)$ decays towards zero, while $\frac{1}{2}(s_1 + s_2)$ approaches the predicted synchronous response. A short segment of the the steady state rotor angular acceleration is shown in the fourth panel of Figure 5, where the thick solid line is the simulated result, the thin solid line is the result reconstructed using the simulated absorber motion and equation (3), and the dashed line is the result using the predicted steady state absorber response in equation (3). It is seen that the analytical prediction, the dashed line, is quite accurate, thus validating the various approximations used. Also, note the rich harmonic content of the rotor vibrations, as described above.

Figure 6 shows similar results for a slightly larger torque level, which takes the system across an NS instability, and thus a nonsynchronous steady state response results. This is seen by the growth of $\frac{1}{2}(s_1 - s_2)$ during the initial transient, as shown. Eventually, one absorber continues to grow in amplitude and reaches the maximum amplitude allowed, a_m. This is shown by the longer time behavior of the individual absorbers, shown in the last panel of Figure 6, where the absorbers are drifting apart, and eventually one absorber will reach the cusp amplitude. Note that the absorber sum coordinate $\frac{1}{2}(s_1 + s_2)$ and w' both settle into steady state responses close to the synchronous response. This indicates that somehow the sum of the two absorbers is close to invariant, even while moving differently, thus leading to an equivalent level of rotor vibration reduction. However, this situation will be destroyed once one of the absorbers reaches its amplitude limits.

With this confidence in the analytical predictions, we focus on the selection of the path parameters to avoid both types of instability. Figure 7 depicts the stability of the synchronous response at $a = a_o = 0.9a_m$ for points in the path parameter space $(\tilde{n}, \gamma)$ near the perfectly tuned tautochronic path $(n, 0)$. White regions represents paths which experience no instabilities for $a < a_o$, gray regions correspond to absorbers that experience symmetry-breaking NS bifurcations, and black regions are for absorbers that undergo saddle-node J instabilities. Note that the most desirable operating point in this space, from the point of view of vibration reduction, is the origin, but an NS instability occurs for this path, as known from previous work (6; 5). Perhaps the simplest way to avoid this NS instability, without

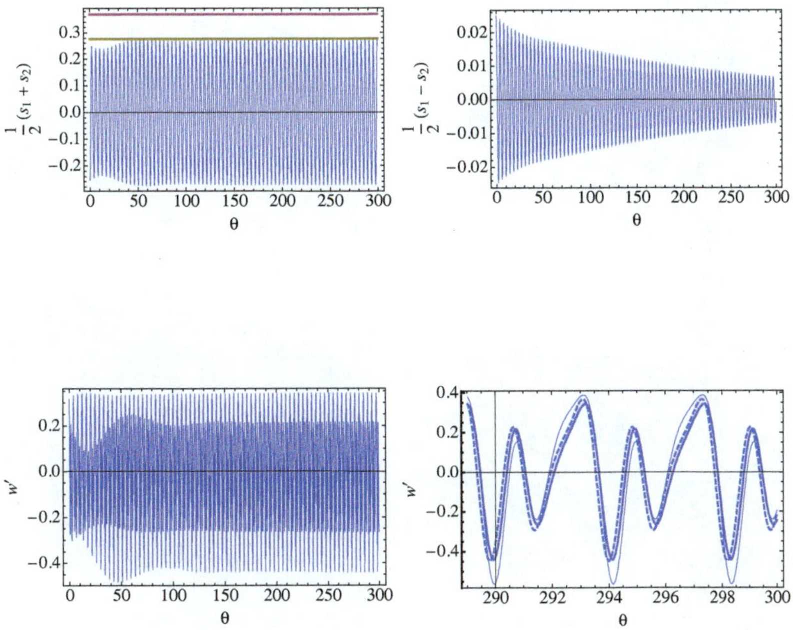

Figure 5. Simulation results for the absorbers and rotor versus the rotor angle θ for system parameter values: $\varepsilon = 0.1$, $n = 1.5$, $\mu = 0.3$, $N = 2$, $\Gamma = 0.679$, $\gamma = 0$. The first panel depicts $\frac{1}{2}(s_1+s_2)$ along with two reference lines: the cusp amplitude and the steady state amplitude predicted by averaging. The second panel shows $\frac{1}{2}(s_1 - s_2)$, which is decaying, indicating the local stability of the synchronous response. The third and fourth panels show w', first over the entire run, and then a segment of the steady state from simulation and reconstructed in the two ways described in the text.

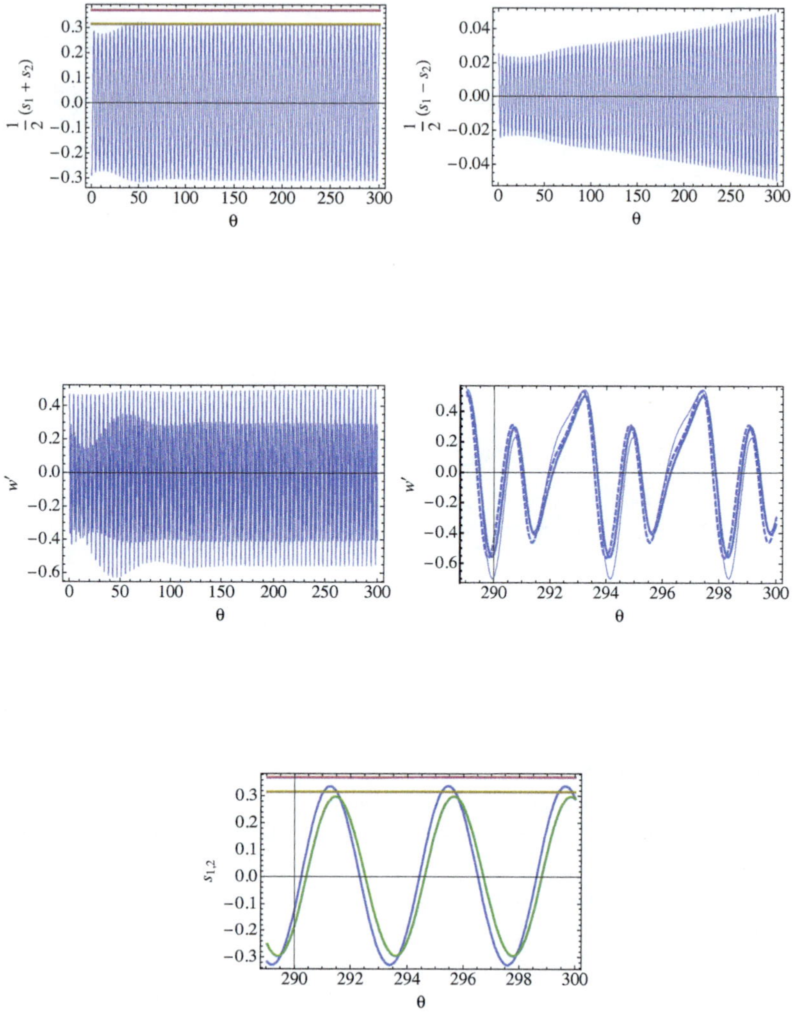

Figure 6. Simulation results versus the rotor angle θ for system parameter values: $\varepsilon = 0.1$, $n = 1.5$, $\mu = 0.3$, $N = 2$, $\Gamma = 0.813$, $\gamma = 0$. The first four panels are similar to those shown in Figure 5, and here the response of $\frac{1}{2}(s_1 - s_2)$ is growing, indicating the instability of the synchronous response. The final panel shows the two individual absorbers near the end of the run, as they drift apart; one absorber eventually approaches the cusp amplitude.

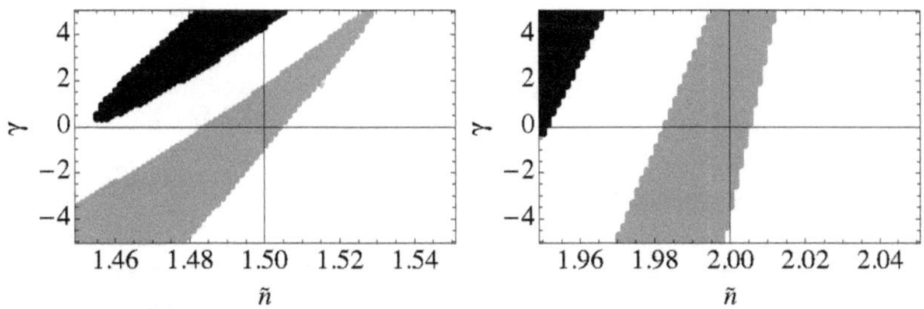

Figure 7. Stability zones in path parameter space, γ vs. $\tilde{n}$, for $a < a_o = \alpha a_{cusp}$ with $\alpha = 0.9$ for excitation orders $n = 3/2$ (left panel) and $n = 2$ (right panel). White regions represent absorber systems with stable synchronous steady state response up to $a = a_o$. Absorbers with paths corresponding to the gray regions experience NS instability to non-synchronous response at some amplitude $a \leq a_o$, and black regions represent absorbers that experience J instabilities, corresponding to jumps, at some amplitude $a \leq a_o$. System parameter values: $\varepsilon = 0.1$, $\mu = 0.3$, $N = 2$.

sacrificing absorber performance, is by slight linear overtuning. As indicated in Figure 4, this approach is seen to provide additional benefits in terms of torque range. From Figure 7 it is observed that one can also avoid the instability by moving γ to a value less than zero, that is, to a hardening path, similar to the cycloid, as used in helicopter applications (15). Thus, in general, it appears that a good strategy for selecting the path is to select a path that is overtuned, either linearly, nonlinearly, or both. For a given design, the amount of overtuning will depend on the inertia ratio ε, the absorber damping μ, and other factors, such as the required torque range. Finally, it is worth noting that if the damping μ is sufficiently large, no NS instabilities will occur near $(\tilde{n}, \gamma) = (n, 0)$, but that, correspondingly, the absorbers may not function as well.

The results for excitation order $n = 2$ are very similar to those for $n = 3/2$, so we show only the path stability result. The right panel of Figure 7 shows the NS and J instability regions in the $(\tilde{n}, \gamma)$ path parameter space, again for $a = a_o = 0.9a_m$. The qualitative structure of the instability zones, and the suggested tuning strategy, are similar to the $n = 3/2$ case. However, note that in this case, nonlinear detuning, that is, moving to negative γ values, is not as effective in this case as it was for $n = 3/2$, since the NS instability region extends further downward. In this case, linear overtuning appears to be the better way to stabilize the synchronous response.

3 Discussion and Closing

The system considered here demonstrates how one can make use of, and even exploit, nonlinear dynamics to push operating limits of a vibratory system of practical relevance. This example is particularly attractive since a reliable mathematical model exists, it experiences interesting bifurcations, it is amenable to analysis, and there exists a clean way to tailor the most important system nonlinearities, in this case by altering the absorber path. While most systems will not offer these features, in the present case a fundamental understanding of nonlinear dynamics plays an important role in the design of the system so that it meets desired specifications.

Based on the results described above, and our experience with these absorbers, the following design guidelines are offered:

- The linear tuning order $\tilde{n}$ should be taken to be at, or slightly above, the excitation order n, and the level of detuning depends on how much absorber inertia mc^2 is available relative to the fluctuating torque that needs to be addressed. The closer $\tilde{n}$ is to n, the better the absorber will perform, but the less torque range it will have. If mc^2 is large enough to keep the absorbers from reaching amplitude limits during transient and steady state responses over the entire torque range encountered, one can tune close to the ideal case.
- The path nonlinearity should be tautochronic, or slightly hardening. As in the case of the linear tuning parameter, one gets the best performance for $\lambda = \lambda_e$, but one can extend the operating range by detuning away from λ_e, towards hardening.
- Systems with multiple absorbers can have rich dynamics, even when the absorbers are identical, and the high level of system symmetry can lead to high sensitivity to imperfections, especially when the system is close to perfect tuning (23; 3; 24). In these cases, non-synchronous responses can occur, a common one being that in which a subset of the absorbers, even one, do all of the vibration reduction. One can avoid these types of responses by detuning away from $\tilde{n} = n$ and $\lambda = \lambda_e$, as described above. However, this is necessary only for absorber systems with very light damping.
- The level of damping, in both the absorber and rotor, also affects the response, and it plays a similar role to detuning, in that it inhibits the absorber response. If the absorbers have substantial damping, even a few percent of critical, many of the nonlinear instabilities described above are avoided. And, since damping is highly varied and difficult to model, one should consider parameter studies that cover the estimated range of damping values in a practical situation.

- Finally, as in any real system, there are unmodeled dynamics not captured by the idealized model. Whether these allow one to be more or less aggressive in terms of tuning is a case-by-case situation, and there is no substitute for physical testing.

Acknowledgements

The author would like to thank Lawrie Virgin and David Wagg for the invitation to present the CISM lectures on which this chapter is based. The work described is part of a long-term collaboration with automotive engineers Bruce Geist, Victor Borowski, and John Brevick, colleagues Alan Haddow and Brian Feeny, and several talented graduate students. Finally, the author is grateful for funding that has helped support this line of work, most recently by grants from U.S. National Science Foundation (#0700307) and the Chrysler Challenge Fund program.

Bibliography

[1] A. S. Alsuwaiyan and S. W. Shaw. Steady-state synchronous and localized responses of tuned pendulum vibration absorbers. In *Proceedings–ASME Design Engineering Technical Conferences*, 1999. Paper DETC99/VIB-8014.

[2] A. S. Alsuwaiyan and S. W. Shaw. Performance and dynamic stability of general-path centrifugal pendulum vibration absorbers. *Journal of Sound and Vibration*, 252(5):791–815, 2002.

[3] A. S. Alsuwaiyan and S. W. Shaw. Steady-state response of systems of nearly-identical torsional vibration absorbers. *Journal of Vibration and Acoustics*, 125:80–87, 2003.

[4] T.K. Caughey and M.E.J. O'Kelly. Classical normal modes in damped linear dynamic systems. *Journal of Applied Mechanics*, 32:583–588, 1965.

[5] C. P. Chao, C. T. Lee, and S. W. Shaw. Non-unison dynamics of multiple centrifugal pendulum vibration absorbers. *Journal of Sound and Vibration*, 204(5):769–794, 1997.

[6] C. P. Chao, S. W. Shaw, and C. T. Lee. Stability of the unison response for a rotating system with multiple tautochronic pendulum vibration absorbers. *Journal of Applied Mechanics*, 64:149–156, 1997.

[7] P. J. Davis. *Circulant Matrices*. Wiley-Interscience, New York, 1979.

[8] H. H. Denman. Tautochronic bifilar pendulum torsion absorbers for reciprocating engines. *Journal of Sound and Vibration*, 159(2):251–277, 1992.

[9] J. Guckenheimer and P. Holmes. *Nonlinear Oscillations, Dynamical Systems, and Bifurcations of Vector Fields.* Springer-Verlag, 1986.

[10] J. P. Den Hartog. *Tuned Pendulums as Torsional Vibration Eliminators*, pages 17–26. Stephen Timoshenko 60th Anniversary Volume. New York: The Macmillan Company, 1938.

[11] J. P. Den Hartog. *Mechanical Vibrations.* Dover, 1985.

[12] B. Jorg, K. Werner, and H. G. Eckel. The speed-adaptive damper dat. *ATZ Magazine*, pages 13–15, 2001.

[13] W. KerWilson. *Practical Solution of Torsional Vibration Problems.* Chapman and Hall Ltd., London, 3rd. edition, 1968.

[14] C. T. Lee and S. W. Shaw. On the counteraction of periodic torques in rotating systems using centrifugally driven vibration absorbers. *Journal of Sound and Vibration*, 191(5):695–719, 1996.

[15] J. F. Madden. Constant frequency bifilar vibration absorber. United States Patent No. 4218187.

[16] Ian Stewart Martin Golubitsky and David G. Schaeffer. *Singularities and Groups in Bifurcation Theory.* Springer Verlag, 1988.

[17] R. J. Monroe and S. W. Shaw. Nonlinear transient dynamics of pendulum torsional vibration absorbers, part i: Theory. Technical report, preprint, Department of Mechanical Engineering, Michigan State University, 2011.

[18] R. J. Monroe and S. W. Shaw. Nonlinear transient dynamics of pendulum torsional vibration absorbers, part ii: Experiment. Technical report, preprint, Department of Mechanical Engineering, Michigan State University, 2011.

[19] R. J. Monroe and S. W. Shaw. On the transient response of forced nonlinear oscillators. Technical report, preprint, Department of Mechanical Engineering, Michigan State University, 2011.

[20] R. J. Monroe, S. W. Shaw, A. G. Haddow, and B. K. Geist. Accounting for roller dynamics in the design of bifilar torsional vibration absorbers. In *Proceedings of the ASME 2009 International Design Engineering Technical Conferences & Computers and Information in Engineering Conference (IDETC/CIE 2009)*, 2009.

[21] A. H. Nayfeh. *Nonlinear Interactions: Analytical, Computational, and Experimental Methods by Ali Hasan Nayfeh.* John Wiley & Sons, 2000.

[22] A. H. Nayfeh and D. T. Mook. *Nonlinear Oscillations.* Wiley, 1995.

[23] T. Nester, A. G. Haddow, and S. W. Shaw. Experimental investigation of a system with multiple nearly identical centrifugal pendulum vibration absorbers. In *Proceedings of the ASME 19th Biennial Conference on Mechanical Vibration and Noise*, Chicago, Illinois, 2003.

[24] T. M. Nester. Experimental investigation of circular path centrifugal pendulum vibration absorbers. M.S. thesis, Michigan State University, East Lansing, MI, 2002.

[25] T. M. Nester, A. G. Haddow, S. W. Shaw, J. E. Brevick, and V. J. Borowski. Vibration reduction in variable displacement engines using pendulum absorbers. In *Proceedings of the SAE Noise and Vibration Conference and Exhibition*, number 2003-01-1484, Traverse City, Michigan, 2003.

[26] D. E. Newland. Nonlinear aspects of the performance of centrifugal pendulum vibration absorbers. *Journal of Engineering for Industry*, 86:257–263, 1964.

[27] B. J. Olson. *Order-Tuned Vibration Absorbers for Systems with Cyclic Symmetry with Applications to Turbomachinery.* Ph.D. dissertation, Michigan State University, East Lansing, MI, 2006.

[28] B. J. Olson and S. W. Shaw. Vibration absorbers for a rotating flexible structure with cyclic symmetry: nonlinear path design. *Nonlinear Dynamics*, 60:149–182, 2010.

[29] P. M. Schmitz. Experimental investigation into epicycloidal centrifugal pendulum vibration absorbers. MS Thesis, Michigan State University, East Lansing, MI, 2003.

[30] S. W. Shaw and B. K. Geist. Tuning for performance and stability in systems of nearly-tautochronic torsional vibration absorbers. *Journal of Vibration and Acoustics*, 132, 2010.

[31] S. W. Shaw, M. B. Orlowski, and A. G. Haddow. Transient dynamics of centrifugal pendulum vibration absorbers. In *The 12th International Symposium on Transport Phenonmenon and Dynamics of Rotating Machinery*, 2008.

[32] S. W. Shaw, P. M. Schmitz, and A. G. Haddow. Tautochronic vibration absorbers for rotating systems. *Journal of Computational and Nonlinear Dynamics*, 1:283–293, 2006.

[33] M. Swank and P. Lindemann. Dynamic absorbers for modern powertrains. In *SAE Transactions*, number 2011-11-1554, 2011.

[34] M. A. Wachs. The main rotor bifilar absorber and its effect on helicopter reliability/maintainability. *SAE Technical Paper Series 730894*, 1973.

Vibrations of Beams in the Elasto-Plastic and Geometrically Nonlinear Regime

Pedro Ribeiro

DEMec/IDMEC, Faculdade de Engenharia, Universidade do Porto, Porto, Portugal

Abstract This chapter presents models for beams vibrating with moderately large displacements and with elasto - plasticity. The beams are initially straight, homogeneous and isotropic, and oscillate always in one plane. A method to solve the equations of motion in the time domain, with computation of plastic strains in a mixed hardening situation, is described. Forced oscillations under harmonic excitations are analysed using this method. A frequency domain procedure to analyse free vibrations with existing plastic strains is also introduced. Free vibration oscillations are then analysed, with particular attention to the combined influence of large displacements, i.e., geometrical nonlinearity, and plastic strains on the shapes assumed during the period of vibration and on the natural frequencies.

1 Introduction

1.1 Chapter Objectives

Often the design of a structural or machine element requires an analysis in order to predict stresses and strains. A major goal of this analysis is to avoid failure in operation. Due to uncertainties - as not anticipated variations in the material properties, slight changes in the geometry, or unexpected loads - designers do not know exactly what stresses and strains the element will endure in practice. In addition, the physical/mathematical models used to represent reality are idealised approximations. The common way that engineers have to account for these uncertainties is to use a factor of safety in design.

There are usually many feasible designs, where the element performs its function without failure, but it is desirable to choose a design that approaches objectives such as minimizing cost, minimizing weight or maximizing the natural frequency without increasing weight. The realization of such a goal generally contradicts a large safety factor. In order to achieve a close to best design, whilst maintaining safety, it is important to reduce uncertainties and therefore it is desirable that the physical/mathematical model of the structural/machine element is close to reality.

Vibrations with large amplitude, which may cause large strains and stresses, occur in a few systems due to large loads or to loads with a frequency component close to a resonance frequency. This is a particular important issue in thin structures. For example in aircraft there is an obvious desire for thin-walled structural components, which are likelier to experience vibrations with amplitude of the order of their thickness (Amabili (2008), Sathyamoorthy (1987), Smith et al. (1961), Teh (1982), White (1978)). Moreover, systems that in current engineering practice are not designed to experience nonlinear vibrations, may, if modelled more accurately, be designed more efficiently and still in such a way that they perform their function safely, taking advantage of nonlinear dynamic analysis.

In spite of the popularity of linear models in engineering, the study of oscillations with large amplitudes requires geometrically nonlinear models (there are numerous publications on this issue that corroborate the former sentence, including Amabili (2008), Bennouna (1982), Kadiri et al. (2002), Han (1993), Mei (1973, 1976), Murphy et al. (1996), Ribeiro (2001, 2004b), Sarma and Varadan (1984a), Sathyamoorthy (1987), Touzé et al. (2004), Wolfe (1995), Zavodney and Nayfeh (1989) and other works quoted in the remainder of this section). The solution of the nonlinear systems of equations of motion is quite often achieved by iterative methods, with repeated update of the nonlinear model. Additional difficulties of nonlinear analyses result from the facts that the superposition principle is not applicable and multiple solutions for the same parameter (e.g., same excitation frequency) can exist.

Even when geometrical type nonlinearity is considered in the literature, it is often assumed that the oscillations occur in the linear elastic regime, with the stresses and strains related by generalised Hooke's law. However, Hooke's law is not valid in all circumstances. In many materials plasticity occurs after a certain stress level. In contrast with elastic deformation, plastic deformation is not reversible and depends on the deformation history (Dieter (1986), Kojić and Bathe (2005)). In this case, an incremental strain computation is advisable.

The knowledge of the natural modes of vibration is of undisputable significance. This is not only due to modal analysis (of particular importance in linear systems), but also because if vibrations are excited at a particular natural frequency, they tend to be of large amplitude. Naturally, larger amplitude vibrations may more easily cause fatigue induced damage and even abrupt failure. Moreover, and related with modal analysis, one can understand a system through its modes.

The goals of this chapter are: to present models that, within certain limits, accurately approach structural elements by considering nonlinear effects; to introduce methods to solve the equations that result from such models; to discuss some of the features that vibrating nonlinear structural elements possess. The text focuses on geometrical nonlinearity and elasto-plasticity. Forced oscillations and free vibrations are analysed, the latter with particular interest on variations of shapes and natural frequencies of vibration. Beams that are initially straight and that oscillate in one plane are used to introduce and discuss the concepts, because with this simple structure one is able to analyse the chief aspects of interest here. Moreover, beams are widely used as structural or machine elements; they can be employed alone, assembled to other beams, or to reinforce other structural elements, as plates or shells, and are found in industries such as aircraft, civil and mechanical. This text is mainly based on references Ribeiro (1998, 2004a, 2010), and Ribeiro and van der Heijden (2009).

1.2 A Review on Geometrically Nonlinear and Elasto-Plastic Vibrations of Beams

Many studies on beam vibrations have been carried out to date and this section cannot provide a complete review, even if a somewhat large number of references is here recalled. We start with large amplitude displacements, where more studies have been published, and proceed in the last paragraphs of the section to elasto-plastic behaviour.

Geometrically nonlinear vibrations. Experimental and theoretical analyses of beams vibrating with large amplitudes, thus, in the geometrically nonlinear regime, have been presented since, at least, the middle of the 20th century (Woinowski-Krieger (1950)). Many of the theoretical approaches assume that the solution is the product of independent time and displacement functions.

Let us first regard free vibration. For a linear conservative system, natural frequencies exist at which the system vibrates in such a way that the ratio between the amplitude of displacement of any two points is constant, thus defining a mode shape of vibration that remains unchanged during the period of vibration and is not amplitude dependent. In linear vibration, modes of vibration are of paramount importance and researchers have been interested in extending the concept to nonlinear systems. The definition of nonlinear mode constitutes a conceptual difficulty of nonlinear analysis and the origin of different approximations.

As the displacement amplitude increases, the stiffness increases due to the effect of the longitudinal forces. Consequently, the shape of most vibrating beams and the natural frequency actually change with the vibration amplitude, and a mode of vibration concept should accommodate for this amplitude dependency (Bennouna and White (1984), Benamar et al. (1991), Ribeiro and Petyt (1999), Szemplinska-Stupnicka (1983)). Another distinction with respect to linear vibrations, is that the natural frequencies of a nonlinear structure may become commensurable at certain vibration amplitudes, creating conditions for the strong interaction of vibration modes with energy interchange. This is a very interesting phenomenon known as internal resonance (Szemplinska-Stupnicka (1990), Nayfeh and Mook (1995)). In Nayfeh and Balachandran (1989) a review on the influence of modal interactions on the nonlinear response of harmonically excited structural systems was carried out. One of the conclusions of that review is that different experiments have shown the existence of internal resonances and that these are responsible for "interesting, unusual and dangerous phenomena".

Rosenberg (1966) introduced the concept of nonlinear normal mode of discrete conservative systems with n degrees of freedom, as a motion where all masses execute periodic vibrations with the same period, reach their maximum amplitudes and pass their static equilibrium points simultaneously.

In two milestone works, Shaw and Pierre (1993, 1994) extended the concept of nonlinear normal modes to continuous, not necessarily conservative, systems and studied the vibration of a linear beam on a nonlinear elastic foundation. The problem was formulated in terms of first order differential equations, including displacements and velocities as dependent variables. A normal mode was defined as a motion which takes place on a two-dimensional invariant manifold in the system's phase space. This manifold has the following properties: it passes through the stable equilibrium point of the system and at this point it is tangent to a plane which is an eigenspace of the system linearized about the equilibrium point. In a normal mode the system behaves like a one degree of freedom system. The validity of the solution procedure proposed is limited to weak nonlinearities; this might explain why Burton and Hamdan (1996) comparison of Shaw and Pierre's technique with a procedure involving the harmonic balance method (HBM) resulted in quantitative disagreement.

It has been assumed that the solution can be expressed as a function of the linear mode shapes. In some cases, a Duffing type equation has been derived, and an exact solution in the form of a cosine elliptic function with the period or frequency of the oscillation given by an elliptic integral has been found (Woinowski-Krieger (1950)). In the absence of exact analytical solutions, perturbation methods, such as the method of multiple scales or the method of averaging, are often used (Nayfeh and Mook (1995)).

Other authors express the time function in the form of a Fourier series and apply the harmonic balance method (Benamar et al. (1991), Ribeiro (1998); Ribeiro and Petyt (1999), Lewandowski (1987, 1991, 1994a,b, 1997a,b, 2003)) or the incremental harmonic balance method (Cheung and Lau (1982)). Approximation methods such as the Rayleigh-Ritz and Galerkin methods, or the finite element method (Cheung and Lau (1982), Ribeiro (1998); Ribeiro and Petyt (1999), Lewandowski (1987, 1991, 1994a,b, 1997a,b), Qaisi (1993)) are commonly used to discretise a continuous structure before applying harmonic balance procedures. This leads to ordinary nonlinear differential equations in the generalized coordinates, from which, via harmonic balance, result the frequency and the shape of vibration. The shapes associated with each harmonic change with the maximum amplitude of vibration and with the period of vibration, but do not change during the period of vibration. However, the shape of the structure is defined by the series containing all harmonics, hence if more than one harmonic is considered in the solution and is not zero, then the vibration shape changes during the period of vibration. By using several harmonics in the time series, bifurcation points and secondary branches due to internal resonances (Szemplinska-Stupnicka (1990)), have been found (see for example Cheung and Lau (1982), Ribeiro and Petyt (1999), Lewandowski (1994a,b, 1997a,b)).

An approach that, although different, has similarities to a one term harmonic balance procedure was followed by Bhashyam and Prathap (1980), Prathap and Varadan (1978), Qaisi (1993) and Sarma and Varadan (1983, 1984b): the problem was reduced to an eigenvalue problem, by the assumption that $\ddot{w}(x,t) = -\omega^2 w(x,t)$ at the point of maximum amplitude of the transverse displacement $w(x,t)$. Parameter $-\omega^2$ was hence interpreted as a coefficient of proportionality between the acceleration $\ddot{w}$ and the displacement w at that point. The solution of the eigenvalue problem defines ω^2 and a shape of vibration.

Yet another alternative is to directly integrate the equations of motion. Runge-Kutta methods are rather popular with first-order-differential equations, and other methods, such as Newmark's method and a symplectic integration scheme, have been applied to second order differential equations (Leung and Mao (1995), Shi and Mei (1996)). Numerical integration demands a large computational effort, particularly in a free vibration and conservative problem, where one wishes to establish the relation between the vibration amplitude and the vibration frequency. Therefore, the equations of motion are usually derived by expressing the displacement as a function of a reduced number of linear modes.

Experimental analysis has also been utilised to verify the variation of shapes and natural frequencies of beams vibrating with large amplitudes. Bennouna (1982); Bennouna and White (1984) experimentally determined the spectral components of the transverse vibration displacements of a clamped-clamped beam. The first harmonic was clearly the most important, but the first three harmonics influenced the response of the beam. It was verified that the shape of a clamped-clamped beam changes during the period of vibration and that its curvature increases near the clamps with increasing amplitude.

A large amount of research has as well been conducted in the study of the nonlinear steady-state oscillation of beams under harmonic transverse excitation. In steady-state nonlinear vibrations, it is common to find more than one solution. Because only the stable solutions are physically meaningful, it is important to carry out a stability study. Usually Floquet's theory is followed for that purpose (Bolotin (1964), Hayashi (1964), Nayfeh and Mook (1995), Ribeiro (2009), Szemplinska-Stupnicka (1990)).

In forced as in free vibration, the beam displacement is generally assumed to be given by a function that is a product of separated functions of space and time, or a sum of such products. Several authors used a limited number of linear modes to express the space dependence of the transverse displacement and applied Galerkin's method. For example, Tseng and Dugundji (1970) applied Galerkin's method and the harmonic balance method to derive the equation of motion of a beam with fixed ends, excited by the periodic motion of its supporting base. Only one linear mode was considered. The stability of the solutions was analysed by solving a variational Hill-type equation. Bennet and Eisley (1970) followed a similar approach, but considering a two and three linear modes expansion, in an investigation of the nonlinear forced response of a clamped-clamped undamped beam subjected to a concentrated harmonic force. Atluri (1973) studied a beam with one end free to move longitudinally. The Galerkin method was used in conjunction with the perturbation procedure of multiple scales. The effect of middleplane stretching was excluded and the effects of large curvature, longitudinal and rotatory inertia included. In these conditions, the nonlinearity tends to be of the softening type. Takahashi (1979) studied the steady-state response of an undamped nonlinear clamped-clamped beam under periodic excitation. Using two-degrees of freedom, the Galerkin and the harmonic balance methods were applied to derive a set of nonlinear algebraic equations. The stability of the solution was studied by investigating the behaviour of a small perturbation to the steady state response.

The finite element method (FEM) is based on approximating the solution of a problem by means of admissible functions and has also been used to construct geometrical nonlinear beam models in forced vibration. One of the first applications of the h-version of the finite element method, version in which better approximations are achieved by refining the mesh, to study nonlinear oscillations of beams was carried out by Mei (1973), who proceeded to an extensive application of the finite element method to investigate nonlinear oscillations of different structures. In another early publication, Busby and Weingarten (1972) applied the FEM to examine the response of simply supported and clamped-clamped beams. The equations of motion were solved by the averaging method. Modal coupling was detected at large amplitudes of vibration. In (Mei and Decha-Umphai (1985)) a finite element method, including a harmonic force matrix, was presented for nonlinear vibrations of undamped beam structures subjected to harmonic excitation. Longitudinal deformation and inertia were included in the formulation. For beams with an axially movable support it was found, as by Atluri (1973), that the nonlinearity can be of the soft spring type. Leung and Fung (1989) applied the finite element method and the harmonic balance method to determine the steady state response of beams and frames. The response curve due to a harmonic excitation was constructed using the Newton method and the phase angle as a parameter. A 1:3 internal resonance was detected in a clamped-hinged beam and resulted in looping characteristics of the response curve. Lewandowski (1994a,b, 1997a,b) applied Newton method but with the arc-length as a parameter, to obtain the relation between the amplitude of vibration and the natural frequency (free vibration studies), and the response to harmonic excitations of diverse beams. 1:3 internal resonances were found and resulted in secondary branches or in an increase of the curvature of the backbone-curves and response curves.

In order to gain a better understanding of the variation of the shape of vibration with the amplitude and during the cycle of vibration, Ma et al. (1995) applied the FEM to derive the equations of motion in an incremental form in time. The equations were integrated in the time domain and, in every integral step, the system was assumed to be linear. Thus, the orthogonality properties of the eigenvectors could be used to decouple the incremental equations of motion. The principal problem with this method is the computational time required, mainly because, for each nonlinear state of the structure, the eigenvalues and eigenvectors have to be calculated.

The p-version of the FEM - in which the mesh is not changed, but the number of generalized coordinates of each element is - has also been applied to geometrically nonlinear vibrations of beams. In Ribeiro and Petyt (1999), a p-version - hierarchical according to the definition of Meirovitch and Baruh (1983) - Bernoulli-Euler finite element was presented and the free and steady-state forced vibrations of simply supported and clamped-

clamped beams were studied in the frequency domain, by the harmonic balance and continuation methods. Internal resonances were found and the variation of the shape of vibration, not only with the vibration amplitude but also along a vibration period, was shown. A similar element, but comparing diverse sets of shape functions, was used for time domain analysis, using Newmark's method, of beams and plane frames in Ribeiro (2001). A Timoshenko, or First Order Shear Deformation (FOSD), type of element for straight and curved beams was developed in Ribeiro (2004a), where nonlinear vibrations were investigated again using Newmark's method. This FOSD element was employed in Ribeiro (2004b) to investigate forced periodic vibrations via shooting method. A more recent study where a p-version FOSD formulation is also employed is presented in Zhu and Leung (2009).

This historical review of works on large amplitude beam vibrations, is concluded with references where experiments are prominent. References Bennouna (1982) and Bennouna and White (1984) were already mentioned in relation with the shape of vibration. Tseng and Dugundji (1970) and Yamamoto et al. (1981) experimentally showed that subharmonic and superharmonic oscillations occur. Yamamoto et al. (1982a,b) experimentally detected so called super-summed and differential harmonic oscillations of beams when the excitation frequency ω satisfies the condition: $2\omega = \omega_i \pm \omega_j$, where ω_i and ω_j are natural frequencies of the beam. Wolfe (1995) analysed a clamped-clamped aluminium alloy beam and a clamped-clamped beam in CFRP (carbon fibre reinforced plastic). The geometrically nonlinear vibration of an aluminium beam hinged at both ends was investigated by Ribeiro and Carneiro (2004). The beam was excited transversely with a harmonic excitation and the amplitudes of the first and higher harmonics analysed. It was demonstrated that internal resonances occur between the first and higher order modes, and between the second and higher order modes. In these works, the fundamental resonance frequency increased as the excitation level increased. Jump phenomena were observed by a large drop in amplitude over a small increase in frequency, when the excitation is increased slowly during sinusoidal excitation. Jump phenomena were also observed when the frequency is above resonance and slowly swept to lower frequencies. Yagasaki (1995) analysed a beam excited by two forces with different frequencies near the first mode frequency. The averaging method was applied to a single-mode Galerkin approximation. The theoretical predictions were in qualitative good agreement with experimental results. It was concluded that for a more precise description, higher order modes must be included in the model.

Vibrations with plastic strains. One of the approaches employed to analyse oscillations with plasticity consist in the application of Shanley-type models. In this type of model a beam is approximated by rigid links (rigid bars) joined by a linear elastic-perfectly plastic element; it is also admitted that the yield stress in tension and compression are equal. Shanley-type models simplify the continuous elasto-plastic beam problem, reducing the system to a low dimensional discrete one. Symonds and Yu (1985) employed a one degree of freedom Shanley-type model to analyse the response of an elasto-plastic beam to a pulse load applied transversely to the beam. The authors found that the beam may rest in the opposite direction of the applied load and achieved similar results with finite element procedures. Lee et al. (1992) investigated the response of a beam in more or less similar lines, but using a two-degree-of-freedom model to represent the beam. The presence of chaos is indicated by positive Lyapunov exponents. Xu and Hasebe (1997) used a Shanley-type model is used, a co-dimension three bifurcation problem is defined and the method of normal forms employed. A three degree of freedom Shanley-type model was used to investigate the dynamic instability of elastic plastic beams by Liu et al. (2004) and Ma et al. (2005). Motions with power spectra, phase space trajectories and Poincaré maps typical of chaotic motions were found.

Naturally, other types of models have been used to investigate elasto-plastic beam vibrations. Manoach and Karagiozova (1993) used the linear modes to discretize the system and integrated numerically the equations of motion. Lepik (1995) discussed vibrations of a buckled beam under harmonic excitation using Galerkin's method. Han and Lu (1999) propose a Space Time FEM scheme for elasto-plastic dynamic analysis of Timoshenko beams, but considering small displacements. Gerstmayr et al. (2001) developed an algorithm to investigate a vibrating thin beam with guided rigid-body motion. Small displacements are again assumed; actually it appears that geometrical nonlinearity is often not considered in analyses that involve plasticity. In Gerstmayr and Irschik (2003) the analysis of Gerstmayr et al. (2001) was extended to a linear elastic/perfectly plastic beam performing rotatory motions about a fixed hinged end.

A first-order-shear deformation theory, p-version finite element with hierarchical basis functions was proposed in (Ribeiro and van der Heijden (2009)). In the later reference, numerical tests are carried out to demonstrate that the p-version may be applied with advantages to analyse elasto-plastic problems; static problems and forced oscillations are addressed. Ribeiro (2010) applies the same element to investigate how plastic strains affect the modes of vibration, i.e. a steady-state periodic free vibration problem is investigated.

1.3 Structure of this Chapter

This chapter has seven main sections. In Section 2 nonlinear equations of motion of vibrating beams, with large displacements and in the presence of elasto-plasticity are developed. The geometrically nonlinear only situation is not presented, since it is a particular case of the former.

Section 3 addresses the solution of the equations of motion in the forced vibration regime and with yielding. The equations are solved in the time domain using a Newmark method and iterating in each time step, in order to compute the plastic strains and to correct the terms that depend upon plastic strains and geometrical nonlinearity.

In Section 4, studies on forced vibrations of beams with geometrical nonlinearity and large amplitude displacements are presented. The influence of the plastic strains on the displacements, velocities, strains and stresses of a thin and a moderately thick beam is discussed.

Section 5 presents the equations of motion of periodic free oscillations with large displacements and with a constant plastic strain field, previously imposed by a force. The equations obtained are algebraic and should ideally be solved by a continuation method.

In Section 6 the formulation previously presented is employed to carry out numerical tests that aim at illustrating the effects of large displacement amplitudes and of plastic strains on the shapes and natural frequencies of vibration.

Section 7 provides, in the guise of conclusions, a short summary of the more important points addressed.

2 Equations of Motion of Beams With Geometrical Nonlinearity and Elasto-Plasticity

In this section the equations of motion of beams vibrating in one plane, that may experience large displacements and elasto-plasticity, are introduced. Reference (Ribeiro and van der Heijden (2009)) and references therein are mostly followed. In what concerns plasticity, the book of Kojić and Bathe (2005) provides a description of the concepts here employed. Nevertheless, readers interested in inelastic problems may wish to additionally consult other books, including (Owen and Hinton, 1980) and (Simo and Hughes, 1998).

2.1 Displacements Field and Strain-Displacement Relations

Displacements are here defined with respect to a stationary reference frame, i.e., in a "total Lagrangian" approach. The displacement field is based upon the following assumptions:

- the beam vibrates only in one plane (we designate this plane as x_1x_3, represented in Figure 1);
- the beam is initially straight;
- the beam cross sections remain plane;
- cross sections are free to rotate about an axis perpendicular to the plane where motion takes place;
- the beam thickness is moderately small in comparison with the length, agreing with the definition of a beam as an elemental structure, but it is not necessarily very small, because rotations of the transverse section are considered.

The assumptions above lead to a first order shear deformation - also known as Timoshenko - model and the displacement field is given by

$$u_1(x_1, x_3, t) = u_1^0(x_1, t) + x_3\theta^0(x_1, t) \tag{1}$$

$$u_3(x_1, x_3, t) = u_3^0(x_1, t) \tag{2}$$

where $u_i(x_1, x_3, t)$ represents the displacement component along axis x_i and $\theta^0(x_1, t)$ represents the cross section rotation about x_2. Superscript ^{0} indicates the longitudinal axis, x_1, which passes through the cross sections centroids, when the beam is straight. The three reference axes and beam dimensions (length ℓ, width b and thickness h) are represented in Figure 1.

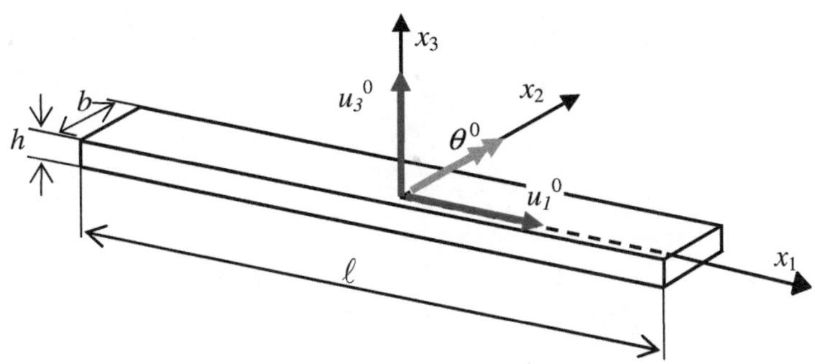

Figure 1. Coordinate axis, reference displacements and beam dimensions.

In addition, it is assumed that the displacements are moderately large and Green strain tensor (Chia (1980), Fung and Tong (2001))

$$\varepsilon_{ij} = \frac{1}{2}\left(\frac{\partial u_i}{\partial x_j} + \frac{\partial u_j}{\partial x_i} + \frac{\partial u_k}{\partial x_i}\frac{\partial u_k}{\partial x_j}\right),\ i,\ j,\ k = 1,2,3 \tag{3}$$

is employed. In the case at hand, the most important nonlinear term is $\left(\frac{\partial u_3}{\partial x_1}\right)^2$ and only this will be considered (a von Kármán approach, Chia (1980)). Hence, the longitudinal strain and the transverse shear engineering strain, $\gamma_{13}(x_1,t)$, which is twice the tensorial shear, are the following

$$\varepsilon_{11}(x_1,x_3,t) = \frac{\partial u_1^0(x_1,t)}{\partial x_1} + \frac{1}{2}\left(\frac{\partial u_3^0(x_1,t)}{\partial x_1}\right)^2 + x_3\frac{\partial \theta^0(x_1,t)}{\partial x_1} \tag{4}$$

$$\gamma_{13}(x_1,t) = 2\varepsilon_{13}(x_1,t) = \frac{\partial u_3^0(x_1,t)}{\partial x_1} + \theta^0(x_1,t) \tag{5}$$

The longitudinal strain can be written in the following form, which is advantageous to define the stiffness matrices (Ribeiro (2001)):

$$\varepsilon_{11}(x_1,x_3,t) = \lfloor\, 1 \quad x_3 \,\rfloor \left(\left\{ \begin{array}{c} \varepsilon_L^\ell(x_1,t) \\ \varepsilon_L^b(x_1,t) \end{array} \right\} + \left\{ \begin{array}{c} \varepsilon_{NL}^\ell(x_1,t) \\ 0 \end{array} \right\}\right) \tag{6}$$

In equation (6) three strain components appear: the linear longitudinal strain, $\varepsilon_L^\ell(x_1,t)$, the bending strain, $\varepsilon_L^b(x_1,t)$, and the geometrically nonlinear longitudinal strain, $\varepsilon_{NL}^\ell(x_1,t)$. These three strain components are given by

$$\varepsilon_L^\ell(x_1,t) = \frac{\partial u_1^0(x_1,t)}{\partial x_1} \tag{7}$$

$$\varepsilon_L^b(x_1,t) = \frac{\partial \theta^0(x_1,t)}{\partial x_1} \tag{8}$$

$$\varepsilon_{NL}^\ell(x_1,t) = \frac{1}{2}\left(\frac{\partial u_3^0(x_1,t)}{\partial x_1}\right)^2 \tag{9}$$

2.2 Constitutive Relation

The relation between stresses and strains adopted is the result of some simplifying assumptions. To start with, and coherently with the planar beam oscillations of interest here, it is assumed that the direct stresses, σ_{22} and σ_{33}, and that the shear stresses, σ_{12} and σ_{23}, are negligible. Moreover, it is assumed that the material is isotropic and homogeneous before plasticity occurs. Finally, a bilinear stress-strain relation is adopted and mixed hardening, with a mixed hardening parameter M, is considered.

The elastic constitutive relation for an isotropic material is adapted to give the stress-strain relation in the presence of plasticity as follows:

$$\left\{ \begin{array}{c} \sigma_{11}\left(x_1, x_3, t\right) \\ \sigma_{13}\left(x_1, x_3, t\right) \end{array} \right\} = \left[\begin{array}{cc} E & 0 \\ 0 & G \end{array} \right] \left\{ \begin{array}{c} \varepsilon_{11}\left(x_1, x_3, t\right) - \varepsilon_{11}^{p}\left(x_1, x_3, t\right) \\ \kappa\gamma_{13}\left(x_1, t\right) - \gamma_{13}^{p}\left(x_1, x_3, t\right) \end{array} \right\} \quad (10)$$

E is the Young modulus and G is the shear modulus of elasticity, which is equal to $E/(2(1+\nu))$. $\varepsilon_{11}^{p}\left(x_1, x_3, t\right)$ represents the longitudinal plastic strain and $\gamma_{13}^{p}\left(x_1, x_3, t\right)$ the shear plastic strain. The letter ν is used to represent the ratio of Poisson and κ the shear correction factor.

Different values have been suggested for the shear correction factor, which actually depends on the beam's cross section (Timoshenko (1922), Kaneko (1975), Hutchinson (2001)). Kaneko (1975) conclude that the value apparently suggested by Timoshenko (reference Timoshenko (1922) is quoted by Kaneko (1975)) gives results closer to experiments. For rectangular beams, this value is $\kappa = (5+5\nu)/(6+5\nu)$ and is the one adopted here. Hutchinson (2001) suggested a more elaborated formula, where κ is a function of the aspect ratio. However, in the particular case of rectangular beams, the same author arrived at the conclusion that "the experimental results neither confirm nor negate the dependence of the new shear coefficient on the aspect ratio".

2.3 Discretization and Equations of Motion

The displacements of points on the centroidal axis of the beam can be written as

$$\left\{ \begin{array}{c} u_1^0\left(\xi,\, t\right) \\ u_3^0\left(\xi,\, t\right) \\ \theta^0\left(\xi,\, t\right) \end{array} \right\} = \left[\begin{array}{ccc} \mathbf{N}^{u_1}\left(\xi\right)^T & 0 & 0 \\ 0 & \mathbf{N}^{u_3}\left(\xi\right)^T & 0 \\ 0 & 0 & \mathbf{N}^{\theta}\left(\xi\right)^T \end{array} \right] \left\{ \begin{array}{c} \mathbf{q}_{u_1}\left(t\right) \\ \mathbf{q}_{u_3}\left(t\right) \\ \mathbf{q}_{\theta}\left(t\right) \end{array} \right\} \quad (11)$$

where the vectors of longitudinal, transverse and rotational basis functions (the names here employed to distinguish the sub-categories of basis functions relate directly with the generalised displacements those functions are connected to. In FEM nomenclature, these are all displacement type "shape functions" or "element displacement functions" - Petyt (1990), Szabó and Babuska (1991)) - are, respectively, given by

$$\mathbf{N}^{u_1}\left(\xi\right)^T = \left\{g_1\left(\xi\right), g_2\left(\xi\right) \;, \cdots\; g_{p_i}\left(\xi\right)\right\} \quad (12)$$

$$\mathbf{N}^{u_3}(\xi)^T = \{f_1(\xi), f_2(\xi), \cdots f_{p_o}(\xi)\} \tag{13}$$

$$\mathbf{N}^{\theta}(\xi)^T = \{\Theta_1(\xi), \Theta_2(\xi), \cdots \Theta_{p\theta}(\xi)\} \tag{14}$$

The displacement shape functions $g_i(\xi)$, $f_i(\xi)$ and $\Theta_i(\xi)$ can be found in (Ribeiro (2001)). The local coordinate,ξ, varies from -1 to 1 and the following relation holds: $x_1 = \xi\ell/2$. Letter $\mathbf{q}$ is employed for time dependent generalised displacement vectors.

This text follows work developed in the area of the p-version of the finite element method (Han (1993), Szabó and Babuska (1991)), so we may designate the results that are presented later as p-version FEM results. One of the characteristics of this method is that the number of elements is defined much more by geometry than by accuracy requirements. Actually, in all case studies of the present text only one element is used and therefore the displacement shape functions (12)-(14) must respect the geometrical boundary conditions of the full beam. Further details on the p-version FEM and its application in nonlinear vibrations of beams in one plane can be found in Ribeiro (1998); Ribeiro and Petyt (1999); Ribeiro (2001, 2004a,b); Ribeiro and van der Heijden (2009); Ribeiro (2010), Zhu and Leung (2009). Applications of this method with more than one element to nonlinear vibrations are shown in Ribeiro (2001) and Ribeiro et al. (2010), although the latter work is on shells, not on beams.

Applying the virtual work principle to our problem, we arrive at the following expression

$$\begin{aligned}&\int_\Omega \delta\varepsilon_{11}\sigma_{11}d\Omega + \int_\Omega \delta\gamma_{13}\sigma_{13}d\Omega - \int_\Omega \rho\left(\delta u_1\, \ddot{u}_1 + \delta u_3 \ddot{u}_3\right) d\Omega \\ &- \int_\Omega \left(\delta u_1\, F_{u_1} + \delta u_3\, F_{u_3} + \delta\theta\, M\right) d\Omega = 0\end{aligned} \tag{15}$$

where F_{u_i} represents external forces with direction x_i and M now represents an external moment about x_2. Letter Ω is used to denote three-dimensional region occupied by the beam. Employing the constitutive relation (10), the first two terms of equation (15) are written as

$$\int_\Omega \delta\varepsilon_{11}\sigma_{11}d\Omega = \int_\Omega \delta\varepsilon_{11}E\varepsilon_{11}d\Omega - \int_\Omega \delta\varepsilon_{11}E\varepsilon_{11}^p d\Omega \tag{16}$$

and

$$\int_\Omega \delta\gamma_{13}\sigma_{13}d\Omega = \int_\Omega \delta\gamma_{13}\kappa G\gamma_{13}d\Omega - \int_\Omega \delta\gamma_{13}G\gamma_{13}^p d\Omega \tag{17}$$

The last term of equation (16) generates the following vectors

$$\mathbf{F}_{u_1}^{plast}\left(\varepsilon_{11}^{p}\left(\xi,\eta,t\right)\right)=E\frac{bh}{2}\int_{-1}^{1}\int_{-1}^{1}\mathbf{N}_{,\xi}^{u_1}\left(\xi\right)\varepsilon_{11}^{p}\left(\xi,\eta,t\right)d\eta d\xi \tag{18}$$

$$\mathbf{F}_{\theta}^{plast}\left(\varepsilon_{11}^{p}\left(\xi,\eta,t\right)\right)=E\frac{bh^2}{4}\int_{-1}^{1}\int_{-1}^{1}\eta\mathbf{N}_{,\xi}^{\theta}\left(\xi\right)\varepsilon_{11}^{p}\left(\xi,\eta,t\right)d\eta d\xi \tag{19}$$

and matrix

$$\mathbf{K}^{plast}\left(\varepsilon_{11}^{p}\left(\xi,\eta,t\right)\right)=E\frac{bh}{\ell}\int_{-1}^{1}\int_{-1}^{1}\mathbf{N}_{,\xi}^{u_3}\left(\xi\right)\mathbf{N}_{,\xi}^{u_3}\left(\xi\right)^{T}\varepsilon_{11}^{p}\left(\xi,\eta,t\right)d\eta d\xi \tag{20}$$

$\mathbf{F}_{u_1}^{plast}$ and $\mathbf{F}_{\theta}^{plast}$ are vectors of generalized forces that exist because of the longitudinal plastic strains; $\mathbf{K}^{plast}$ is a matrix that appears due to the interaction between the plastic strains and the large displacements. Equations (18)-(20) were written for the particular, but quite common, case where E, b and h are constant. Letter η represents the dimensionless coordinate in direction x_3: $x_3 = \eta h/2$. The plastic shear strains originate the following force vectors:

$$\mathbf{F}_{\gamma u_3}^{plast}\left(\gamma_{13}^{p}\left(\xi,\eta,t\right)\right)=\frac{bh}{2}G\int_{-1}^{1}\int_{-1}^{1}\mathbf{N}_{,\xi}^{u_3}\left(\xi\right)\gamma_{13}^{p}\left(\xi,\eta,t\right)d\eta d\xi \tag{21}$$

$$\mathbf{F}_{\gamma\theta}^{plast}\left(\gamma_{13}^{p}\left(\xi,\eta,t\right)\right)=\frac{bh\ell}{4}G\int_{-1}^{1}\int_{-1}^{1}\mathbf{N}^{\theta}\left(\xi\right)\gamma_{13}^{p}\left(\xi,\eta,t\right)d\eta d\xi \tag{22}$$

Forces and matrix defined by equations (18)-(22) involve plastic strains, and numerical integration (Davis and Polonsky (1972)) is employed to compute them.

The remaining terms in equation (15) are not related with plasticity and originate: (1) mass matrices that will be represented by $\mathbf{M}_i$, $i = u_1, u_3, \theta$; (2) the often called linear - linear because they appear in a linear system, the matrices are actually constant - stiffness matrices, that will be written as $\mathbf{K}_{\ell ij}^{k}$, where i, j are 1, 2 or 3, and k can be p, γ or b (respectively standing for longitudinal, shear and bending); (3) the stiffness matrices $\mathbf{K}_{n\ell ij}$, i, j=1,2, that represent the isolated effect of geometrical nonlinearity; and the vectors of generalized external forces $\mathbf{F}_i, i = u_1, u_3, \theta$. These terms are computed resorting to analytical, exact, integration. Within the actual formulation, these matrices were defined in Ribeiro (2004a) and their expressions are given in the following paragraphs.

In the case of the beams analysed in this text, which have constant rectangular cross sections, the longitudinal stiffness matrix $\mathbf{K}^p_{\ell 11}$ can be written as:

$$\mathbf{K}^p_{\ell 11} = E\, bh \frac{2}{\ell} \int_{-1}^{1} \mathbf{N}^{u_1}_{,\xi} \mathbf{N}^{u_1 T}_{,\xi} d\xi \tag{23}$$

and the bending matrix $\mathbf{K}^b_{\ell 33}$ is given by

$$\mathbf{K}^b_{\ell 33} = E \frac{bh^3}{12} \frac{2}{\ell} \int_{-1}^{1} \mathbf{N}^{\theta}_{,\xi} \mathbf{N}^{\theta T}_{,\xi} d\xi \tag{24}$$

The shear stiffness matrix is

$$\begin{bmatrix} \mathbf{K}^{\gamma}_{\ell 22} & \mathbf{K}^{\gamma}_{\ell 23} \\ \mathbf{K}^{\gamma}_{\ell 32} & \mathbf{K}^{\gamma}_{\ell 33} \end{bmatrix} = \lambda G bh \int_{-1}^{1} \begin{bmatrix} \frac{4}{\ell^2} \mathbf{N}^{u_3}_{,\xi} \mathbf{N}^{u_3 T}_{,\xi} & \frac{2}{\ell} \mathbf{N}^{u_3}_{,\xi} \mathbf{N}^{\theta T} \\ \frac{2}{\ell} \mathbf{N}^{\theta} \mathbf{N}^{u_3 T}_{,\xi} & \mathbf{N}^{\theta} \mathbf{N}^{\theta T} \end{bmatrix} \frac{\ell}{2} d\xi \tag{25}$$

where the shear modulus is $G = \frac{E}{2(1+\nu)}$.

There are three nonlinear stiffness matrices equal to the thin straight beam ones, which are given in Ribeiro and Petyt (1999). These are matrix $\mathbf{K}_{n\ell_{12}}$, which depends linearly on the generalised transverse displacements $\mathbf{q}_{u_3}$, matrix $\mathbf{K}_{n\ell_{22}}$, which is a quadratic function of $\mathbf{q}_{u_3}$, and matrix $\mathbf{K}_{n\ell_{21}}$, which is equal to $2\mathbf{K}^T_{n\ell_{12}}$.

$\mathbf{K}_{n\ell_{12}}$ is given by

$$\mathbf{K}_{n\ell_{12}} = Ebh \frac{2}{\ell^2} \int_{-1}^{1} u_{3,\xi} \mathbf{N}^{u_1}_{,\xi} \mathbf{N}^{u_3 T}_{,\xi} d\xi \tag{26}$$

Matrix $\mathbf{K}_{n\ell_{22}}$ is a quadratic function of the transverse displacement u_3. Unlike $\mathbf{K}_{n\ell_{12}}$, $\mathbf{K}_{n\ell_{22}}$ is not affected by the longitudinal displacement of the reference line; it is defined as:

$$\mathbf{K}_{n\ell_{22}} = Ebh \frac{4}{\ell^3} \int_{-1}^{1} \mathbf{N}^{u_3 T}_{,\xi} \mathbf{N}^{u_3}_{,\xi} u^2_{3,\xi} d\xi \tag{27}$$

A consistent mass matrix (Ribeiro (2004a)) is employed.

Using the just defined force vectors and matrices, the equations of motion are written in the following form (not all function arguments are represented):

$$\begin{bmatrix} \mathbf{M}_{u_1} & 0 & 0 \\ 0 & \mathbf{M}_{u_3} & 0 \\ 0 & 0 & \mathbf{M}_\theta \end{bmatrix} \begin{Bmatrix} \ddot{\mathbf{q}}_{u_1}(t) \\ \ddot{\mathbf{q}}_{u_3}(t) \\ \ddot{\mathbf{q}}_\theta(t) \end{Bmatrix} + \begin{bmatrix} \mathbf{K}^p_{\ell_{11}} & 0 & 0 \\ 0 & \mathbf{K}^\gamma_{\ell_{22}} & \mathbf{K}^\gamma_{\ell_{23}} \\ 0 & \mathbf{K}^\gamma_{\ell_{32}} & \mathbf{K}^\gamma_{\ell_{33}} + \mathbf{K}^b_{\ell_{33}} \end{bmatrix} \times$$

$$\times \begin{Bmatrix} \mathbf{q}_{u_1}(t) \\ \mathbf{q}_{u_3}(t) \\ \mathbf{q}_\theta(t) \end{Bmatrix} + \begin{bmatrix} 0 & \mathbf{K}_{n\ell_{12}} & 0 \\ \mathbf{K}_{n\ell_{21}} & \mathbf{K}_{n\ell_{22}} - \mathbf{K}_{plast} & 0 \\ 0 & 0 & 0 \end{bmatrix} \begin{Bmatrix} \mathbf{q}_{u_1}(t) \\ \mathbf{q}_{u_3}(t) \\ \mathbf{q}_\theta(t) \end{Bmatrix} =$$

$$= \begin{Bmatrix} \mathbf{F}_{u_1}(t) \\ \mathbf{F}_{u_3}(t) \\ \mathbf{F}_\theta(t) \end{Bmatrix} + \begin{Bmatrix} \mathbf{F}^{plast}_{u_1}(\varepsilon^p_{11}) \\ \mathbf{F}^{plast}_{\gamma u_3}(\gamma^p_{13}) \\ \mathbf{F}^{plast}_{\theta}(\varepsilon^p_{11}) + \mathbf{F}^{plast}_{\gamma\theta}(\gamma^p_{13}) \end{Bmatrix} \tag{28}$$

3 Time Domain Solution of the Equations of Motion and Computation of Plastic Strains

In the presence of external forces and yielding, it is natural to solve the equations of motion in the time domain. If the plastic strains are constant, Equations (28) are simpler ordinary differential equations and there are a number of established methods that can be used for the numerical solution of these ODEs. These methods can be adapted to a situation where the plastic strains change. Here a procedure based on Newmark method with Newmark's parameters Bathe (1996), which has proven to be reliable and efficient in a wide number of tests, will be employed. The method is implicit, a feature that is understood to be important, as it allows correcting the nonlinear terms.

Because the plastic terms change whenever there is yielding and due to the geometrical nonlinearity, some stiffness matrices and vectors of equation (28) are unknown in each time step. This combined problem will be solved in a rather simple way using two cycles. Instead of using the "consistent elastic-plastic tangent matrix" (Kojić and Bathe (2005)), direct substitution and iteration on the original equations are employed. A method that employs a tangent matrix likely requires fewer iterations for convergence and may, eventually, converge in cases where the method here proposed fails. But since the determination of a consistent elastic-plastic tangent matrix requires an additional effort and the simper method here proposed worked out quite effectively in our case studies, we will pass without the tangent matrix.

In a first cycle of the solution algorithm, the plastic strains from the previous time step - either zero or not - are used to compute the forces and matrix given by Equations (18)-(22). The generalized displacements are computed by solving equation (28) iteratively with correction of the geometrically nonlinear stiffness matrix. When convergence is achieved in the former cycle (i.e., convergence has been achieved in the computation of generalized displacements without updating the plastic strains. It is also always verified that the equation of motion (28) is satisfied), the stresses are computed at the Gauss points using Equations (10) and a yield criterion employed to verify if yielding occurred at any point. Here von Mises yield criterion with mixed hardening is used. It has been confirmed that von Mises criterion provides results close to experimental results in metals (Kojić and Bathe (2005)). The yield function f is thus defined as

$$f_y = \frac{1}{2}\left({}^{t+\Delta t}S_{ij}\,{}^{t+\Delta t}S_{ij} - {}^{t+\Delta t}\alpha_{ij}\,{}^{t+\Delta t}\alpha_{ij}\right) - \frac{1}{3}\,{}^{t+\Delta t}\sigma_y^2 \qquad i,\, j = 1,\, 2,\, 3 \tag{29}$$

where the summation convention applies. α_{ij} are the components of the back stress tensor $\boldsymbol{\alpha}$ which are zero until plasticity takes place and define the position of the yield surface; ${}^{t+\Delta t}\sigma_y$ is the yield stress, which depends upon the degree of plasticity of the particular point within the domain. S_{ij} are the deviatoric stress components which are given by

$$S_{ij} = \sigma_{ij} - \sigma_m \delta_{ij} \tag{30}$$

with σ_m the mean stress and δ_{ij} Kronecker delta. In the case of beams that obey the assumptions stated in the beginning of sub-section 2.2 we have: $S_{11} = 2\sigma_{11}/3$, $S_{22} = S_{33} = -\sigma_{11}/3$, $S_{13} = S_{31} = \sigma_{13}$, $S_{12} = S_{21} = S_{23} = S_{32} = 0$.

The nine deviatoric stress components - a second order tensor - are here written in vector form, i.e., $\mathbf{S} = \{S_{11},\, S_{12},\, S_{13},\, S_{21},\, \ldots S_{33}\}$. This vector form is also employed here for the other second order tensors related with stresses and strains.

When yielding occurs at any point, the plastic strains and the plastic dependent variables, including the yield stress, must be updated. It is as well necessary to re-calculate the generalized displacements. These are the tasks of a second cycle. The procedure here suggested to compute the plastic strains is the governing parameter method with the increment of effective plastic strain, Δe^P , as governing parameter (Kojić and Bathe (2005)). This procedure is described in the following paragraphs.

The von Mises yield function, our governing function, cannot be greater than zero. A bilinear stress-strain relation with mixed hardening is here assumed, Figure 2, and, therefore, the yield function can be written as

$$f_y\left(\Delta e^P\right) = \frac{{}^{t+\Delta t}\sigma^E}{{}^{t+\Delta t}\sigma_y + \left[3G + (1-M)\,E_P\right]\Delta e^P} - 1 \tag{31}$$

When yielding takes place, Δe^P is calculated such that the governing function $f_y\left(\Delta e^P\right)$ is zero.

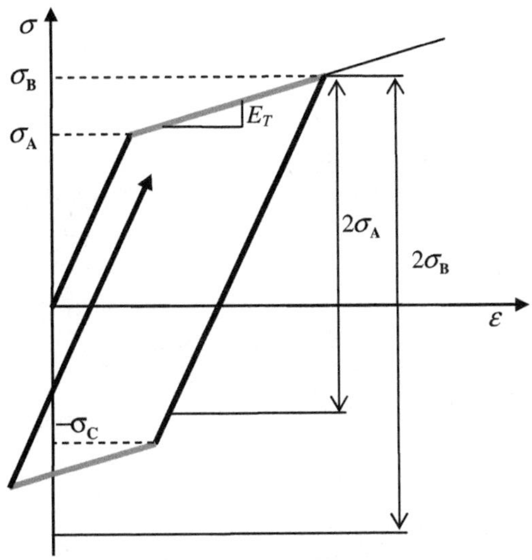

Figure 2. Bilinear uniaxial stress-strain curve. In mixed hardening the yield stress in compression, $-\sigma_C$, is smaller than $\sigma_B - 2\sigma_A$ and larger than $-\sigma_B$.

M in equation (31) is a mixed hardening parameter, which is a characteristic of the material and quantifies Bauschinger effect (Dieter (1986)). A mixed hardening material model ($0< M <1$) is placed between isotropic (M=0) and kinematic hardening material models (M=1). Only the isotropic part of the effective plastic stress affects the size of the yield surface (Kojić and Bathe (2005)). Another consequence of Bauschinger effect is the movement of the yield surface, represented by the back stress, $\boldsymbol{\alpha}$, that was introduced in equation (29).

The plastic modulus E_P can be written as

$$E_P = \frac{E E_T}{E - E_T} \tag{32}$$

where E_T is the tangent modulus and E Young modulus. Because we assume that the stress strain relation is bilinear, E_T and E_P are constants.

${}^{t+\Delta t}\sigma^E$, in equation (31), is the effective plastic stress when $\Delta e^P = 0$, defined as

$$ {}^{t+\Delta t}\sigma^E = \left(\frac{3}{2} {}^{t+\Delta t}\hat{\mathbf{S}}^E \cdot {}^{t+\Delta t}\hat{\mathbf{S}}^E \right)^{1/2} \tag{33} $$

with ${}^{t+\Delta t}\hat{\mathbf{S}}^E$ the radius of the elastic stress surface, given by

$$ {}^{t+\Delta t}\hat{\mathbf{S}}^E = {}^{t+\Delta t}\mathbf{S}^E - {}^{t}\boldsymbol{\alpha} \tag{34} $$

${}^{t+\Delta t}\mathbf{S}^E$ represents the deviatoric stress of the elastic solution, i.e., the deviatoric stress of a solution with no plastic deformation in the current step. To compute ${}^{t+\Delta t}\mathbf{S}^E$ one uses the following four expressions:

$$ {}^{t+\Delta t}e'_{ij} = {}^{t+\Delta t}\varepsilon_{ij} - {}^{t+\Delta t}\varepsilon_m \qquad i = j \tag{35} $$

$$ {}^{t+\Delta t}e'_{13} = \frac{1}{2} {}^{t+\Delta t}\gamma_{13} \tag{36} $$

$$ {}^{t+\Delta t}e''_{ij} = {}^{t+\Delta t}e'_{ij} - {}^{t}\varepsilon^P_{ij} \tag{37} $$

$$ {}^{t+\Delta t}\mathbf{S}^E = 2G\,{}^{t+\Delta t}\mathbf{e}'' \tag{38} $$

Terms ${}^{t+\Delta t}e'_{ij}$ and ${}^{t}\varepsilon^P_{ij}$ are, respectively, the components of the deviatoric strain tensor in the actual time step and of the plastic strain tensor in the previous step. ${}^{t+\Delta t}\varepsilon_m$ is the mean strain.

Since σ_{22} and σ_{33} are neglected, the following relations hold

$$ {}^{t+\Delta t}\varepsilon_{11} = {}^{t+\Delta t}\frac{\sigma_{11}}{E} + {}^{t+\Delta t}\varepsilon^P_{11} \tag{39} $$

$$ {}^{t+\Delta t}\varepsilon_{22} = -\frac{\nu}{E} {}^{t+\Delta t}\sigma_{11} + {}^{t+\Delta t}\varepsilon^P_{22} \tag{40} $$

$$ {}^{t+\Delta t}\varepsilon_{33} = -\frac{\nu}{E} {}^{t+\Delta t}\sigma_{11} + {}^{t+\Delta t}\varepsilon^P_{33} \tag{41} $$

Moreover, the volumetric plastic strain is equal to zero ($\varepsilon^P_V = {}^{t+\Delta t}\varepsilon^P_{11} + {}^{t+\Delta t}\varepsilon^P_{22} + {}^{t+\Delta t}\varepsilon^P_{33} = 0$) and, consequently, the mean strain is given by

$$ {}^{t+\Delta t}\varepsilon_m = \frac{1 - 2\nu}{3} \frac{\sigma_{11}}{E} \tag{42} $$

Next we define adimensional parameter $\Delta\lambda$ which relates the increments of plastic strain and the total deviatoric stresses as

$$\Delta\lambda = \frac{3}{2}\frac{\Delta e^P}{{}^{t+\Delta t}\sigma_y} \tag{43}$$

${}^{t+\Delta t}\sigma_y$ is the yield stress at time step $t + \Delta t$ and is given by

$${}^{t+\Delta t}\sigma_y = \sigma_{yv} + M\,E_P\left(e^P + \Delta e^P\right) \tag{44}$$

where σ_{yv} represents the initial yield stress, meaning the yield stress before any plastic hardening occurred.

The deviatoric stress ${}^{t+\Delta t}\mathbf{S}$ minus the back stress ${}^{t+\Delta t}\boldsymbol{\alpha}$ is designated as stress radius and represented by ${}^{t+\Delta t}\hat{\mathbf{S}}$. In a bilinear stress-strain relation, it can be obtained using the following expression

$${}^{t+\Delta t}\hat{\mathbf{S}} = \frac{{}^{t+\Delta t}\hat{\mathbf{S}}^E}{1 + 2\left[G + (1-M)\,E_P/3\right]\Delta\lambda} \tag{45}$$

It is easier to deduct equation (45) if the deviatoric stresses (30) are written in the following form:

$${}^{t+\Delta t}\mathbf{S} = {}^{t+\Delta t}\mathbf{S}^E - 2G\Delta\lambda\;{}^{t+\Delta t}\hat{\mathbf{S}} \tag{46}$$

The increment of plastic strains is given by

$$\Delta^{t+\Delta t}\boldsymbol{\varepsilon}^P = \Delta\lambda\;{}^{t+\Delta t}\hat{\mathbf{S}} \tag{47}$$

which are the Prandtl-Reuss equations (3.2.28) of reference (Kojić and Bathe (2005)), but where ${}^{t+\Delta t}\hat{\mathbf{S}}$ are the deviatoric stresses, defined in equation (30), minus the back stresses ${}^{t+\Delta t}\boldsymbol{\alpha}$ (in equations (3.2.28) of (Kojić and Bathe (2005)) there is no back stress). The total plastic strain at time $t + \Delta t$ is

$${}^{t+\Delta t}\boldsymbol{\varepsilon}^P = {}^{\Delta t}\boldsymbol{\varepsilon}^P + \Delta\boldsymbol{\varepsilon}^P \tag{48}$$

and the back stress is given by

$${}^{t+\Delta t}\boldsymbol{\alpha} = {}^{\Delta t}\boldsymbol{\alpha} + \Delta\boldsymbol{\alpha} \tag{49}$$

$$\Delta\boldsymbol{\alpha} = 2\,(1-M)\,E_P\Delta\boldsymbol{\varepsilon}^P/3 \tag{50}$$

Equation (50) is a form of Prager's hardening rule (Kojić and Bathe (2005)) where $2E_P/3$ is the kinematic hardening modulus.

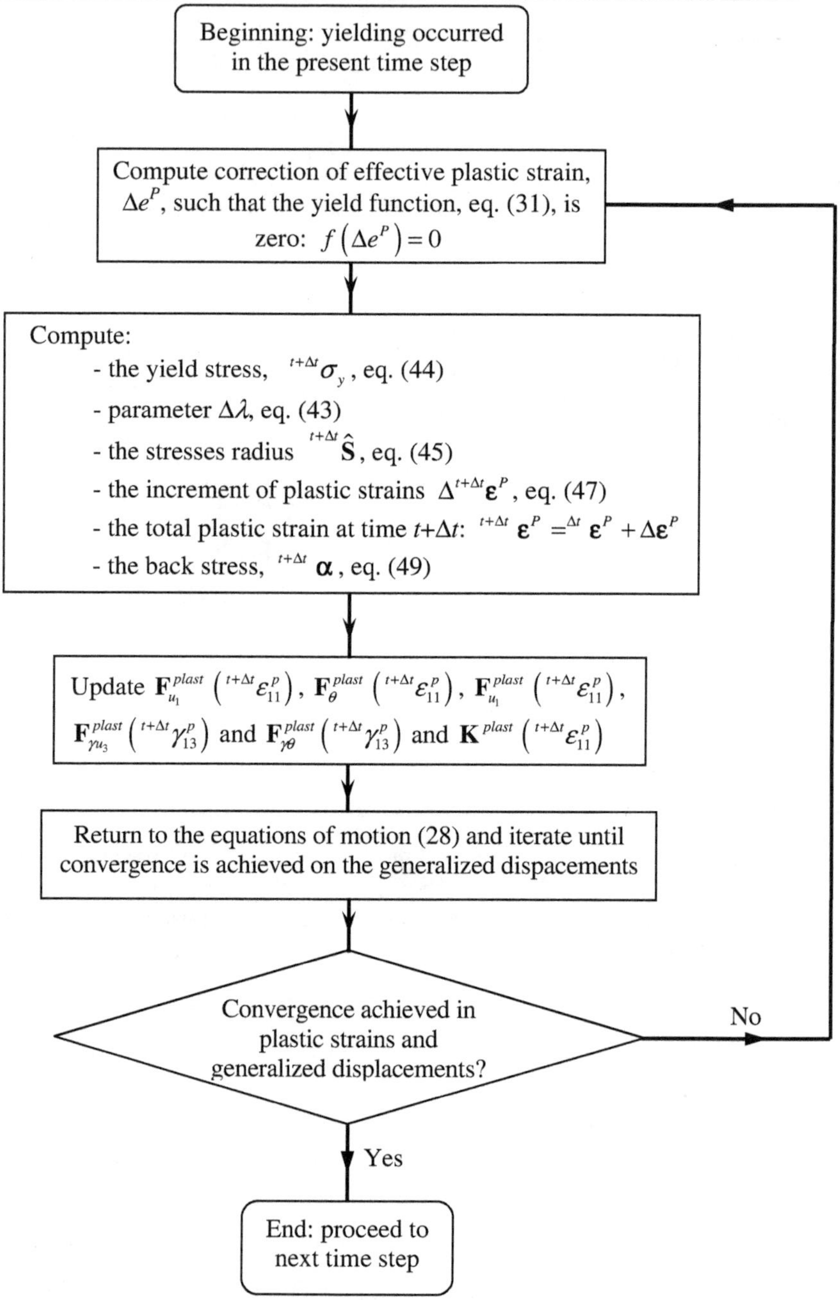

Figure 3. Flow chart showing procedure for computation of plastic strains with mixed hardening.

When the plastic strains are obtained, the forces and matrix given by Equations (18)-(22) can be updated. After, one returns to the equations of motion (28) and corrects the generalized displacements in an inner iterative procedure where the geometrically nonlinear terms are corrected. After convergence has been achieved in this inner iterative procedure, the new generalized displacements are used to compute the strains and stresses. Then it is verified if convergence has also been achieved in the computation of the plastic strains and back stresses. If it has, one proceeds to the next time step, otherwise one returns to the computation of the plastic strains using Equations (29) to (50) and the updated generalised displacements. Hence, the iterations within a particular time step only stop when neither the generalized displacements, nor the plastic strains vary more than predefined threshold values. It is always enforced that the solution is inside or on the yield surface.

The just presented algorithm for computation of plastic strains is further illustrated in the flow chart of Figure 3.

4 Examples of Forced Elasto-Plastic Vibrations

This section employs the time domain procedure described in Section 3 to investigate the oscillations of beams in the elasto-plastic regime and with geometrical nonlinearity. The section starts by addressing the validation of the formulation and computational codes employed, and proceeds to an illustration of the dynamic behaviour of elasto-plastic and geometrically nonlinear beams.

The approach described in the previous section has been put to test in different examples, in order to verify the consistency of its results. The geometrical nonlinear, linear elastic model has been often used with success and some examples of this are given in (Ribeiro (1998)) and (Ribeiro and Petyt (1999)). It was in addition tested by comparing displacements and velocities with the ones computed by McEwan et al. (2001).

The elasto-plastic and geometrically nonlinear code has been more recently developed and tested (Ribeiro and van der Heijden (2009)) and appears to be correct. In fact, it provided plastic strain distributions due to a distributed, transverse, step force that agree (visual comparison) with data published by Manoach and Karagiozova (1993). It also gave results that agree with the ones of finite element software ANSYS (ANSYS (2007)). Two ANSYS elements were employed in the comparisons, namely a beam element with shear deflection, BEAM23, and a shell element, SHELL43. Both ANSYS elements have plasticity and large deflection capabilities. The ANSYS beam element obviously follows a beam theory and therefore provides a fair term of comparison in what concerns the number of degrees of freedom necessary for convergence. The shell element was used to obtain reference results. It was verified that the p-version element provides linear natural frequencies, displacements in elastic and elasto-plastic regimes closer to the ones of the shell element and with less degrees of freedom than the beam h-version element (Ribeiro and van der Heijden (2009)). Moreover, the plastic strains are computed by the p-version beam finite element within the whole beam domain with a distribution quite similar to detailed shell models. Nevertheless, to obtain this detailed picture of plastic strains a few Gauss points are required (remember: it is just one element for the whole beam). Quite expectedly, more degrees of freedom are needed in the presence of plasticity than when this is absent.

In the tests shown in (Ribeiro and van der Heijden (2009)) using the p-version FEM, the same numbers of transverse, longitudinal and rotational shape functions were employed, a restriction implemented to limit the number of numerical tests. Naturally, the number of shape functions could have been chosen independently, and that would allow further reducing the number of degrees of freedom.

The following test example is taken from McEwan et al. (2001). The beam material properties are typical of steel; these are: $\nu = 0.3, E = 2.00 \times 10^{11}$ N m^{-2}, $\rho = 7.8 \times 10^3$ kg m^{-3}. The geometric properties are b=0.03 m, ℓ=1 m and h=0.01 ℓ. The yield stress, $\sigma_{yv} = 2.0 \times 10^8$ N m^{-2} is as well appropriate for some steels and taken from reference (Kojić and Bathe (2005)). The tangent modulus is either $E_T = 10^8$ N m^{-2}, which is somewhat small, approaching the popular elastic-perfectly plastic model (Kojić and Bathe (2005), Simo and Hughes (1998)), or $E_T = 10^9$ N m^{-2}, a value from reference (Kojić and Bathe (2005)). M is in these tests equal to 1, therefore Bauschinger effect is neglected.

Mass proportional viscous damping with a proportionality factor β equal to one is assumed, as in (McEwan et al. (2001)). This means that the equations of motion (28) take the form (function arguments are not represented)

$$\mathbf{M}\ddot{\mathbf{q}} + \beta\mathbf{M}\dot{\mathbf{q}} + \mathbf{K}_\ell\mathbf{q} + [\mathbf{K}_{n\ell} - \mathbf{K}_{plast}]\,\mathbf{q} = \mathbf{F} + \mathbf{F}^{plast} \tag{51}$$

Figure 4 [1] and Figure 5 show the steady state transverse displacements as a function of time and displacements versus velocities of two points (ξ=0 and ξ =0,875). Only data computed with tangent modulus $E_T = 10^8$ N

[1] The data shown in Figure 4 (a) was published in (Ribeiro and van der Heijden (2009)), Copyright (2009), and is reprinted with permission from Elsevier. Figures 9, 10 and 13 on the present text are also similar to figures presented in the same reference, although they use slightly different data here.

m^{-2} is shown; computations with $E_T = 10^9$ N m^{-2} were also carried out, but there is no visible difference and hence are not included in the figures. The oscillations are due to a transverse force, uniformly distributed in space with amplitude 1000 N m^{-1} and sinusoidal in time, with frequency 117.8097 rad/s (18.75 Hz). The oscillations are periodic with the same period as the excitation, but they are not harmonic. Data computed with different numbers of shape functions and Gauss points agrees. It is clear that a model with a small number of degrees of freedom (for example 15 DOF) is sufficient. Actually, an exhaustive convergence study was not carried out, so it is possible that fewer degrees of freedom would suffice in this example. As said, this example is taken from reference (McEwan et al. (2001)), where a linear elastic, geometrically nonlinear model is applied. The displacements and velocities in direction x_3 of the middle point (Figure 4) are similar to those portrayed in Figure 7 of that reference.

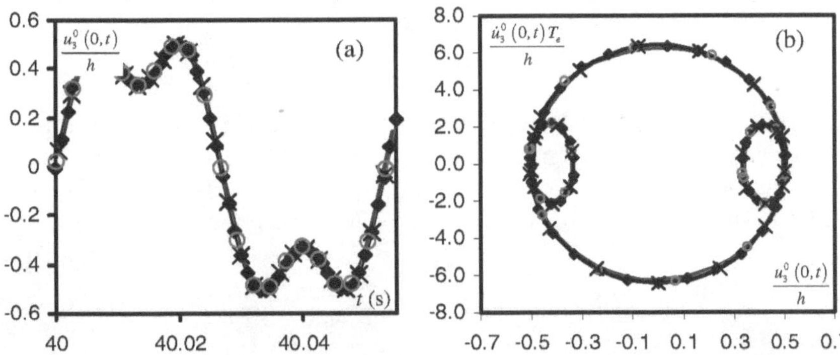

Figure 4. (a) Transverse displacement; (b) displacement versus velocity. Data computed at middle point of the beam with: — 15 DOF, 20 × 10 Gauss points; –o– 27 DOF, 40 × 20 Gauss points; –×– 75 DOF, 40 × 20 Gauss points; ♦ 75 DOF 64 × 64 Gauss points.

The direct, $\sigma_{11}(x_1, x_3, t)$, and transverse shear, $\sigma_{13}(x_1, x_3, t)$, stresses at t=0.1786 s computed with the geometrically nonlinear only model and with the model that considers both geometrical nonlinearity and plasticity are shown in Figure 6 and Figure 7. 75 degrees of freedom and a grid of 64 × 64 Gauss points were employed to compute the results shown in these figures and in the following paragraphs. Because, in this example, plastic strains are really only of importance closer to the boundaries, they are of small consequence in displacements and velocities, and that is the reason why displacements and velocities are similar to the ones of reference (McEwan

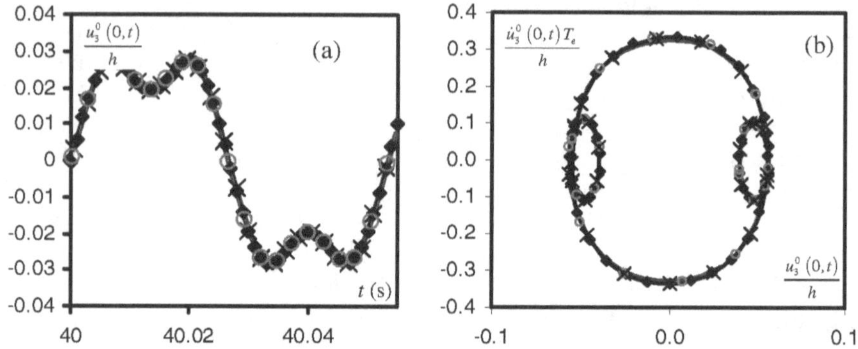

Figure 5. (a) Transverse displacement; (b) displacement versus velocity. Data computed at point ξ =0,875 of the beam with: — 15 DOF, 20 × 10 Gauss points; - ○ - 27 DOF, 40 × 20 Gauss points; -×- 75 DOF, 40 × 20 Gauss points; ♦ 75 DOF 64 × 64 Gauss points.

et al. (2001)). On the other hand, it is apparent that it is needed to consider them for accurate stress computation, since near the beam limits, which are the critical areas in terms of stresses, the plastic strains experience steep variations that lead to substantial changes in the stresses.

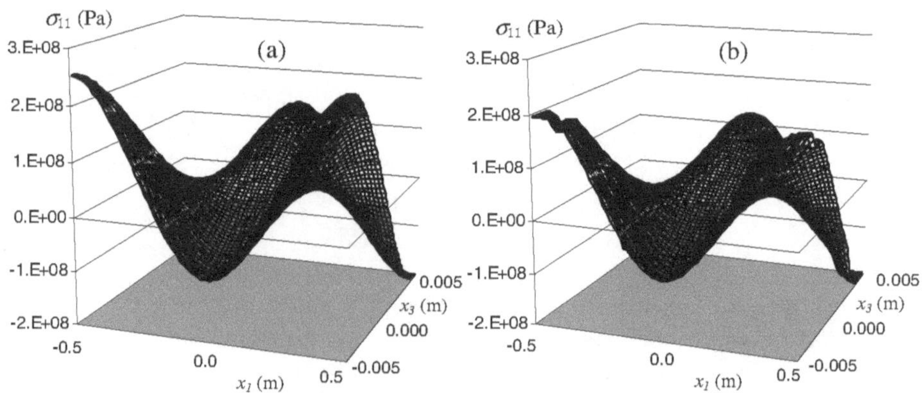

Figure 6. Direct stresses $\sigma_{11}(x_1, x_3, t)$ at t=0.1786 s: (a) geometrically nonlinear only (b) elasto-plastic and geometrically nonlinear model.

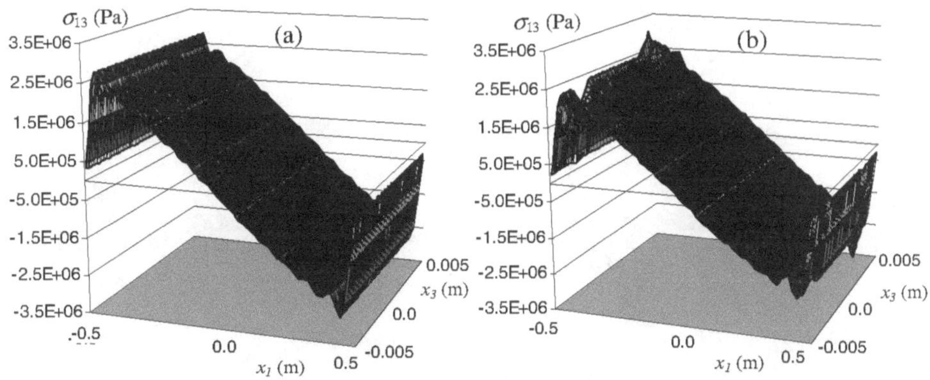

Figure 7. Direct stresses $\sigma_{13}(x_1, x_3, t)$ at t=0.1786 s: (a) geometrically nonlinear only (b) elasto-plastic and geometrically nonlinear model.

In this example, transverse shear stresses are about two orders of magnitude lower than direct stresses.

Figure 8 shows plastic strains computed with $E_T = 10^8$ N m^{-2} (close to perfect plasticity) and with $E_T = 10^9$ N m^{-2}. The plastic strains computed with $E_T = 10^9$ N m^{-2} differ only slightly from the ones computed with $E_T = 10^8$ N m^{-2} in this case. It is curious to notice the irregular pattern that appears with sinusoidal excitations.

By increasing the excitation amplitude and/or changing the excitation frequency, we can lead the same beam to oscillations where not only strains and stresses, but also displacements and velocities that result from elastic only and from elasto-plastic models differ visibly. An example where the later occurs is shown in Figure 9. The excitation frequency is equal to the first linear natural frequency, 326.827 rad/s, and the amplitude of the distributed transverse force is 500 Nm $^{-1}$. The figure shows the transverse displacement of the middle beam point along two excitation cycles and the projection of the trajectory on the phase plane defined by the transverse displacement and velocity. T_e represents the excitation period. In this case, both models predict periodic oscillations dominated by the excitation frequency but the elasto-plastic model predicts smaller displacement and velocity amplitudes than the elastic model.

Figure 10 and Figure 11 show the longitudinal and shear stresses, σ_{11} and σ_{13}, computed with the linear elastic model and with the elasto-plastic

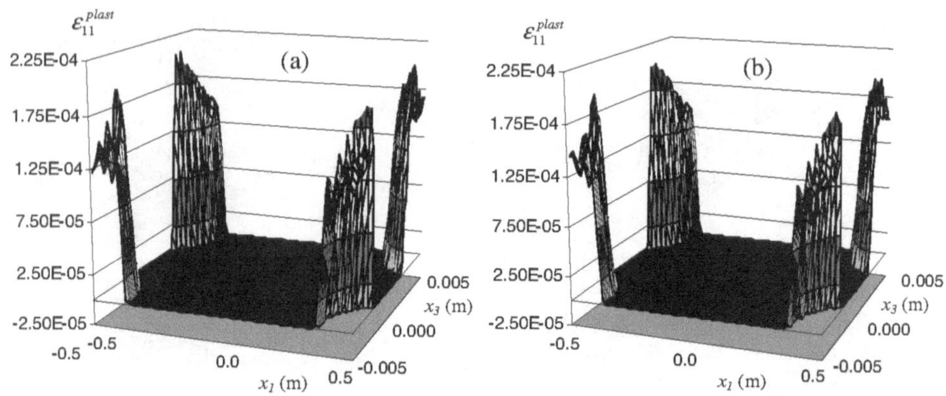

Figure 8. Plastic strains $\varepsilon_{11}^{plast}(x_1, x_3, t)$ at t=30.0 s: (a) $E_T = 10^8$ N m^{-2} (b) $E_T = 10^9$ N m^{-2}.

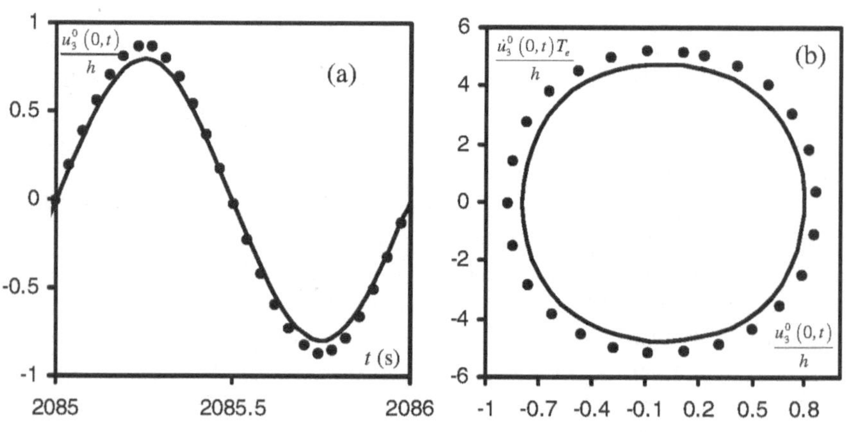

Figure 9. (a) Transverse displacement of middle point and (b) projection on a phase plane, excitation at first linear natural frequency with 500 N m^{-1} amplitude. — elasto-plastic and geometrically nonlinear; • geometrically nonlinear.

model, both models geometrically nonlinear, at $t = 0.7547$ s. The difference between values computed with the different models is higher near the clamps and in the middle of the beam. Due to plasticity, the surfaces defined by σ_{11} and σ_{13} become uneven near the clamps (as occurred in the former example, see Figure 8). Although plastic strains also alter the stresses in statics, this unevenness effect was only found in dynamics and is due to the oscillations. The initial yield stress of this particular steel is not high, but it is noted that the stresses computed without considering plastic strains attain values over 400 MPa and that, therefore, plasticity would as well take place in a higher strength steel.

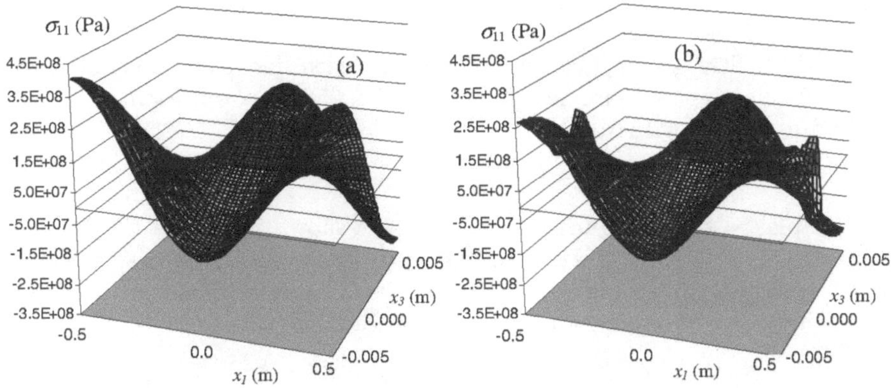

Figure 10. Stresses $\sigma_{11}(x_1, x_3, t)$ computed with linear elastic model (a) and with elasto-plastic model (b), when t=0.7547 s, excitation at linear natural frequency with 500 N m^{-1} amplitude.

Transverse shear stresses are still about two orders of magnitude lower than direct stresses, as occurred in the former example. Also as before, but now in a rather more pronounced way, localised areas exist where transverse shear stresses increase or decrease very sharply.

In the following lines, results on a thicker beam, h/ℓ=0.05, are shown. The force amplitude was increased to 10 000 N m^{-1} in order to achieve large displacements in this thicker beam. The damping factor β was chosen so that the nondimensional damping ratio, which is equal to the viscous damping coefficient divided by the critical damping coefficient (Kelly (1993)), of a one degree of freedom system representing the beam does not change with respect to the previous thin beam. The value hence obtained for β is 4.92206.

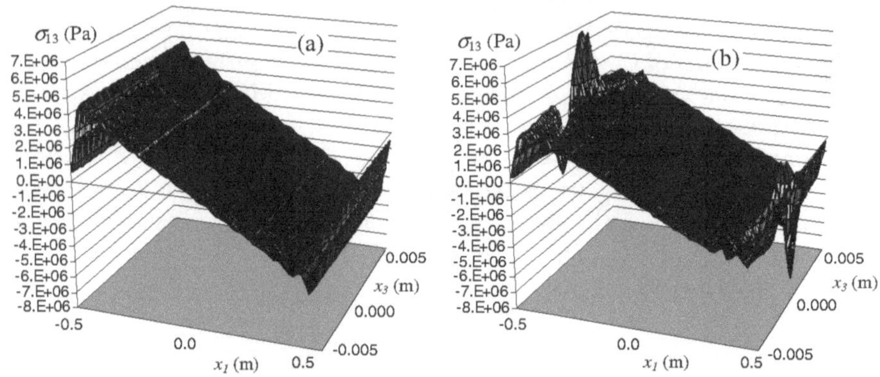

Figure 11. Stresses $\sigma_{13}(x_1, x_3, t)$ computed with linear elastic model (a) and with elasto-plastic model (b), when t=0.7547 s, excitation at linear natural frequency with 500 N m^{-1} amplitude.

The plastic strains at instant 0.1518 s are shown in Figure 12. In this case, although those strains are large close to the boundaries, they are also quite large in areas around the beam centre. The shear plastic strains are about one order of magnitude smaller than the direct strains.

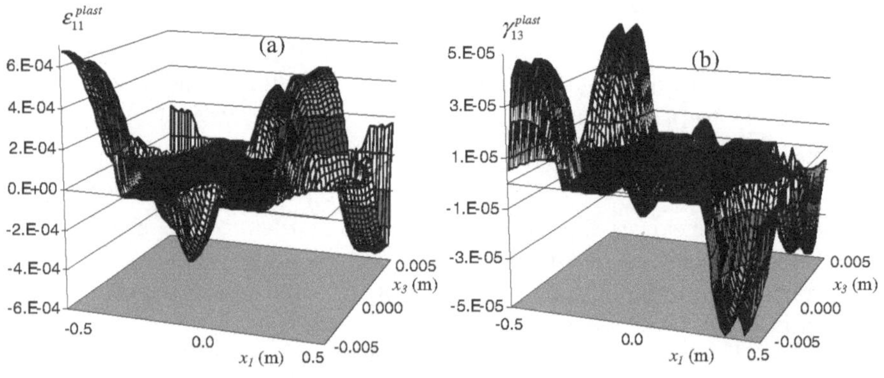

Figure 12. (a) Direct, $\varepsilon_{11}^{plast}(x_1, x_3, t)$, and shear plastic strains (b)$\gamma_{13}^{plast}(x_1, x_3, t)$, at t=0.7547 s.

Figure 13 shows the transverse displacement and the projection of the trajectory on a phase plane at later cycles. The geometrically nonlinear,

linear elastic model over predicts the displacement and velocity amplitudes. An explanation for this is that the plastic strains contribute to the variation of the natural frequencies of the beam (illustrated in Section 6), therefore the fundamental frequency becomes more distant from the excitation frequency.

In the cases here shown the displacements computed by the model that included plasticity were almost equal or larger than the ones computed with the geometrically nonlinear only model. It can also happen that the displacements obtained by the elastic-plastic geometrical nonlinear model are larger than the ones obtained by using only the elastic model, due to softening induced by plasticity. Reference (Ribeiro and van der Heijden (2009)) shows an example of this occurrence. We again recall that in a dynamic problem the frequency content of the external excitation and its relation with the resonance frequencies (generally close to the natural frequencies) of the excited system are of paramount importance and, as we shall see in an ensuing section, plastic strains affect the natural frequencies.

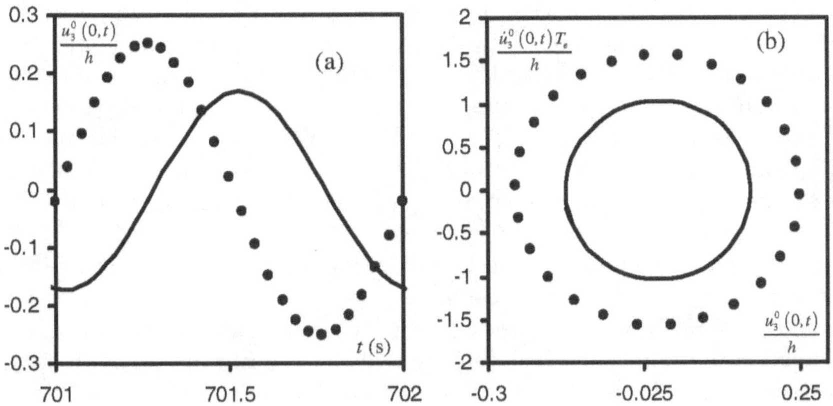

Figure 13. (a) Transverse displacement of middle point and (b) projection on a phase plane, excitation at first linear natural frequency of beam h/ℓ=0.05, with 10 000 N m^{-1} amplitude. — elasto-plastic and geometrically nonlinear; • geometrically nonlinear.

5 Vibration Modes in Large Amplitude Vibrations and With Plastic Strains - Equations of Motion

5.1 Mode of Vibration

The importance of natural modes of vibration in linear systems is widely known (e.g. Ewins (2000), Kelly (1993), Thomsen (2003)), but, as an introduction to this section, we recall the essential features of these modes. Linear conservative systems have natural frequencies of vibration, where they vibrate in such a way that the ratio between the vibration amplitude of any two points is constant, thus defining a form of vibration of the system, the natural mode shape of vibration. Hence, a natural mode of vibration of a linear conservative system is as an oscillation at a natural frequency (i.e., a harmonic oscillation) and with a natural mode shape.

Three reasons make the linear modes very important: (1) the equations of motion can be uncoupled using modal coordinates and the orthogonality of the natural mode shapes with respect to the mass and to the stiffness matrices; (2) the vibrations of a system often are defined by a reduced number of modes, therefore, modal reduction, which consists in expressing the motion as a superposition of a selected number of modes, can be employed to analyse vibrating systems with a reduced cost, good accuracy and allowing for an easier interpretation of the system dynamic behaviour; (3) when damping is small, resonance occurs close to the natural frequencies.

Whilst in a linear conservative system the natural frequencies and mode shapes are constants that do not change with the vibration amplitude, in a system experiencing geometrically nonlinear vibrations this is not true. Periodic - therefore steady-state - oscillations in free regime can be obtained that tend to the linear modes as the vibration amplitude decreases, but the shape assumed by the system along a vibration period is generally not constant and the oscillations are not exactly harmonic. A few studies have been carried out on the variation of mode shape and vibration period and on definitions of mode shapes for nonlinear systems. Rosenberg (1966) suggested that a nonlinear natural mode of a discrete system could be interpreted as a motion where the masses execute periodic, but not necessarily harmonic, vibrations, all masses vibrating with the same period, achieving the maximum amplitude displacement and the equilibrium position simultaneously. The co-ordinates of the system are determined by the position of any of the masses, as in linear modes, but the ratio between the vibration amplitudes of any two points is not constant. This concept can be extended to continuous systems, by considering and infinite number of infinitesimal masses.

The natural modes of discrete (or discretized) linear systems are computed by solving eigenvalue problems. In discrete models of nonlinear systems an eigenvalue problem related with steady-state free vibration can - as shown in the next section - also be defined, but the stiffness matrix depends upon the unknown eigenvector. Hence, the eigenvalues and the eigenvectors are, generally, amplitude dependent. We defend that these define frequencies and shapes of vibration, which depend on the amplitude of vibration displacement and that approximately represent nonlinear modes of vibration. Accordingly, the designation "natural frequency" will be attributed to the square root of the eigenvalue that solves the nonlinear eigenvalue problem and the "mode shape" of linear systems will be replaced by the shape defined by the eigenvectors. In the presence of internal resonance, at least two modes are involved.

5.2 Reduced Model Equations for Free Vibrations

In this text we will show the effect of plastic strains on the free vibration modes and proceed now to the presentation of the model employed for this purpose. In the case here considered the beam has been deformed quasi-statically by an external force, in such a way that the yield stress is passed at least in some points of the domain. After a desired deflection has been achieved, the force is removed and free vibrations with displacement amplitudes of the order of the static displacement that originated the plastic strain, or lower, are analysed. This restriction on the vibration amplitude is used to justify the assumption, here taken, that the stresses achieved during oscillations are small enough for the plastic strains to remain constant. Accordingly, in the following equations, the plastic strains $\varepsilon_{11}^{plast}(\xi,\eta,t)$ and $\gamma_{13}^{plast}(\xi,\eta,t)$ become $\varepsilon_{11}^{plast}(\xi,\eta)$ and $\gamma_{13}^{plast}(\xi,\eta)$. If the former assumption is not true, then the plastic strain field should be re-calculated.

We will only be concerned with beams where both ends are fixed and analyse vibration modes that in the linear regime are bending modes. In this case, the longitudinal inertia is generally of small significance (Ribeiro (2001)) and can be neglected. The following two equations (Equations (52) and (53)) are hence derived from Equation (28). Notice that the longitudinal displacements were not neglected. They are important if the vibration amplitude is not small (Ribeiro (1998)) and neglecting them leads to a too stiff model.

$$\mathbf{q}_{u_1}(t) = \mathbf{K}_{\ell_{11}}^{p^{-1}} \mathbf{F}_{u_1}^{plast}\left(\varepsilon_{11}^{p}(\xi,\eta)\right) - \mathbf{K}_{\ell_{11}}^{p^{-1}} \mathbf{K}_{n\ell_{12}}\left(\mathbf{q}_{u_3}(t)\right) \mathbf{q}_{u_3}(t) \tag{52}$$

$$\begin{bmatrix} \mathbf{M}_{u_3} & \mathbf{0} \\ \mathbf{0} & \mathbf{M}_\theta \end{bmatrix} \left\{ \begin{array}{c} \ddot{\mathbf{q}}_{u_3}(t) \\ \ddot{\mathbf{q}}_\theta(t) \end{array} \right\} + \begin{bmatrix} \mathbf{K}^\gamma_{\ell_{22}} & \mathbf{K}^\gamma_{\ell_{23}} \\ \mathbf{K}^\gamma_{\ell_{32}} & \mathbf{K}^\gamma_{\ell_{33}} + \mathbf{K}^b_{\ell_{33}} \end{bmatrix} \left\{ \begin{array}{c} \mathbf{q}_{u_3}(t) \\ \mathbf{q}_\theta(t) \end{array} \right\} -$$
$$\begin{bmatrix} \mathbf{K}_{plast}\left(\varepsilon^{plast}_{11}(\xi,\eta)\right) & \mathbf{0} \\ \mathbf{0} & \mathbf{0} \end{bmatrix} \left\{ \begin{array}{c} \mathbf{q}_{u_3}(t) \\ \mathbf{q}_\theta(t) \end{array} \right\} +$$
$$\begin{bmatrix} \mathbf{K}_{n\ell_{22}}(\mathbf{q}_{u_3}(t)) - \mathbf{K}_{n\ell_{21}}(\mathbf{q}_{u_3}(t))\mathbf{K}^{p^{-1}}_{\ell_{11}}\mathbf{K}_{n\ell_{12}}(\mathbf{q}_{u_3}(t)) & \mathbf{0} \\ \mathbf{0} & \mathbf{0} \end{bmatrix} \left\{ \begin{array}{c} \mathbf{q}_{u_3}(t) \\ \mathbf{q}_\theta(t) \end{array} \right\} +$$
$$\left\{ \begin{array}{c} \mathbf{K}_{n\ell_{21}}(\mathbf{q}_{u_3}(t))\mathbf{K}^{p^{-1}}_{\ell_{11}}\mathbf{F}^{plast}_{u_1}\left(\varepsilon^{plast}_{11}(\xi,\eta)\right) \\ \mathbf{0} \end{array} \right\} =$$
$$= \left\{ \begin{array}{c} \mathbf{F}^{plast}_{\gamma u_3}\left(\gamma^{plast}_{13}(\xi,\eta)\right) \\ \mathbf{F}^{plast}_{\theta}\left(\varepsilon^{plast}_{11}(\xi,\eta)\right) + \mathbf{F}^{plast}_{\gamma\theta}\left(\gamma^{plast}_{13}(\xi,\eta)\right) \end{array} \right\} \tag{53}$$

Equations (53) can be otherwise written as follows

$$\begin{bmatrix} \mathbf{M}_{u_3} & \mathbf{0} \\ \mathbf{0} & \mathbf{M}_\theta \end{bmatrix} \left\{ \begin{array}{c} \ddot{\mathbf{q}}_{u_3}(t) \\ \ddot{\mathbf{q}}_\theta(t) \end{array} \right\} + \begin{bmatrix} \mathbf{K}^\gamma_{\ell_{22}} & \mathbf{K}^\gamma_{\ell_{23}} \\ \mathbf{K}^\gamma_{\ell_{32}} & \mathbf{K}^\gamma_{\ell_{33}} + \mathbf{K}^b_{\ell_{33}} \end{bmatrix} \left\{ \begin{array}{c} \mathbf{q}_{u_3}(t) \\ \mathbf{q}_\theta(t) \end{array} \right\}$$
$$+ \begin{bmatrix} \mathbf{K}^{plast}_{Gn\ell}\left(\varepsilon^{plast}_{11}(\xi,\eta)\right) - \mathbf{K}^{plast}\left(\varepsilon^{plast}_{11}(\xi,\eta)\right) & \mathbf{0} \\ \mathbf{0} & \mathbf{0} \end{bmatrix} \left\{ \begin{array}{c} \mathbf{q}_{u_3}(t) \\ \mathbf{q}_\theta(t) \end{array} \right\}$$
$$+ \begin{bmatrix} \mathbf{K}_{n\ell_{22}}(\mathbf{q}_{u_3}(t)) - \mathbf{K}_{n\ell_{21}}(\mathbf{q}_{u_3}(t))\mathbf{K}^{p^{-1}}_{\ell_{11}}\mathbf{K}_{n\ell_{12}}(\mathbf{q}_{u_3}(t)) & \mathbf{0} \\ \mathbf{0} & \mathbf{0} \end{bmatrix} \left\{ \begin{array}{c} \mathbf{q}_{u_3}(t) \\ \mathbf{q}_\theta(t) \end{array} \right\}$$
$$= \left\{ \begin{array}{c} \mathbf{F}^{plast}_{\gamma u_3}\left(\gamma^{plast}_{13}(\xi,\eta)\right) \\ \mathbf{F}^{plast}_{\theta}\left(\varepsilon^{p}_{11}(\xi,\eta)\right) + \mathbf{F}^{plast}_{\gamma\theta}\left(\gamma^{plast}_{13}(\xi,\eta)\right) \end{array} \right\} \tag{54}$$

where matrix $\mathbf{K}^{plast}_{Gn\ell}\left(\varepsilon^{plast}_{11}(\xi,\eta)\right)$ obeys the following relation

$$\mathbf{K}^{plast}_{Gn\ell}\left(\varepsilon^{plast}_{11}(\xi,\eta)\right)\mathbf{q}_{u_3}(t) = \mathbf{K}_{n\ell_{21}}(\mathbf{q}_{u_3}(t))\mathbf{K}^{p^{-1}}_{\ell_{11}}\mathbf{F}^{plast}_{u_1}\left(\varepsilon^{plast}_{11}(\xi,\eta)\right) \tag{55}$$

It is noted that although $\mathbf{K}^{plast}_{Gn\ell}\left(\varepsilon^{plast}_{11}(\xi,\eta)\right)$ is a constant matrix that remains in a linearized version of Equations (54) (Ribeiro (2010)), it only arises due to the nonlinearity of the system, since it is a consequence of plastic strains and geometrical nonlinearity.

5.3 Harmonic Balance Method

The ordinary differential equations of motion (54), or (53), will be transformed into algebraic equations by the harmonic balance method (HBM). In the HBM the solution is expanded in a truncated Fourier series. Since only undamped periodic motions are here analysed, a series only with cosine (sine) terms is used. Equations (54) contain linear and cubic terms in the generalised displacements, and the beam is plastically deformed from its straight configuration. Consequently, the Fourier series should contain the constant term and, at least, the first two harmonics. In addition, previous experiences indicate that the third harmonic can be very important in these oscillations (Ribeiro and Petyt (1999), Ribeiro (2004b), Lewandowski (1994a), Lewandowski (1997b)). Therefore, the third harmonic will be also included in the Fourier series and it is assumed that the solution of Equations (54) has the form:

$$\begin{aligned} &\left\{ \begin{array}{c} \mathbf{q}_{u_3}(t) \\ \mathbf{q}_{\theta}(t) \end{array} \right\} = \frac{1}{2}\left\{ \begin{array}{c} \mathbf{w}_0 \\ \theta_0 \end{array} \right\} + \sum_{i=1}^{3}\left\{ \begin{array}{c} \mathbf{w}_i \\ \theta_i \end{array} \right\} \cos(i\omega t) \\ &\Leftrightarrow \mathbf{q}(t) = \frac{1}{2}\mathbf{Q}_0 + \sum_{i=1}^{3} \mathbf{Q}_i \cos(i\omega t) \end{aligned} \tag{56}$$

If the consideration of more harmonics may eventually lead to different results, series (56) contains terms of all types (constant, odd and even) and allows us to analyse the essential features of the beam vibrations. The fact that the longitudinal inertia was neglected brings the benefit that the longitudinal displacements are automatically expressed by a truncated Fourier series with the constant term and six harmonics. To verify that one just needs to introduce series (56) in equation (52).

Inserting the truncated series (56) into Eqs. (54) and applying the harmonic balance method (Lewandowski (1997a), Nayfeh and Mook (1995), Ribeiro and Petyt (1999)), a set of algebraic equations which depend on the fundamental frequency of vibration and on the coefficients of each harmonic is obtained. The procedure to solve these algebraic equations is described - in brief, but giving references where the missing details can be found - in the next section. To facilitate the description that follows, Equations (54) are written in a simplified manner as follows:

$$
\begin{aligned}
&\bar{\mathbf{M}}\ddot{\mathbf{q}}(t)+\bar{\mathbf{K}}_{\ell}\mathbf{q}(t)+\bar{\mathbf{K}}^{plast}\left(\varepsilon_{11}^{plast}(x,y)\right)\mathbf{q}(t)+\bar{\mathbf{K}}_{Gn\ell}\left(\mathbf{q}_{u_3}(t)\right)\mathbf{q}(t)\\
&=\bar{\mathbf{F}}^{plast}\left(\varepsilon_{11}^{plast}(x,y),\gamma_{13}^{plast}(x,y)\right)
\end{aligned}
\tag{57}
$$

Most matrices and the force vector in equation (57) are constant. Only the nonlinear matrix $\bar{\mathbf{K}}_{Gn\ell}\left(\mathbf{q}_{u_3}(t)\right)$ is a quadratic function of $\mathbf{q}_{u_3}(t)$.

5.4 Eigenvalue Problem

When the solution is given by equation (56), one can transform the equations of motion into the following:

$$
\begin{aligned}
&\left(-\omega^2\overline{\mathbf{M}}\begin{bmatrix}\mathbf{0}&\mathbf{0}&\mathbf{0}&\mathbf{0}\\ \mathbf{0}&\mathbf{I}&\mathbf{0}&\mathbf{0}\\ \mathbf{0}&\mathbf{0}&4\mathbf{I}&\mathbf{0}\\ \mathbf{0}&\mathbf{0}&\mathbf{0}&9\mathbf{I}\end{bmatrix}+\left(\overline{\mathbf{K}}_{\ell}+\overline{\mathbf{K}}^{plast}\right)\begin{bmatrix}\frac{1}{2}\mathbf{I}&\mathbf{0}&\mathbf{0}&\mathbf{0}\\ \mathbf{0}&\mathbf{I}&\mathbf{0}&\mathbf{0}\\ \mathbf{0}&\mathbf{0}&\mathbf{I}&\mathbf{0}\\ \mathbf{0}&\mathbf{0}&\mathbf{0}&\mathbf{I}\end{bmatrix}\right)\\
&\qquad\times\begin{Bmatrix}\mathbf{Q}_0\\ \mathbf{Q}_1\\ \mathbf{Q}_2\\ \mathbf{Q}_3\end{Bmatrix}+\begin{Bmatrix}\mathbf{F}_{c0}\\ \mathbf{F}_{c1}\\ \mathbf{F}_{c2}\\ \mathbf{F}_{c3}\end{Bmatrix}=\begin{Bmatrix}\bar{\mathbf{F}}^{plast}\\ \mathbf{0}\\ \mathbf{0}\\ \mathbf{0}\end{Bmatrix}
\end{aligned}
\tag{58}
$$

Time is not present in Equations (58), that are algebraic and have as unknowns the fundamental frequency, ω, of Fourier series (56), and the coefficients of the harmonics $\mathbf{Q}_0$, $\mathbf{Q}_1$, $\mathbf{Q}_2$, $\mathbf{Q}_3$. Although, for the sake of simplicity, this is not explicitly written in Equations (58), $\bar{\mathbf{K}}^{plast}$ and $\bar{\mathbf{F}}^{plast}$ are functions of the plastic strains and $\mathbf{F}_{c0}$, $\mathbf{F}_{c1}$, $\mathbf{F}_{c2}$ and $\mathbf{F}_{c3}$ are functions of the coefficients of the harmonics. Vectors $\mathbf{F}_{ci}$, i=0-3, are given by

$$
\mathbf{F}_{ci}=\frac{2}{T}\int_0^T\overline{\mathbf{K}}_{Gn\ell}\left(\mathbf{q}_{u_3}(t)\right)\,\mathbf{q}_{u_3}(t)\cos(i\omega t)\,\mathrm{dt}\quad i=0,1,2,3 \tag{59}
$$

where sub-vectors $\mathbf{F}_{ci}$ involve all harmonic coefficients; for instance $\mathbf{F}_{c0}$ is a function of $\mathbf{Q}_0$, $\mathbf{Q}_1$, $\mathbf{Q}_2$ and$\mathbf{Q}_3$.

Equation (58) is equivalent to

$$
\mathbf{R}=\mathbf{0} \tag{60}
$$

where vector $\mathbf{R}$ is defined as follows

$$\mathbf{R} = \left(-\omega^2\overline{\mathbf{M}} \begin{bmatrix} \mathbf{0} & \mathbf{0} & \mathbf{0} & \mathbf{0} \\ \mathbf{0} & \mathbf{I} & \mathbf{0} & \mathbf{0} \\ \mathbf{0} & \mathbf{0} & 4\mathbf{I} & \mathbf{0} \\ \mathbf{0} & \mathbf{0} & \mathbf{0} & 9\mathbf{I} \end{bmatrix} + \left(\overline{\mathbf{K}}_\ell + \overline{\mathbf{K}}^{plast}\right) \begin{bmatrix} \frac{1}{2}\mathbf{I} & \mathbf{0} & \mathbf{0} & \mathbf{0} \\ \mathbf{0} & \mathbf{I} & \mathbf{0} & \mathbf{0} \\ \mathbf{0} & \mathbf{0} & \mathbf{I} & \mathbf{0} \\ \mathbf{0} & \mathbf{0} & \mathbf{0} & \mathbf{I} \end{bmatrix} \right) \times \begin{Bmatrix} \mathbf{Q}_0 \\ \mathbf{Q}_1 \\ \mathbf{Q}_2 \\ \mathbf{Q}_3 \end{Bmatrix} + \begin{Bmatrix} \mathbf{F}_{c0} \\ \mathbf{F}_{c1} \\ \mathbf{F}_{c2} \\ \mathbf{F}_{c3} \end{Bmatrix} - \begin{Bmatrix} \bar{\mathbf{F}}^{plast} \\ \mathbf{0} \\ \mathbf{0} \\ \mathbf{0} \end{Bmatrix} \tag{61}$$

Equations (60) and (61) define an eigenvalue problem where one of the matrices depends on the eigenvector $\mathbf{Q}_0$, $\mathbf{Q}_1$, $\mathbf{Q}_2$, $\mathbf{Q}_3$ and the eigenvalue is the fundamental frequency of the Fourier series squared ω^2. This problem may be solved by an arc-length continuation method (Lewandowski (1994a, 1997a,b), Ribeiro (1998, 2010), Ribeiro and Petyt (1999)), which is basically a Newton (also known as Newton-Raphson) procedure complemented by a constraint equation. In a direct application of Newton method to this problem the frequency or the amplitude of vibration would be used as a parameter and turning points not passed. We employ instead the distance between two points of the backbone curve as a parameter. The continuation procedure converges in cases where a simpler Newton algorithm experiences difficulties, but, with the aforementioned exception of turning points, if Newton is coupled with a secant predictor it is also rather effective.

Other more or less efficient methods have been implemented to solve the present problem. AUTO (Doedel (1981, 2006)) is a popular continuation software. B. Cochelin and co-workers made available a procedure based on the so-called asymptotic numerical method (ANM) continuation technique (Arquier et al. (2006), Cochelin and Vergez (2009)).

6 Vibration Modes in Large Amplitude Vibrations and With Plastic Strains - Illustrative Test Cases

In this section, numerical tests are presented in order to illustrate the effects of large displacement amplitudes and of plastic strains on the shapes and natural frequencies of vibration.

To achieve the goal of illustrating the effect of plasticity, beams with different plastic strain fields are required. These will be obtained by applying the same force, quasi-statically, to beams similar in everything except in the yield stress. Appendix E of reference (Thomsen (2003)) states that the yield stress, σ_{yv}, of steel ranges from 2×10^8 N m^{-2} to 16 $\times10^8$ Nm^{-2}. Values within these limits can be found elsewhere in the literature (Kojić and Bathe (2005), Manoach and Karagiozova (1993)) and in material data sheets. With this in mind, the four (academic but possible) steels with yield stresses as written in Table 1 were considered.

Table 1. Initial yield stresses of steels 1-4 (in Nm^{-2}).

Steel	1	2	3	4
Initial yield stress	2×10^8	3.5×10^8	5×10^8	6.5×10^8

The remaining material properties are the ones of Section 3: ν =0.3, E=2.00 $\times10^{11}$Nm^{-2}, ρ = 7.8 $\times10^3$kg m^{-3}. The geometric properties are the ones of the thinner beam of that section: b=0.03 m, ℓ =1 m and h=0.01 ℓ. The shear correction factor is still κ =(5+5ν)/(6+5ν) and the tangent modulus E_T = 10^9 Nm^{-2}, taken from an example of (Kojić and Bathe (2005)) is once again employed.

The first six linear natural frequencies of vibration of the beam in the plane x_1x_3 and without plastic strains, computed with the full model (i.e., not neglecting the longitudinal inertia, are shown in Table 2. The model employed uses eleven symmetric bending shape functions and twenty one longitudinal and rotational shape functions (p_o=11, p_i=p_θ=21).

Table 2. Linear natural frequencies of vibration, $\omega_\ell(rad/s)$, of beam without plastic deformation. SB - symmetric bending mode; L - longitudinal mode.

Mode number	1 (SB)	2 (SB)	3 (SB)	4 (SB)	5 (SB)	6 (L)
ω_ℓ	326.83	1762.5	4337.9	8029.7	12809	15908

As written in a previous section, the convergence and accuracy of the p-version model in elasto-plastic, geometrically nonlinear vibrations, were investigated in (Ribeiro and van der Heijden (2009)). Further validating analysis were developed in the preparation of (Ribeiro (2010)). The convergence analysis demonstrates that 12 symmetric transverse, 23 longitudinal and 23 rotational shape functions provide very accurate displacement and sufficiently accurate plastic strains in the cases considered here where plastic strains are larger. This set of shape functions will hence be used when the plastic strains are not small. For the cases with smaller plastic strains, a model with p_o=11, p_i=p_θ=21 is enough and will be employed. In the problems at hand we can, and did, use only symmetric out-of-plane functions. A reduction using the symmetry properties was some times also implemented in the longitudinal and rotational generalised coordinates, but reduction of these coordinates is not so important because the elastic nonlinear terms are, in this model, only related with transverse generalized coordinates. A set of 75×75 Gauss points, which is actually more dense than necessary in most instances (Ribeiro and van der Heijden (2009)), was employed.

In the following case study, transverse, uniformly distributed forces are first applied in a quasi-static fashion to beams in steels 1 to 4. The amplitudes of the forces are such that the deflection of each beam in its middle point is 1.50000 h. After the static force is removed the beams partially recover, but plastic strains remain. In this way the plastic strain distributions shown in Figure 14 were implemented. As would be expected, the plastic strains are larger and the force amplitude required is smaller in steels with lower yield stress. As we have seen in previous cases, the plastic strains develop closer to the clamped ends, where stresses are larger. Plastic strains also appear in the middle of the beam, although they are difficult to see in the picture. The direct plastic strains, ε_{11}^{plast}, are much larger than the shear plastic strains, γ_{13}^{plast}, in this thin beam.

Free vibration studies were carried out in these four beams assuming that the plastic strains do not change and following the procedure described in the previous section (Section 5). As explained, an eigenvalue problem is actually solved, and eigenvectors and eigenvalues that provide approximations to the vibration modes in the nonlinear regime are computed. Since the yield stress does not have any implication now, the beams are similar in everything except the plastic strain field.

Curves that relate the frequency of each harmonic with the vibration displacement amplitude (backbone curves) are shown in Figure 15, where the vibration amplitudes were computed in the middle of the beam ($\xi = x_1{=}0$). A first noticeable difference with respect to elastic beams is that due to the plastic strains there is a non-zero constant term. The amplitude of this constant term decreases as the amplitudes of other harmonics grow, most probably because an augmentation of the harmonics amplitudes corresponds to larger vibration amplitudes and results in an increase of stiffness. Since on the other hand, the constant force $\bar{\mathbf{F}}_{\mathbf{plast}}$ does not vary and, obviously, there is no inertia related with the constant term, the amplitude of the constant term decreases.

In an elastic system with no plastic strains the natural frequency at which the first harmonic amplitude is approximately zero is the linear natural frequency, $\omega_{\ell 1}$. But Figure 15 shows that the frequency at which the oscillation amplitude is near zero varies significantly with the installed plastic strain

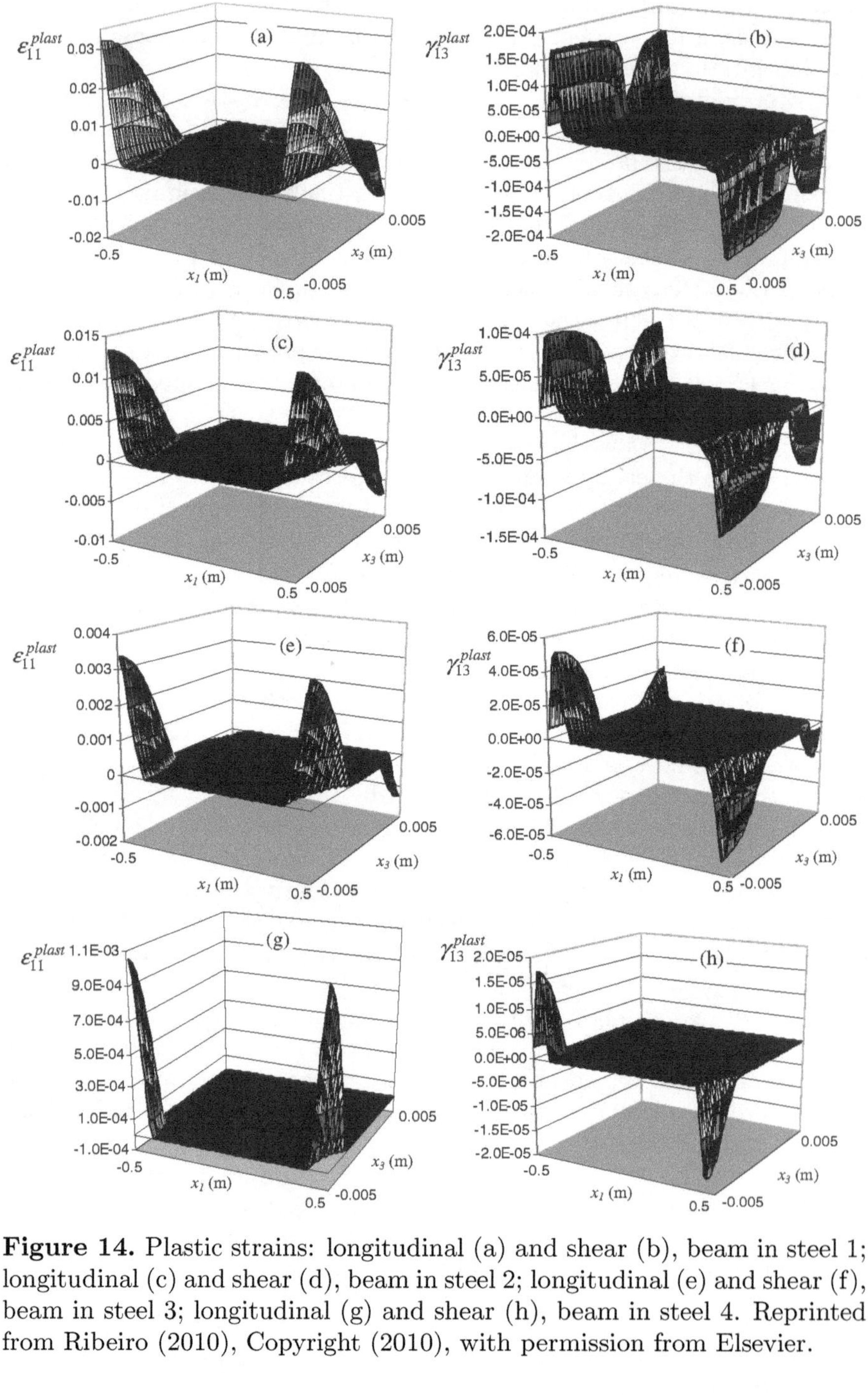

Figure 14. Plastic strains: longitudinal (a) and shear (b), beam in steel 1; longitudinal (c) and shear (d), beam in steel 2; longitudinal (e) and shear (f), beam in steel 3; longitudinal (g) and shear (h), beam in steel 4. Reprinted from Ribeiro (2010), Copyright (2010), with permission from Elsevier.

field. In the case of beam in steel 1 (the beam where larger plastic strains were achieved), this frequency is about 1.278 $\omega_{\ell 1}$; in the case of beam in steel 2 (second larger plastic strains), the starting frequency is about 1.049 $\omega_{\ell 1}$. In these cases the significant deformation of the beam due to plastic forces leads to large amplitude displacements with respect to the initial, straight, configuration and stiffen the beam. On the other hand, the plastic strains have a softening effect on the beam, represented in the present model by the minus sign before matrix $\mathbf{K}^{plast}\left(\varepsilon_{11}^{plast}(\xi,\eta)\right)$ in equation (54). This softening effect competes with the hardening effect of the geometrical nonlinearity. In the case of the initial point of beam in steel 3, the plastic strains are more important than the geometrically nonlinear terms, because the amplitudes of the initial displacements are smaller. Therefore the initial natural frequency of vibration is lower than the one of the elastic beam (0.813 $\omega_{\ell 1}$). The plastic strains and initial displacements of beam in steel 4 are very small; therefore, the starting frequency is approximately equal to the natural frequency of the linear elastic beam.

All oscillations related with Figure 15 are periodic and the larger harmonic is the first, features that are common with oscillation of geometrically nonlinear beams on the main branch (Ribeiro and Petyt (1999)), which is the branch here analysed. What is not common with geometrically nonlinear beams is the softening that beam in steel 1 shows, and the presence of a constant term and second harmonic in all beams with plastic strains.

After the initial softening of beam in steel 1, the geometrical nonlinear terms become preponderant and hardening appears. Beams in steels 2, 3 and 4 solely experience hardening spring effect: the frequency increases as the amplitude of the first harmonic increases. In the beam in steel 4 the constant term and the second harmonic are very small because the plastic strains are also small ((g) and (h)). This means that the solutions are very close to symmetric about an equilibrium configuration that almost does not differ from the one of elastic beams (a configuration where the beam is straight) because the plastic forces are approximately zero.

The normalized shapes of each harmonic of the nonlinear beam in steel 1 at some frequencies of vibration are shown in Figure 16. The shape of harmonic i defined by

$$w_i(\xi) = \mathbf{N}^{u_3}(\xi)^T \mathbf{Q}_i \tag{62}$$

and the shapes represented in Figure 16 were normalized such that the displacement amplitude in the middle of the beam, represented by wm_i, i=0-3, is 1.

The shape of the constant term, with an apparent rotation near the

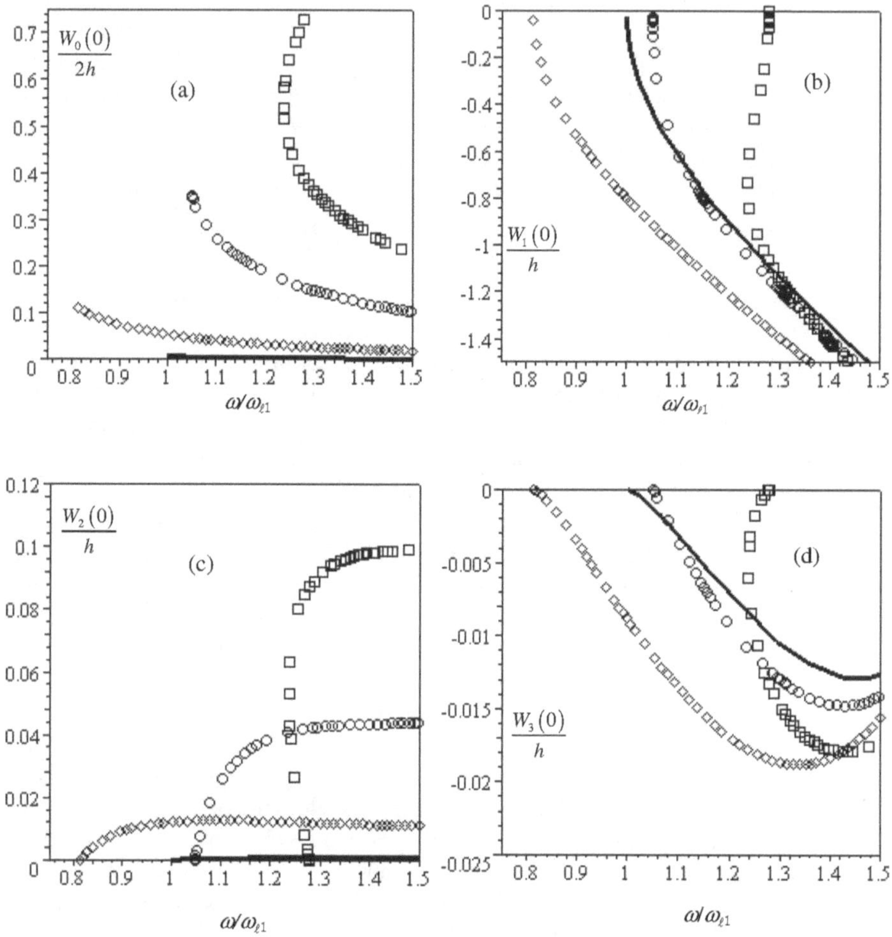

Figure 15. Backbone curves computed in the middle of the beam (point ξ =0.0); □ beam in steel 1; ○ beam in steel 2; ◊ beam in steel 3; — beam in steel 4. Reprinted from Ribeiro (2010), Copyright (2010), with permission from Elsevier.

clamped ends, is initially similar to the first mode shape of a simply supported beam. This is due to the large plastic strains near the clamped ends, which particularly soften the beam in these regions. Proceeding along the backbone curve, the shape of the constant term changes, with a relative increase of the displacement of points away from the central point. This may be a consequence of stiffness increase in the centre of the beam, itself a consequence of the larger amplitude of higher harmonics.

The shape of the first harmonic is very similar to the first linear bending mode shape and changes very little with the fundamental frequency (i.e. with the vibration amplitude). The second harmonic shape also remains similar to the one of the first linear mode shape of vibration, but changes more than the shape of the first harmonic. The third harmonic shape is essentially a combination of the first and second symmetric mode shapes, where the second symmetric mode becomes more important as we proceed along the backbone curve.

In Figure 16 we can see normalised shapes assumed by each harmonic of the beam in steel 3, at different fundamental frequencies of vibration. This example is shown because the backbone curve of the beam in steel 3 is qualitatively different from the backbone curve of beam in steel 1 (in the case of the beam in steel 3, there is no softening and the initial frequency of vibration is smaller than the linear fundamental frequency). Nevertheless, Figure 17 shows that the evolution of each harmonic shape is rather similar to the one of beam 1. One should recall that for the same nondimensional frequency the relative importance of each harmonic is different in the two beams, and therefore the shape assumed by each beam at any particular instant along the vibration period is actually quite different.

The trajectories on the phase plane defined by $u_3^0\left(0,t\right)/h$ and $\dot{u}_3^0\left(0,t\right)T/h$, where T represents the vibration period, are portrayed in Figure 18. The oscillations occur about centres that do not coincide with the origin of the phase plane, since these centres are defined by the plastic strains. The deviation is greater in beam 1, because it has larger plastic strains. Beam in steel 1 also presents phase plots that are more visibly asymmetric with respect to a vertical line that crosses the centre. This is a consequence of the higher excitation of the second harmonic in beam in steel 1, and the latter is another result of the asymmetry introduced by the plastic strains.

7 Conclusion

This text addressed the nonlinear vibrations of beams in one plane. An overview of the large body of work carried to date on geometrically nonlinear, planar, vibrations of beams was presented. A shorter review on

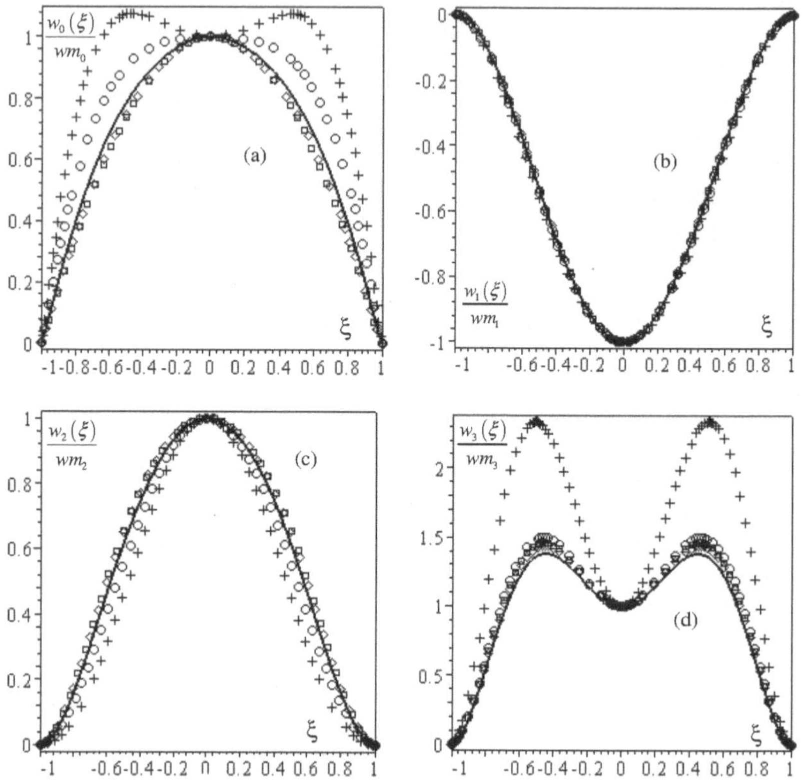

Figure 16. Shapes of constant term (a), first harmonic (b), second harmonic (c) and third harmonic (c) of the response of beam in steel 1 at the following fundamental vibration frequencies ($\omega/\omega_{\ell 1}$): □ 1.2782, ◊ 1.2681, — 1.2412, ○ 1.2692, + 1.4250. Reprinted from Ribeiro (2010), Copyright (2010), with permission from Elsevier.

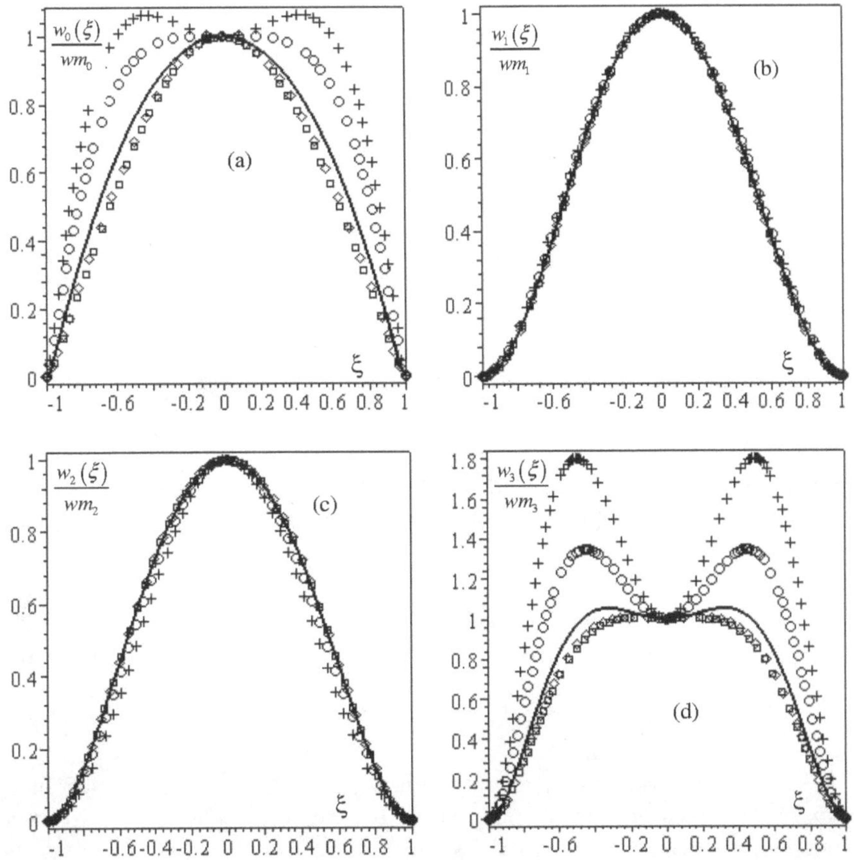

Figure 17. Shapes of constant term (a), first harmonic (b), second harmonic (c) and third harmonic (c) of the response of beam in steel 3 at the following fundamental vibration frequencies ($\omega/\omega_{\ell 1}$): □ .81324, ◇ .82753, — .90822, ∘ 1.1077, + 1.2566. Reprinted from Ribeiro (2010), Copyright (2010), with permission from Elsevier.

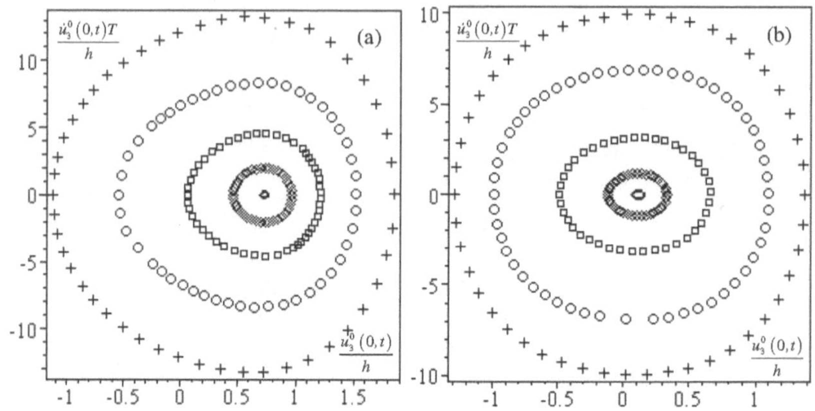

Figure 18. Projections of trajectories on a phase plane at different frequencies and vibration amplitudes (phase portraits): (a) beam in steel 1; (b) beam in steel 3. Reprinted from Ribeiro (2010), Copyright (2010), with permission from Elsevier.

the not so large number of papers on oscillations of the same structural element with plasticity was also written. Equations of motion that model the vibrations of beams with moderately large displacements and plastic deformations were given.

Forced, with large displacements and with plastic strains, vibrations of beams under harmonic excitations were investigated. The procedure used to determine plastic strains is based on a "governing parameter" and the solution of the equations of motion entails the method of Newmark with Newmark parameters. The latter, which can be replaced by other ordinary differential equations of motion solver, performed rather well. Even in cases where the plastic strains were not large on a particular beam, they had a significant influence on the stresses. Moreover, by increasing the excitation amplitude and/or changing the excitation frequency, a beam can experience oscillations where not only strains and stresses, but also displacements and velocities that result from elastic only and from elasto-plastic models significantly differ.

Free, large displacement, vibrations of beams with constant plastic strains were also analysed. The harmonic balance and continuation procedures were now employed to find periodic oscillations, which were interpreted as nonlinear modes of vibration. These modes change with the vibration amplitude and are not harmonic. Moreover, it is important to note that, due to plastic strains, the constant term and the second harmonic appear in the main branch of solutions.

Models for nonlinear oscillations are expected to allow for a greater understanding of the behaviour of beams, an understanding that may be used in engineering design. Even in structure or machine elements that do not in today's practice experience large amplitude oscillations, it is possible to change the design to a more efficient/economic one and without compromising safety, if the engineer employs formulations that better approach nature by taking into account nonlinear effects. Knowledge on how a structure will behave in the plastic regime, perhaps before failure in a limit situation, is also rather important. Finally, better understanding of the structure dynamics, including knowledge of its modes in free vibration, increases the potential of vibration based structural health monitoring methodologies.

Bibliography

M. Amabili. *Nonlinear Vibrations and Stability of Shells and Plates.* Cambridge University Press, 2008.

ANSYS. *ANSYS Academic Research, Release 11.0*, 2007.

R. Arquier, S. Bellizzi, R. Bouc, and B. Cochelin. Two methods for the computation of nonlinear modes of vibrating systems at large amplitudes. *Computers and Structures*, 24-25:1565–1576, 2006.

S. Atluri. Nonlinear vibrations of a hinged beam including nonlinear effects. *Transactions of the ASME, Journal of Applied Mechanics*, 40:121–126, 1973.

K. J. Bathe. *Finite Element Procedures.* Prentice Hall, 1996.

R. Benamar, M. M. K. Bennouna, and R. G. White. The effects of large vibration amplitudes on the mode shapes and natural frequencies of thin elastic structures. part i: Simply supported and clamped-clamped beams. *Journal of Sound and Vibration*, 149:179–195, 1991.

J. A. Bennet and J. G. Eisley. A multiple-degree-of freedom approach to nonlinear beam vibrations. *AIAA Journal*, 8:734–739, 1970.

M. M. K. Bennouna. *Nonlinear Dynamic Behaviour of a Clamped-clamped Beam with Consideration of Fatigue Life, Ph.D. Thesis.* University of Southampton, 1982.

M. M. K. Bennouna and R. G. White. The effects of large vibration amplitudes on the fundamental mode shape of a clamped-clamped uniform beam. *Journal of Sound and Vibration*, 96:309–331, 1984.

G. R. Bhashyam and G. Prathap. Galerkin finite element method for nonlinear vibrations. *Journal of Sound and Vibration*, 72:191–203, 1980.

V. V. Bolotin. *The Dynamic Stability of Elastic Systems.* Holden Day, 1964.

T. D. Burton and M. N. Hamdan. On the calculation of non-linear normal modes in continuous systems. *Journal of Sound and Vibration*, 197:117–130, 1996.

H. R. Busby and V. I. Weingarten. Nonlinear response of a beam to periodic loading. *International Journal of Non-linear Mechanics*, 7:289–303, 1972.

Y. K. Cheung and S. L. Lau. Incremental time-space finite strip method for non-linear structural vibrations. *Earthquake Engineering and Structural Dynamics*, 10:239–253, 1982.

C. Y. Chia. *Nonlinear Analysis of Plates*. McGraw-Hill, 1980.

B. Cochelin and C. Vergez. A high order purely frequency-based harmonic balance formulation for continuation of periodic solutions. *Journal of Sound and Vibration*, 324:243–262, 2009.

P. J. Davis and I. Polonsky. *Handbook of Mathematical Functions With Formulas, Graphs, and Mathematical Tables, Series 55*, 1972.

G. E. Dieter. *Mechanical Metallurgy*. McGraw-Hill, 1986.

D. J. Ewins. *Modal Testing: Theory, Pratice and Application.* Research Studies Press, 2000.

Y. C. Fung and P. Tong. *Classical and Computational Solid Mechanics.* World Scientific, 2001.

J. Gerstmayr and H. Irschik. Vibrations of the elasto-plastic pendulum. *International Journal of Non-Linear Mechanics*, 38:111–122, 2003.

J. Gerstmayr, H. J. Holl, and H. Irschik. Development of plasticity and damage in vibrating structural elements performing guided rigid-body motions. *Archive of Applied Mechanics*, 71:135–145, 2001.

R. P. S. Han and J. Lu. A space-time finite element method for elasto-plastic shock dynamics. *Journal of Sound and Vibration*, 222:65–84, 1999.

W. Han. *The analysis of isotropic and laminated rectangular plates including geometrical non-linearity using the p-version finite element method, Ph.D. Thesis.* University of Southampton, 1993.

C. Hayashi. *Nonlinear Oscillations in Physical Systems.* McGraw-Hill, 1964.

J. R. Hutchinson. Shear coefficients for timoshenko beam theory. *Transactions of the ASME, Journal of Applied Mechanics*, 68:87–92, 2001.

M. El Kadiri, R. Benamar, and R. G. White. Improvement of the semi-analytical method, for determining the geometrically non-linear response of thin straight structures. part 1: Application to clamped-clamped and simply supported-clamped beams. *Journal of Sound and Vibration*, 249: 263–305, 2002.

T. Kaneko. On timoshenko's correction for shear in vibrating beams. *Journal of Physics D*, 8:1927–1936, 1975.

S. G. Kelly. *Mechanical Vibrations.* McGraw-Hill, 1993.

M. Kojić and K.-J. Bathe. *Inelastic Analysis of Solids and Structures.* Springer-Verlag, 2005.

J.-Y. Lee, P. S. Symonds, and G. Borino. Chaotic responses of a two-degree-of freedom elastic-plastic beam model to short pulse loading. *Transactions of the ASME, Journal of Applied Mechanics*, 59:711–721, 1992.

U. Lepik. Elastic-plastic vibrations of a buckled beam. *International Journal of Non-Linear Mechanics*, 30:129–139, 1995.

A. Y. T. Leung and T. C. Fung. Non-linear steady state vibration of frames by finite element method. *International Journal for Numerical Methods in Engineering*, 28:1599–1618, 1989.

A. Y. T. Leung and S. G. Mao. A symplectic galerkin method for non-linear vibration of beams and plates. *Journal of Sound and Vibration*, 183:475–491, 1995.

R. Lewandowski. Application of the ritz method to the analysis of non-linear free vibrations of beams. *Journal of Sound and Vibration*, 114: 91–101, 1987.

R. Lewandowski. Non-linear, steady-state analysis of multispan beams by the finite element method. *Computers and Structures*, 39:83–93, 1991.

R. Lewandowski. Non-linear free vibrations of beams by the finite element and continuation methods. *Journal of Sound and Vibration*, 170:577–593, 1994a.

R. Lewandowski. Solutions with bifurcation points for free vibration of beams: an analytical approach. *Journal of Sound and Vibration*, 177: 239–249, 1994b.

R. Lewandowski. Computational formulation for periodic vibration of geometrically nonlinear structures - part 1: Theoretical background. *International Journal Solids Structures*, 34:1925–1947, 1997a.

R. Lewandowski. Computational formulation for periodic vibration of geometrically nonlinear structures - part 2: Numerical strategy and examples. *International Journal Solids Structures*, 34:1949–1964, 1997b.

R. Lewandowski. Free vibration of structures with cubic non-linearity - remarks on amplitude equation and rayleigh quotient. *Computer Methods in Applied Mechanics and Engineering*, 192:1681–1709, 2003.

Y. M. Liu, G. W. Ma, and Q. M. Li. Chaotic and asymmetrical beam response to impulsive load. *International Journal of Solids and Structures*, 41:765–784, 2004.

A. J. Ma, S. H. Chen, and D. T. Song. A new method of nonlinear analysis for large deflection forced vibration of beams. *Finite Elements in Analysis and Design*, 20:39–46, 1995.

G. W. Ma, Y. M. Liu, J. Zhao, and Q. M. Li. Dynamic asymmetrical instability of elastic-plastic beams. *International Journal of Mechanical Sciences*, 47:43–62, 2005.

E. Manoach and D. Karagiozova. Dynamic response of thick elastic-plastic beams. *International Journal of Mechanical Sciences*, 35:909–919, 1993.

M. I. McEwan, J. R. Wright, J. E. Cooper, and A. Y. T. Leung. A combined modal/finite element analysis technique for the dynamic response of a non-linear beam to harmonic excitation. *Journal of Sound and Vibration*, 243:601–624, 2001.

C. Mei. Finite element displacement method for large amplitude free flexural vibrations of beams and plates. *Computers and Structures*, 3:163–174, 1973.

C. Mei. Nonlinear vibration of beams by matrix displacement method. *AIAA Journal*, 10:355–357, 1976.

C. Mei and K. Decha-Umphai. A finite element method for non-linear forced vibrations of beams. *Journal of Sound and Vibration*, 102:369–380, 1985.

L. Meirovitch and H. Baruh. On the inclusion principle for the hierarchical finite element method. *International Journal for Numerical Methods in Engineering*, 19:281–291, 1983.

K. D. Murphy, L. N. Virgin, and S. A. Rizzi. Characterizing the dynamic response of a thermally loaded, acoustically excited plate. *Journal of Sound and Vibration*, 196:635–658, 1996.

A. H. Nayfeh and B. Balachandran. Modal interactions in dynamical and structural systems. *Applied Mechanics Review*, 42:S175–S201, 1989.

A. H. Nayfeh and D. T. Mook. *Nonlinear Oscillations*. John Wiley and Sons, 1995.

M. Petyt. *Introduction to Finite Element Vibration Analysis*. Cambridge University Press, 1990.

B. S. Prathap and T. K. Varadan. The large amplitude vibrations of hinged beams. *Computers and Structures*, 9:219–222, 1978.

M. I. Qaisi. Application of the harmonic balance principle to the nonlinear free vibration of beams. *Applied Acoustics*, 40:141–151, 1993.

P. Ribeiro. *Geometrical Nonlinear Vibration of Beams and Plates by the Hierarchical Finite Element Method, Ph.D. Thesis*. University of Southampton, 1998.

P. Ribeiro. Hierarchical finite element analyses of geometrically non-linear vibration of beams and plane frames. *Journal of Sound and Vibration*, 246:225–244, 2001.

P. Ribeiro. A p-version, first order shear deformation, finite element for geometrically non-linear vibration of curved beams. *International Journal for Numerical Methods in Engineering*, 61:2696–2715, 2004a.

P. Ribeiro. Non-linear forced vibrations of thin/thick beams and plates by the finite element and shooting methods. *Computers and Structures*, 82: 1413–1423, 2004b.

P. Ribeiro. Free periodic vibrations of beams with large displacements and initial plastic strains. *International Journal of Mechanical Sciences*, 52: 1407–1418, 2010.

P. Ribeiro. Stability of multi-degree-of-freedom Duffing oscillators. *Engineering and Computational Mechanics, Proceedings of the Institution of Civil Engineers*, 2:87–97, 2009.

P. Ribeiro and R. Carneiro. Experimental detection of modal interaction in the non-linear analysis of a hinged-hinged beam. *Journal of Sound and Vibration*, 277:943–954, 2004.

P. Ribeiro and M. Petyt. Non-linear vibration of beams with internal resonance by the hierarchical finite element method. *Journal of Sound and Vibration*, 224:591–624, 1999.

P. Ribeiro and G. H. M. van der Heijden. Elasto-plastic and geometrically nonlinear vibrations of beams by the p-version finite element method. *Journal of Sound and Vibration*, 325:321–337, 2009.

P. Ribeiro, B. Cochelin, and S. Bellizzi. Non-linear vibrations of deep cylindrical shells by the p-version finite element method. *Shock and Vibration*, 17:21–37, 2010.

R. M. Rosenberg. On non-linear vibrations of systems with many degrees of freedom. *Advances Applied Mechanics*, 9:155–242, 1966.

B. S. Sarma and T. K. Varadan. Ritz finite element approach to non-linear vibrations of beams. *International Journal for Numerical Methods in Engineering*, 20:353–367, 1984a.

B. S. Sarma and T. K. Varadan. Lagrange-type formulation for finite element analysis of non-linear beam vibrations. *Journal of Sound and Vibration*, 86:61–70, 1983.

B. S. Sarma and T. K. Varadan. Ritz finite element approach to non-linear vibrations of beams. *International Journal for Numerical Methods in Engineering*, 20:353–367, 1984b.

M. Sathyamoorthy. Nonlinear vibration analysis of plates: a review and survey of current developments. *Applied Mechanics Review*, 40:1553–1561, 1987.

S. W. Shaw and C. Pierre. Normal modes for non-linear vibratory systems. *Journal of Sound and Vibration*, 164:85–121, 1993.

S. W. Shaw and C. Pierre. Normal modes of vibration for non-linear continuous systems. *Journal of Sound and Vibration*, 169:319–347, 1994.

Y. Shi and C. Mei. A finite element time domain modal formulation for large amplitude free vibration of beams and plates. *Journal of Sound and Vibration*, 193:453–464, 1996.

J. C. Simo and T. J. R. Hughes. *Computational Inelasticity.* Springer-Verlag, 1998.

P. W. Smith, C. I. Malme, and C. M. Gogos. Nonlinear response of a simple clamped panel. *The Journal of the Acoustical Society of America*, 33: 1476–1482, 1961.

P. S. Symonds and T. X. Yu. Counter-intuitive behavior in a problem of elastic-plastic beam dynamics. *Transactions of the ASME, Journal of Applied Mechanics*, 52:517–522, 1985.

B. Szabó and I. Babuska. *Finite Element Analysis.* John Wiley and Sons, 1991.

W. Szemplinska-Stupnicka. *The Behaviour of Non-linear Vibrating Systems.* Kluwer Academic Publishers, 1990.

W. Szemplinska-Stupnicka. "non-linear normal modes" and the generalized ritz method in the problems of vibrations of non-linear elastic continuous systems. *International Journal of Non-Linear Mechanics*, 18:149–165, 1983.

K. Takahashi. A method of stability analysis for non-linear vibration of beams. *Journal of Sound and Vibration*, 67:43–54, 1979.

C. E. Teh. *Dynamic Behaviour and Acoustic Fatigue of Isotropic and Anisotropic Panels Under Combined Acoustic Excitation and Static In-plane Compression, Ph.D. Thesis.* University of Southampton, 1982.

J. J. Thomsen. *Vibrations and Stability.* Springer, 2003.

S. P. Timoshenko. On the transverse vibrations of bars of uniform cross section. *Philosophy Magazine*, 43:125–131, 1922.

C. Touzé, O. Thomas, and A. Huberdeau. Asymptotic non-linear normal modes for large-amplitude vibrations of continuous structures. *Computers and Structures*, 82:31–32, 2004.

W. Y. Tseng and J. Dugundji. Nonlinear vibrations of a beam under harmonic excitation. *Transactions of the ASME, Journal of Applied Mechanics*, 37:292–297, 1970.

R. G. White. A comparison of some statistical properties of the responses of aluminium alloy and cfrp plates to acoustic excitation. *Composites*, 9: 251–258, 1978.

S. Woinowski-Krieger. The effect of an axial force on the vibration of hinged bars. *Journal of Applied Mechanics*, 17:35–36, 1950.

H. Wolfe. *An Experimental Investigation of Nonlinear Behaviour of Beams and Plates Excited to High Levels of Dynamic Response, Ph.D. Thesis.* University of Southampton, 1995.

J.-X. Xu and N. Hasebe. The problem of an elastic-plastic beam dynamics and an incomplete co-dimension two bifurcation. *International Journal of Non-Linear Mechanics*, 32:127–143, 1997.

K. Yagasaki. Bifurcations and chaos in a quasi-periodically forced beam: theory, simulation and experiment. *Journal of Sound and Vibration*, 183: 1–31, 1995.

T. Yamamoto, K. Yasuda, and K. Aoki. Subharmonic oscillations of a slender beam. *Bulletin of the JSME*, 24:1011–1020, 1981.

T. Yamamoto, K. Yasuda, and K. Aoki. Summed and differential harmonic oscillations in a slender beam. *Bulletin of the JSME*, 24:1214–1222, 1982a.

T. Yamamoto, K. Yasuda, and K. Aoki. Super summed and differential harmonic oscillations in a slender beam. *Bulletin of the JSME*, 25:959–968, 1982b.

L.D. Zavodney and A.H. Nayfeh. The non-linear response of a slender beam carrying a lumped mass to a principal parametric excitation: Theory and experiments. *International Journal of Non-Linear Mechanics*, 24: 105–125, 1989.

B. Zhu and A. Y. T. Leung. Linear and nonlinear vibration of non-uniform beams on two-parameter foundations using p-elements. *Computers and Geotechnics*, 36:743–750, 2009.

Control and exploitation of nonlinearity in smart structures

David Wagg*
Department of Mechanical Engineering, University of Bristol, Bristol, UK

Abstract In this Chapter the control and exploitation of nonlinearity are considered when applied to so called 'smart structures'. Here we consider how active control can be applied to structures particularly in the presence of nonlinearity. We also consider how snap-through structures can potentially be exploited as hinges in morphing applications.

1 Introduction

The phrase "smart structure" is increasingly used to describe structural systems which have characteristics which go beyond the traditional definition of a structure (sometimes also called adaptive or intelligent structures). These new "smart" characteristics include things such as: (i) monitoring the structure for signs of damage, (ii) reducing unwanted vibrations, or (iii) changing the shape of the structure. The general trend in structural design is towards lighter structures, which typically leads to increased flexibility. In order for structures to carry out the smart functions it is now possible for structural elements to have actuator and sensor networks.

Nonlinear behaviour in structural dynamics arises naturally from a range of common nonlinearities. In some cases nonlinearities either cannot be avoided, or add some potential benefit, which leads to designing in the presence of nonlinearity. The most common form of nonlinearity for structural dynamics is geometric nonlinearity. For example, in the design and construction of bridges there has always been the desire to build longer spans, and therefore more flexible bridges. Figure 1 shows the Sutong Bridge which is a cable-stayed bridge that spans the Yangtze River in China. It is currently the cable stay bridge with the longest main span in the world, with

*The author would also like to acknowledge the work of Simon Neild, Peter Gawthrop, Jack Potter, Andres Arrieta Diaz, Nihal Malik and Lin Yang, who contributed to the results shown in this chapter. JP and NM were supported for PhD studies by EPSRC studentships; AAD and LY were supported by ORS scholarships.

Figure 1. The Sutong Bridge is currently (in 2011) the world's longest span cable-stay bridge. It spans the Yangtze River between Nantong and Changshu in the People's Republic of China. It has a main span of 1,088 metres, and two side spans are 300 metres each, and there are also four small cable spans. Photo credit: Wikipedia.

a span of 1,088 metres (3,570 ft). In fact suspension bridges can be even longer, and the Akashi Kaikyo bridge in Japan has a main span of 1.9 kilometres, which is more than a mile long. Some of the structural elements for these bridges, especially the cables, have very low damping, and as a result large deflections become unavoidable.

There are also structures where flexibility occurs primarily as a result of the requirement of low weight. For example, Figure 2 shows the NASA Helios which was designed as an unmanned long-term, high-altitude aircraft powered by solar and fuel cells. It suffered a structural failure and crashed during a flight across the pacific on June 26, 2003. The cause of the failure was due to higher than expected wind loads which led to an unexpected, persistent, high deformation of the wing. This in turn led to

Figure 2. The Helios was designed as a long duration solar powered aircraft. An unexpected, persistent, high dihedral configuration during poor weather on a test flight during June 26, 2003, led to unstable oscillations in a pitch mode and structural failure. Photo credit NASA.

unstable oscillations in one of the vibration modes which grew in amplitude until structural failure occurred. More recent aircraft being developed for the same purposes include the European built Solar Impulse project.

All the structures mentioned so far are large, but smart structures can also be very small. In fact there has been increasing interest in small micro and nano scale structures. Figure 3, for example, shows the tip of an atomic force microscope. In fact this is just a very small cantilever beam made from silicon, and about as wide as a human hair, approximately 30 microns in this case. The radius of curvature of the tip is of the order of nanometres. The microscope works by dragging the tip across the surface of the material, "feeling" the surface as it goes in order to create an image at the atomic scale. However, vibrations caused by contact with the surface can distort the image if the process is not adequately controlled. And getting rid of unwanted vibrations is one of the most basic tasks of a smart structure. For an atomic force microscope it is usually done using piezoelectric materials.

Space structures are another important area of structural design. Figure

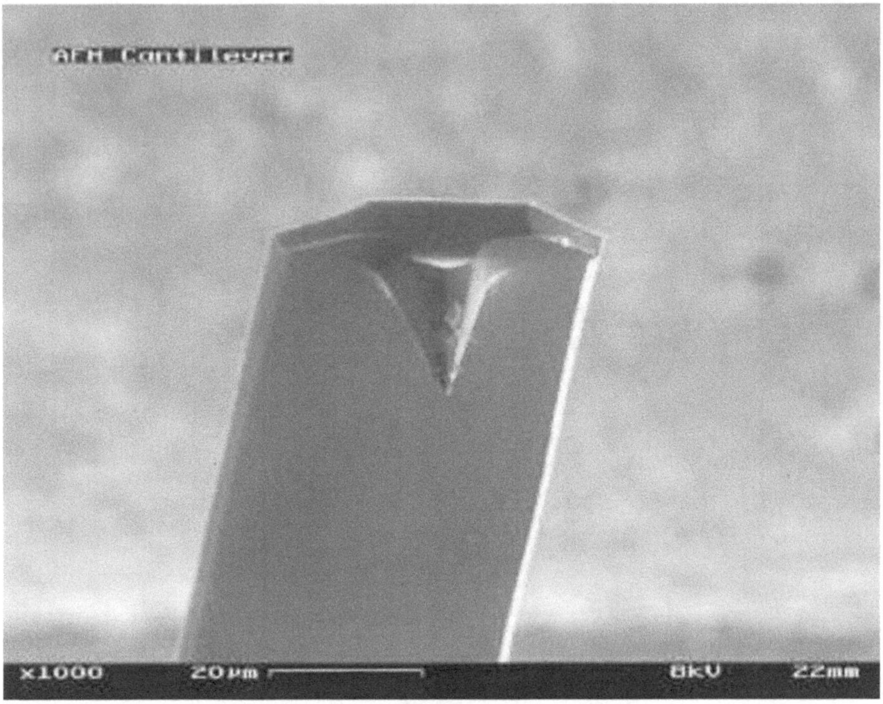

Figure 3. The atomic force microscope (AFM) consists of a cantilever with a sharp tip (probe) at its end that is used to scan the specimen surface. The cantilever is typically silicon or silicon nitride with a tip radius of curvature on the order of nanometres. Photo credit Wikipedia

4 shows the concept of a solar sail, which is a futuristic idea for propelling spacecraft using solar photon radiation onto large, highly reflecting sails. This is a form of passive propulsion, very like wind pushing a sailing ship along on earth. This idea is in the prototype stage, and several solar sails are currently being designed and tested around the world.

In addition to large deformations, geometric nonlinearity also includes the effects of combined stretching/compressing with vibration and nonlinear alignment of structural elements. Geometric nonlinearity can be exploited in some smart structures applications. For example, to design high performing spring elements such as bi-stable structures with snap-through behaviour, which will be discussed later in this Chapter.

Control forces can be added to a structural system in order to control

Figure 4. Solar Sails are spacecraft which utilize the momentum transfer of solar photons onto large, highly reflecting sails for passive propulsion. This sail is being developed at the German Aerospace Centre (DLR). Photo credit: DLR.

the behaviour in some way and make it an adaptive structure. However, control of structural vibrations is different from the majority of control problems, because there are typically multiple lightly damped resonances in the system response. In addition, when an actuator is attached to the structure, its effect will be coupled to some resonances much more strongly than others. As a result, careful design is required to reduce particular resonant responses.

Using feedback can induce instability in the system, and so ensuring any control design is stable is of primary importance. The underlying ideas of stability for nonlinear systems have been introduced in Chapter 1. In this Chapter, these ideas are extended to include systems with feedback control, and the stability analysis is carried out using a particular type of potential function, called a Lyapunov function. The basic ideas of Lyapunov based control design can be extended to a range of other approaches. The main control method described here is how to effectively linearise the system

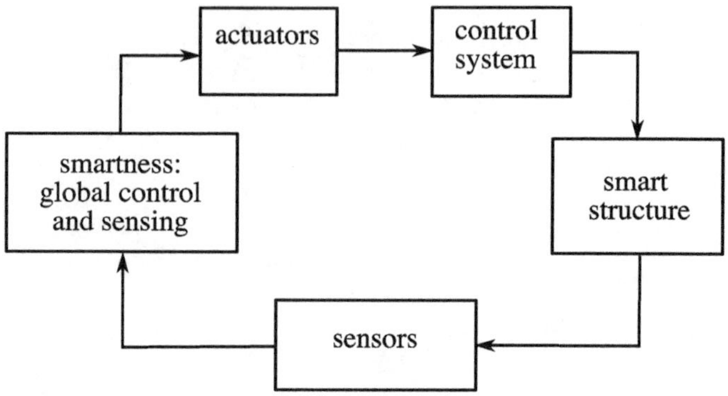

Figure 5. Schematic diagram of a smart structure control system.

using feedback. Adaptive control, which can also be a useful method for nonlinear or uncertain systems was introduced in Chapter 1, and will be discussed briefly later in this Chapter.

In this Chapter we assume that the ultimate objective of adding control capability to structures is to create smart structures. The key elements of a smart structure are shown in Figure 5. The most important starting point is that the structure needs to have some awareness of its condition and/or its environment. This is typically achieved by having a series of measurement sensors mounted on (or integrated into) the structure. Information from the sensors is then used by the global control system. This is where the smart behaviour is applied. The global control system will monitor the condition of the structure and when required give command signals to a series of actuators which act on the structure.

Understanding the nonlinear structural dynamics is very important for many smart structures, because they typically have one or more of the following characteristics: (i) the ability to have large deformations, (ii) non-homogeneous material properties, (iii) material parameters which vary (or can be varied), (iv) multiple stable states, (v) highly flexible elements, (vi) very light damping, and (vii) need to be operated in a dynamic environment.

For the interested reader wishing to explore the background to this area, there are a number of texts discuss linear vibration with control, such as Beards (1981), Fuller et al. (1996), Moheimani et al. (2003) and Inman (2006). There are also texts which discuss the vibration and control of smart structures such as Preumont (1997), Clark et al. (1998), Srinivasan and McFarland (2001), Worden et al. (2003), Leo (2007), Wagg and Neild (2009) and Vepa (2010). A good overview of linear control is given by

Goodwin et al. (2000). A good introduction to nonlinear control can be found in Khalil (1992) and for adaptive control see Åström and Wittenmark (1995). An excellent starting point for consulting background literature on structural control in general is the review by Housner et al. (1997).

2 Control design for smart structures

One of the most common requirements in a smart structure is to reduce unwanted vibrations. The traditional way to reduce vibrations is to design the system with additional damping, by using special materials or adding physical damping devices. This approach is called passive vibration control (or sometimes redesign) and it is a well developed subject area for linear vibration problems, see for example Soong and Dargush (1997), Mottershead and Ram (2006) and Inman (2006). Passive techniques, such as the tuned mass damper can be extended to structures with nonlinear behaviour — see for example Alexander and Schilder (2009) and references therein. An example of a tuned mass damper is shown in Figure 6. In some applications, the nonlinear characteristics can be exploited to improve the vibration isolation, see for example Semercigil et al. (2002), Vyas et al. (2003), Shoeybi and Ghorashi (2006), Mikhlin and Reshetnikova (2006) and Milovanovic et al. (2009).

Passive vibration solutions are normally preferred in practice as they can be built into the system and there is no control element, which eliminates any issues with control stability or robustness. Engineering applications include helicopters (Panda et al., 1996), space structures (Nair and Keane, 2001), buildings/structures (Kasai et al., 1998; Wang et al., 2003; Ghosh and Basu, 2008) and automotive applications (Cole and Cebon, 1996; Tamboli and Joshi, 1999). However, for a growing class of structures for which reduced weight and flexibility are important features, passive vibration control is not an effective solution.

In recent years there has been growing interest in passive actuation methods, which have received considerable attention for use in morphing aircraft structures (Bharti et al., 2004; Lucato et al., 2004; Spadoni and Ruzzene, 2007; Wang et al., 2007; Thill et al., 2008; Baker and Friswell, 2009; Daynes et al., 2010). This type of application is typically related to flight control surfaces which are designed such that they can morph (i.e. change shape) in response to specific aerodynamic loads. In some of these designs techniques the nonlinear characteristics of particular structures are deliberately exploited, such as bi-stable shells see Chapters 4, 5 and 7 of Wagg et al. (2007). This is still an active area of research, and in many cases the passive actuation method is designed to work in conjunction with an active control

Figure 6. Passive vibration reduction using a tuned mass damper (TMD). This is the TMD for the Taipei 101 which is currently the world's second tallest building. The 660-tonne TMD acts like a giant pendulum to counteract the building's movement–reducing sway due to wind by 30 to 40 percent. Photo copyright J Aaron; Wikipiedia alternatives.

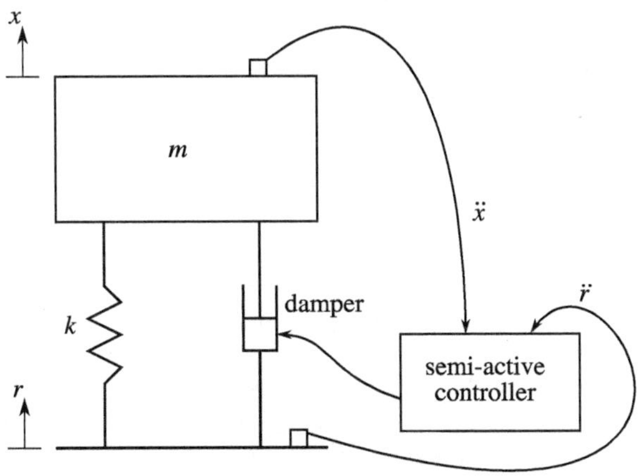

Figure 7. Single-degree-of-freedom oscillator with semi-active vibration control.

element Daynes et al. (2010).

The alternatives to passive vibration control are to use either active control or semi-active vibration reduction techniques, which we consider next.

2.1 Semi-active vibration control

Semi-active control is a technique for controlling a vibrating system without using control actuators. Instead, a semi-active element, such as a damper, is used to effect change in the system (Karnopp, 1995; Ahmadian, 1999). Within the semi-active element it is possible to vary one or sometimes more system parameters. An important difference between semi-active and active control is that, semi-active cannot add energy to the system, and therefore is normally an unconditionally stable form of control.

Semi-active control methods have been used extensively in many structural engineering applications. A review of these techniques as applied to structures such as bridges and tall buildings is given by Spencer and Nagarajaiah (2003). Other comprehensive discussions of semi-active control techniques can be found in Casciati et al. (2006) and Preumont and Seto (2008)

An example of a single-degree-of-freedom oscillator with semi-active vibration control and excitation by a moving support input of $r(t)$, is shown in Figure 7. Semi-active control design is based around how to select c_v, the damping value of the variable damper, to give the best vibration suppression. To do this, information is needed about the relative velocity of the mass and the input. This can be obtained by using accelerometers to measure the acceleration of the mass and the support, $\ddot{x}$ and $\ddot{r}$ respectively, which can be integrated to give velocities $\dot{x}$ and $\dot{r}$.

A commonly used semi-active control strategy is sky-hook control (Karnopp, 1995; Preumont et al., 2002; Hong et al., 2002). In this approach the control objective is that the mass be isolated from the support input by getting the semi-active damper to behave as if it is a grounded passive damper. When this is achieved, a damping force which resists the absolute velocity of the mass will be provided. To implement this in a semi-active element a typical method is to switch between a high and a low damping value — see for example Potter et al. (2010) and references therein. The high damping is selected when the damper force is resisting the direction of motion of the mass, and the low damping force is used when this is not the case. In practice, this can be achieved in various ways, for example by switching between high and low viscosity in a magneto-rheological damper.

Now consider an example of designing a semi-active vibration control strategy, for the mass-spring-damper system shown in Figure 7. In this case, the feedback to the controller is the acceleration of the mass, $\ddot{x}$, and input, $\ddot{r}$, measured by accelerometers. We will assume that the semi-active damper can be switched between two values, c_{high} and c_{low}.

The acceleration signals are fed to a semi-active controller, which then needs to decide when to switch between c_{high} and c_{low}. The governing equation of motion for the oscillator is given by

$$m\ddot{x} + c_v(\dot{x} - \dot{r}) + k(x - r) = 0,$$

where the displacement of the mass, m, is given by x, k is the spring stiffness and c_v is a variable damping parameter which can be controlled by the semi-active controller. The control objective is to reduce vibration in the system as much as possible, or in other words, to isolate the mass (or minimise the absolute acceleration of the mass $\ddot{x}$).

Using the sky-hook strategy to achieve this, the damper needs to be in the high damping state when opposing the motion of the mass, and in the low damping state when aiding the motion of the mass. This can be defined by noting that when the relative velocity $(\dot{x} - \dot{r})$ has the same sign as the absolute velocity $\dot{x}$ then the damper is opposing the mass. So the semi-active control law applied to the single degree-of-freedom system can be written as

$$c_v = \begin{cases} c_{high} & (\dot{x} - \dot{r})\dot{x} > 0, \\ c_{low} & \text{otherwise.} \end{cases}$$

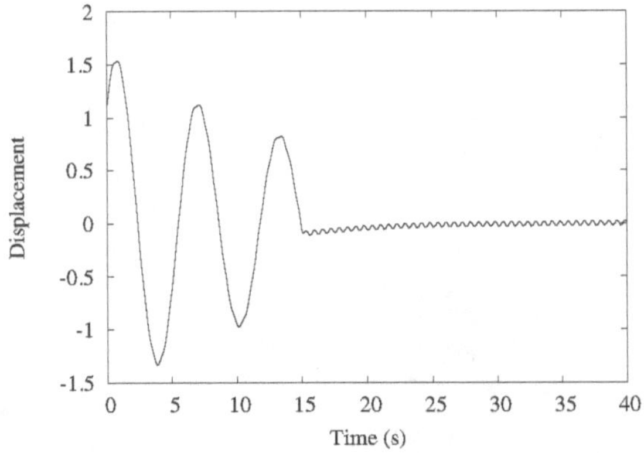

Figure 8. Time simulation of on-off sky hook control.

The semi-active control will act like additional damping in the linear oscillator, which in turn will reduce the height of the resonance peak.

A time simulation for the example when $r = \sin(10t)$ is shown in Figure 8 with $x(0) = 1.1$ and $\dot{x}(0) = 1.0$, $m = 1$, $k = 1$ and $c_v = 0.1$. Initially the sky hook control is switched off, and then at time $t = 15s$ the control is switched on, with $c_{high} = 0.7$, $c_{low} = 0.1$ and excitation signal $r = \sin(15t)$. A dramatic reduction in vibration amplitude can be seen as soon as the control is switched on.

That said, in some cases, the skyhook approach has been shown to be non-optimal across a wide frequency range, and in fact it appears to work best a lower frequencies (Potter et al., 2010). The switching relationship for a skyhook damper is shown in Figure 9 (a), and this can be compared with the more general switching surface controller proposed by Potter et al. (2010), which is shown in Figure 9 (b). The control law for the switching surface controller is given by

$$c_v = \begin{cases} c_{high} & (\dot{x} + \alpha\dot{r})(\dot{x} - \beta\dot{r}) > 0, \\ c_{low} & \text{otherwise.} \end{cases}$$

In this context, the sky-hook controller is a special case of the switching surface controller with $\beta = 1$ and $\alpha = 0$. In the study carried out by Potter et al. (2010) choosing $\alpha = 1$ was shown to improve the damping effect over standard skyhook for some parameter values.

Although on-off approaches offer simplicity, there are situations when a variable level of damping is more advantageous. This can still be achieved

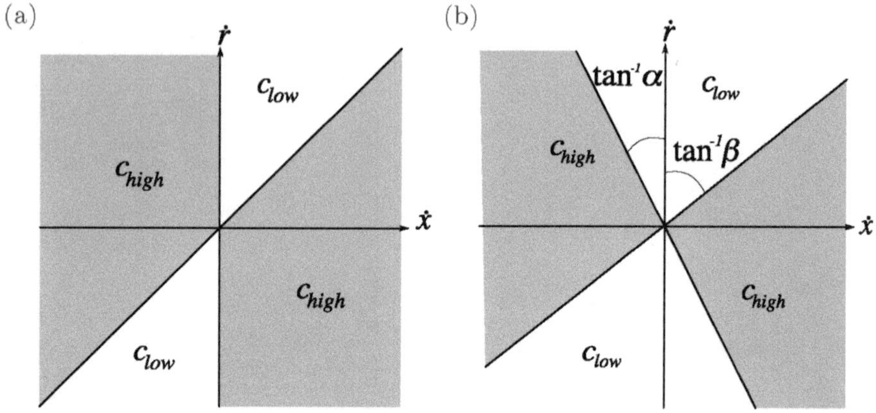

Figure 9. Switching relationships for semi-active control. (a) skyhook, and (b) switching surface controller.

in the context of semi-active control by applying a control law of the type given by

$$c_v = \begin{cases} \max\left[c_{low}, \min\left[\dfrac{c_{high}\dot{x}}{\dot{x}-\dot{r}}, c_{high}\right]\right], & (\dot{x}-\dot{r})\dot{x} > 0, \\ c_{low}, & \text{otherwise.} \end{cases}$$

This allows the semi-active controller to apply a damping value between the low and high values, see Liu et al. (2005) and references therein for further details.

Sky-hook, and other semi-active approaches, can produce significantly improved vibration isolation compared to passively damped systems. The most common application is in automotive suspension systems, especially low degree-of-freedom quarter-car models (Besinger et al., 1992; Hrovat, 1997; Zaremba et al., 1997; Fialho and Balas, 2000; Kitching et al., 2000; Hong et al., 2002; Jalili, 2002; Sammier et al., 2003; Verros et al., 2005; Shen et al., 2006; Giorgetti et al., 2006). They can also be applied to other types of base isolation systems (Barbat et al., 1995; Papoulia and Kelly, 1997; Yoshioka et al., 2002; Ramallo et al., 2002; Bani-Hani and Sheban, 2006; Nagarajaiah and Narasimhan, 2007; Bahar et al., 2010). An example of semi-active vibration control for reducing cable vibration in the Dongting Lake bridge is shown in Figure 10

It is common for semi-active strategies to involve switching off the control action when the conditions are not favourable for control. As a result the system behaviour is uncontrolled for significant portions of time. We also

Figure 10. Semi-active vibration reduction on the Dongting Lake Bridge, which is a cable-stayed bridge crossing the Dongting Lake where it meets the Yangtze River in China. The cables were observed to have rain-wind-induced vibration, especially under adverse weather conditions and so a magnetorheological (MR) damping system was introduced to reduce cable vibration. Here you can see how the pair of MR dampers, at an inclined angle are attached to the cable. Photo credit: Wikipedia.

note that systems with multiple degrees of freedom, such as continuous structural elements it becomes increasingly difficult to apply semi-active control methods like sky-hook, unless the behaviour is limited to a very low number (usually one) of modes of vibration. Usually, active vibration control is required to tackle these types of application.

2.2 Active control

An introduction to the basic ideas of feedback control has been given in Chapter 1. The interested reader may also want to consult Preumont (1997), or Vepa (2010) for an discussion on linear control methods applied to smart structures or Wagg and Neild (2009) for a discussion of nonlinear

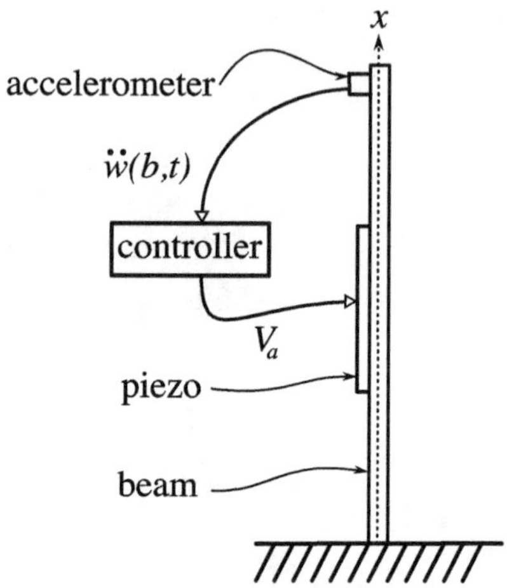

Figure 11. A vibration control configuration for a cantilever beam with a piezoelectric actuator. The acceleration is measured near the tip at point $x = b$, and the control signal is a voltage, V_a sent to the piezo from the controller.

control methods for smart structure applications. The general governing equation for a nonlinear control system is given by

$$\dot{x} = f(x) + g(x)u, \tag{1}$$

where f is the nonlinear system function and g is the nonlinear controller function, x is the state vector $x = [\mathbf{x}^T, \dot{\mathbf{x}}^T]^T$ and $\mathbf{x}$ is the displacement vector. Note that to align with notation conventions of both nonlinear and structural dynamics communities, there is a subtle but significant difference between x which is the $2N \times 1$ *state* vector and $\mathbf{x}$ is the $N \times 1$ *displacement* vector. The control output is defined as $y = h(x)$, where h is the nonlinear output function.

To specify the governing equations of the system, a model of the smart structure is required. A typical scenario is that the displacements and velocities in the state vector represent an approximate modal model of a continuous structural element such as beam, cable or plate etc. Although the idealised modal model is infinite, in practice it must be truncated to the sum of N modal contributions.

An example of a structural control configurations for a cantilever beam is shown schematically in Figure 11. The cantilever beam is being controlled by a piezoelectric actuator. In addition the beam acceleration at a second point along the cantilever is measured by an accelerometer. In this case the underlying vibrating system (i.e. the cantilever beam) is infinite dimensional, but is acted on by only by a single actuator, and measured with a single sensor. Piezoceramic materials have been used extensively in structural control and other smart structure applications (Preumont et al., 1992; Khajepour and Golnaraghi, 1997; Ashour and Nayfeh, 2002; Preumont et al., 2003; Schultz and Hyer, 2003; Chen and Chen, 2004; Zhou and Wang, 2004; Moheimani and Vautier, 2005; Preumont et al., 2005; Song et al., 2006; Moheimani and Fleming, 2006; de Marneffe and Preumont, 2008; Harari et al., 2009). Of particular note is the increasing development and use of Macro Fibre Composite actuators (MFC) (Song et al., 2006; Deraemaeker et al., 2009) which allow large curvature deflections to be measured and actuated.

2.3 Observability and controllability

Part of the control design is to determine the system observability and controllability. For continuous structures where the control actuators are located at a limited number of discrete points on the structure, when a modal decomposition is carried out, the effect of the control actuators appear on the right-hand side of the modal equations multiplied by the modal participation factor. As a result, in this type of modal model of a structural system controllability of a particular mode will depend directly on the associated modal participation factor.

Using a sensor to measure at a discrete point has a similar effect on observability, because the transverse displacement, $w(x,t)$, at a point a is typically approximated as $w(a,t) \approx \sum_{j=1}^{N} \phi_j(a) q_j(t)$, where the $\phi_j(a)$ values are the mode shapes evaluated at point a and the $q_j(t)$ values are the modal coordinates. Although N modes are taken in a truncated model of the continuous system, observability and controllability is related only to the *controlled* part of the system. If, for example, the control objective is just to control the first mode of vibration of the structure, then it is only of interest to know if this mode is controllable or observable. However, the modes other than the controlled modes may still have significant dynamics.

For example, if the controlled modes run from $1, 2, ..., N_c$ and the uncontrolled modes from $N_c + 1, ...\infty$, then when making a measurement at point a, the response is actually given by $w(a,t) \approx \sum_{j=1}^{N_c} \phi_j(a) q_j(t) + \sum_{j=N_c+1}^{\infty} \phi_j(a) q_j(t)$. As a result if the response of the uncontrolled modes is significant the measurement of $w(a,t)$ will be corrupted by their contribution to the measurement. This effect is called observation spillover. A similar effect occurs when the control force is applied at a single point, because in general the modal participation factors are non-zero for the uncontrolled

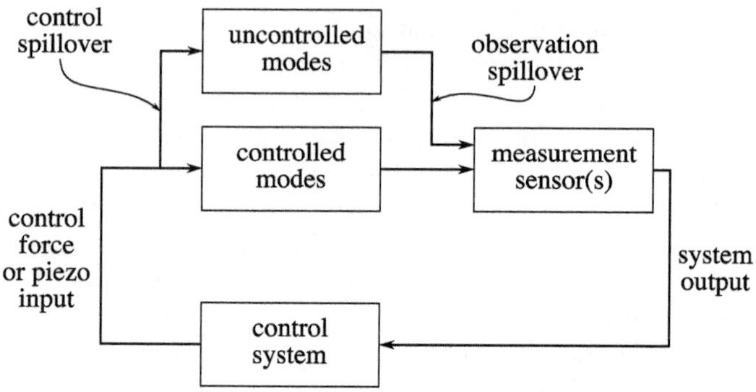

Figure 12. Modal control block diagram.

modes, so the effect of the control force is to excite the uncontrolled modes. This is called control spillover (Balas, 1978, 1998). The scenario is shown as a control block diagram in Figure 12.

The position of the sensors and actuators is important, because for modes with node points, both the mode-shape function, ϕ, and the modal participation factor will be zero at some points along the beam. For linear modal systems, in-depth analysis of the effect of actuator and sensor placement has been developed, see for example Gawronski (2004).

Assume the state vector, $x = [\mathbf{x}^T, \dot{\mathbf{x}}^T]^T$, consists of an equal number of system displacements and velocities, for the controlled modes, and so the vector length is $2N_c$. Then for most systems it will be sufficient to consider the observability and controllability of the underlying linearized system. This can be derived from equation (1), by using the same approach described in Chapter 1 for linearizing about equilibrium points and will typically give a linearized system of the form

$$\dot{x} \approx Ax + Bu,$$

with an output $y \approx Cx$. Then the normal linear conditions for observability and controllability that matrices $\mathcal{O}$ and $\mathcal{R}$ have full rank can be applied Goodwin et al. (2000), where

$$\mathcal{O} = \begin{bmatrix} C \\ CA \\ \vdots \\ CA^{2N-1} \end{bmatrix}, \quad \mathcal{R} = [B \quad AB \quad A^2B \quad \ldots \quad A^{2N-1}B].$$

A more general discussion on controllability and observability in nonlinear

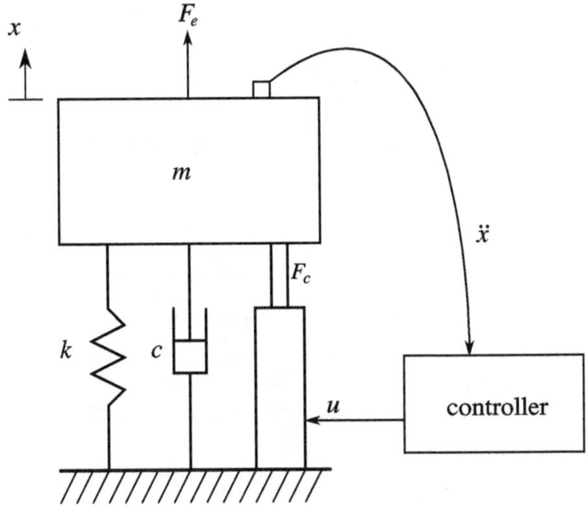

Figure 13. Single-degree-of-freedom oscillator with active vibration control.

systems, particularly those in which the underlying linear system is not necessarily controllable or observable can be found in Nijmeijer and van der Schaft (1990), Isidori (1995), Sastry (1999) and Vidyasagar (2002). The interested reader may also want to investigate the related property of flatness for nonlinear systems, see Fliess et al. (1995) and references therein.

2.4 Control law design

Control law design effectively means how to choose u. Control objectives fall broadly into the two main classes of stabilization or tracking. A stabilization problem is concerned with finding u such that the state vector tends to a stable equilibrium point (typically zero) for any initial conditions and parameter values in the required range. Tracking means getting a state, or output variable, to follow a predefined reference trajectory.

For tracking control, accuracy in replicating the required reference signal and the speed of response are of primary interest. For active vibration control, the amount of vibration reduction is the main performance measure. Cost is important both in terms of implementation and also limiting excessive actuation which will quickly lead to wear and high maintenance costs.

Now consider the system shown in Figure 13. The only feedback to the controller is the acceleration of the mass, measured by an accelerometer. The mass is also acted on by an external excitation force, F_e, which is a disturbance signal. The control force produced by the actuator can be assumed to be $F_c = bu(t)$, where b is a scalar constant. To design an active vibration control law, u, for the mass-spring-damper system we can use direct velocity feedback.

To do this we note that the governing equation of motion for the oscillator is given by

$$m\ddot{x}(t) + c\dot{x}(t) + kx(t) = F_e(t) + bu(t). \tag{2}$$

The control objective is to reduce vibration in the system as much as possible, or in other words, increase the effective damping of the system. To add damping, the control can be directly related to an extra velocity term by integrating the acceleration feedback signal

$$u(t) = -\kappa \int_0^t \ddot{x} \mathrm{d}t,$$

where κ is a control gain parameter. Now equation (2) becomes

$$m\ddot{x}(t) + (c + b\kappa)\dot{x}(t) + kx(t) = F_e(t), \tag{3}$$

which increases the damping by $b\kappa$ and providing both $b > 0$ and $\kappa > 0$ the control will act like additional damping in the linear oscillator, which in turn will reduce the height of the resonance peak.

This type of velocity feedback is sometimes called integral acceleration feedback and needs careful implementation in practice to avoid noise being amplified at higher frequencies, see for example Preumont (1997). It can be directly applied to nonlinear oscillators with linear damping, such as this

$$m\ddot{x}(t) + c\dot{x}(t) + k_1x(t) + k_2x^2 + k_3x^3 = F_e(t) + F_c(t),$$

to give

$$m\ddot{x}(t) + (c + b\kappa)\dot{x}(t) + k_1x(t) + k_2x^2 + k_3x^3 = F_e(t), \tag{4}$$

which is a forced nonlinear oscillator with increased damping.

3 Stability theory

The basic ideas of dynamic stability were introduced in Chapter 1. This section extends the stability concepts to the case where a system, is subject to a control signal. For more detailed discussion of these topics the interested

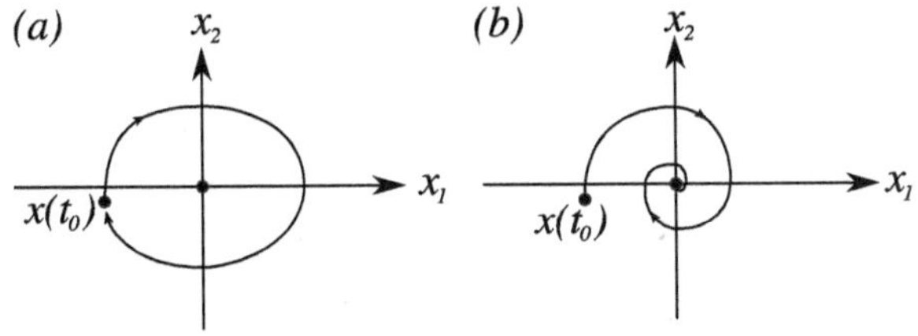

Figure 14. Stability in the phase plane (a) neutral stability, and (b) asymptotic stability.

reader should consult, Slotine and Li (1991), Khalil (1992), Isidori (1995), Krstić et al. (1995), Sastry (1999), Fradkov et al. (1999). An equilibrium point, x^*, is stable if a solution, $x(t)$, starting close to the equilibrium point stays close for all time t, and there are two specific cases of interest: an equilibrium point, x^*, is (i) Lyapunov or neutrally stable if trajectories stay close to it, and (ii) asymptotically stable if nearby trajectories are attracted to it. For systems with two states, such that $x = [x_1, x_2]^T$, where $x_1 = x$ is displacement and $x_2 = \dot{x}$ is velocity, the two types of stability are shown in the phase plane in Figure 14. In both cases the equilibrium point is at the origin.

3.1 Lyapunov functions

A technique for analysing the stability of the system with a single equilibrium point at the origin, is to use a Lyapunov function. To do this first select a Lyapunov function, $V(x,t)$, and then find the time derivative $\dot{V}$. Then $\dot{V}$ gives the rate of change of energy with time and there are three possible cases which relate directly to the stability of the equilibrium point at the origin:

- If $\dot{V}$ is increasing $\rightarrow$ energy increasing $\rightarrow$ unstable
- If $\dot{V}$ zero $\rightarrow$ energy stays the same $\rightarrow$neutrally stable
- If $\dot{V}$ decreasing $\rightarrow$ energy decreasing $\rightarrow$ asymptotically stable

It is important to note that this is only true if V is a positive definite function, such that for $x^* = 0$, $V(0,t) = 0$ and then $V(x,t) > 0$ must hold for all x other than $x = 0$. Schematic phase portraits for the cases of neutral and asymptotic stability are shown in Figure 14 with initial condition $x(t_0)$.

Note also that V is not unique for any particular system. In fact, the main difficulty with using Lyapunov function analysis is deciding which function to select. For oscillators with two states, $x = [x_1, x_2]^T$, a parabolic potential function of the form $V(x) = \frac{x_1^2}{2} + \frac{x_2^2}{2}$, can often be used as a first-

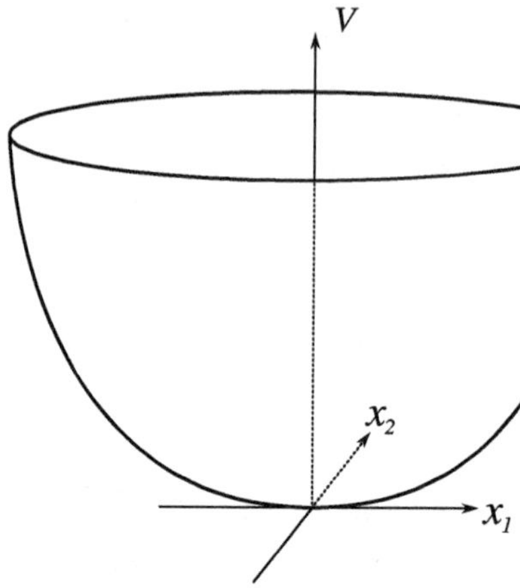

Figure 15. Parabolic Lyapunov function.

guess Lyapunov function. The parabolic function is shown in Figure 15 and, if the energy in the system is represented as a ball rolling on the inside surface of the function, it is easy to see that the ball can only come to rest at the origin. This is because V is positive definite, and therefore has a single unique minimum at the origin.

Consider using a Lyapunov function to assess the stability of the controlled Duffing equation, given by equation (4), for the case when only a single equilibrium point at the origin exists in the system and F_e is zero. The first step is to write the governing equations of motion for the oscillator in first-order form

$$\begin{aligned} \dot{x}_1 &= x_2 \\ \dot{x}_2 &= -\left(\frac{c+b\kappa}{m}\right)x_2 - \frac{k_1}{m}x_1 - \frac{k_2}{m}x_1^2 - \frac{k_3}{m}x_1^3. \end{aligned} \tag{5}$$

The state vector is $x = [x_1, x_2]^T$, and it will be assumed that the Lyapunov function is

$$V(x) = \frac{x_2^2}{2} + \frac{k_1}{m}\frac{x_1^2}{2} + \frac{k_2}{m}\frac{x_1^3}{3} + \frac{k_3}{m}\frac{x_1^4}{4}.$$

differentiating via the chain rule gives

$$\dot{V} = x_2\dot{x}_2 + \frac{k_1}{m}x_1\dot{x}_1 + \frac{k_2}{m}x_1^2\dot{x}_1 + \frac{k_3}{m}x_1^3\dot{x}_1.$$

Then substituting for $\dot{x}_1$ and $\dot{x}_2$ from equation (5) gives

$$\dot{V} = x_2\left(-\left(\frac{c+b\kappa}{m}\right)x_2 - \frac{k_1}{m}x_1 - \frac{k_2}{m}x_1^2 - \frac{k_3}{m}x_1^3\right) + \frac{k_1}{m}x_1x_2 + \frac{k_2}{m}x_1^2x_2 + \frac{k_3}{m}x_1^3x_2,$$

which reduces to

$$\dot{V} = -\left(\frac{c+b\kappa}{m}\right)x_2^2,$$

and providing $(c+b\kappa)/m$ is positive then $\dot{V}$ is always negative definite and the equilibrium point at the origin is asymptotically stable. Note this assumes that the new Lyapunov function is positive definite, which in this example is true.

This example demonstrates how a Lyapunov function analysis can be carried out using an initial guess for V. However, notice that the form of the Lyapunov function, V, is actually the sum of the kinetic energy and the potential energy, all divided by the mass, m, i.e. $V = E_t/m$ where E_t is the kinetic plus potential energy. It is often the case that using energy as a guess for the Lyapunov function analysis is often a good choice for mechanical oscillators. It also possible to guess numerous functions and still not be able to determine the sign of $\dot{V}$, a problem that increases with the complexity of the system. It is also difficult to extend this technique to multi-degree-of-freedom systems.

4 Linearization using feedback

Feedback linearization techniques are designed to linearise the nonlinear system using the feedback control signal. Typically displacement control is required for this approach and signals can be taken from transducers such as an LVDT (linear variable differential transformer). A typical scenario is shown in Figure 16 for a single-degree-of-freedom oscillator with a nonlinear spring.

The governing equations for the system in Figure 16, with $F_e = 0$, can be written as

$$\begin{aligned} \dot{x}_1 &= x_2, \\ \dot{x}_2 &= -\frac{c}{m}x_2 - \frac{k_1}{m}x_1 - \frac{k_3}{m}x_1^3 + \frac{p}{m}u(t), \end{aligned} \tag{6}$$

where $F_c = pu(t)$. This can be written in the matrix form

$$\begin{bmatrix} \dot{x}_1 \\ \dot{x}_2 \end{bmatrix} = \begin{bmatrix} 0 & 1 \\ -\frac{k_1}{m} & -\frac{c}{m} \end{bmatrix}\begin{bmatrix} x_1 \\ x_2 \end{bmatrix} + \begin{bmatrix} 0 \\ -\frac{k_3}{m}x_1^3 \end{bmatrix} + \begin{bmatrix} 0 \\ \frac{p}{m} \end{bmatrix}u(t), \tag{7}$$

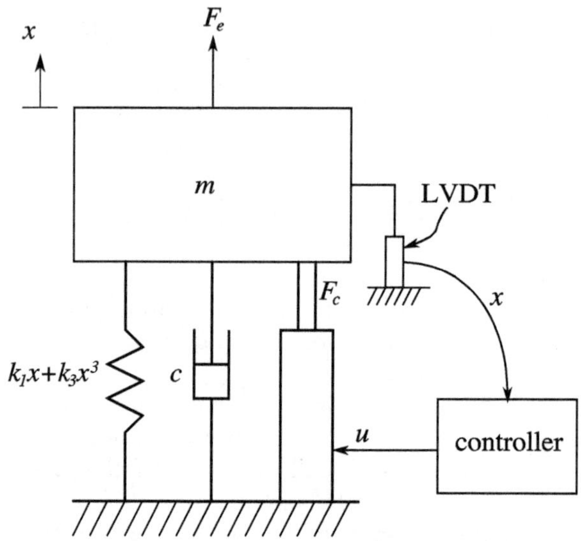

Figure 16. Single-degree-of-freedom oscillator with active vibration control using accelerometer and LVDT.

or in general

$$\dot{x} = Ax + \mathcal{N}(x) + Bu. \tag{8}$$

By inspection of either equation (6) or (7), it can be seen that, by setting $u = \frac{k_3}{p}x_1^3$, then $\mathcal{N}(x) + Bu(t) \to 0$ as $t \to \infty$. The system thereby reduces to $\dot{x} = Ax$, which is linear and, providing A has stable eigenvalues, it is also stable. For the general case, $\mathcal{N}(x)$ can be rewritten as $\mathcal{N}(x) = B\mathcal{N}^*(x)$ and equation (8) becomes

$$\dot{x} = Ax + B\mathcal{N}^*(x) + Bu = Ax + B(\mathcal{N}^*(x) + u). \tag{9}$$

Setting $u = -\mathcal{N}^*$ gives the feedback linearization control signal for the system in the case where A is a stable matrix. This assumes that the system states in the expression $-\mathcal{N}^*$ can be readily accessed for use in the control signal u. So systems which can be expressed in the form of equation (9) can be linearized using the feedback control signal.

The control signal can also include an additional control task, like adding damping. For example, $u = \frac{k_3}{p}x_1^3 - \frac{\kappa}{m}x_2$ means that $\mathcal{N}(x) + Bu(t) \to [0, -\frac{\kappa}{m}x_2]^T$ as $t \to \infty$, and the damping increases by $\frac{\kappa}{m}$. In general, the control input is chosen as $u = -\mathcal{N}^*(x) + c(x)$, where $c(\bullet)$ is the desired control function.

If the underlying linear system happens to be unstable, then the control function needs to be designed to provide a stable linear control after the nonlinear terms have been removed. Now consider an example of an oscillator with nonlinear damping. We will use feedback linearization to remove the nonlinear damping terms in the following nonlinear oscillator

$$m\ddot{x} + c\dot{x}(1 + \delta x) + kx = pu(t),$$

where $u(t)$ is the control input.

First, write the governing equations of motion for the oscillator in the form $\dot{x} = A(x,t) + \mathcal{N}(x) + Bu$, to give

$$\begin{bmatrix} \dot{x}_1 \\ \dot{x}_2 \end{bmatrix} = \begin{bmatrix} 0 & 1 \\ -\dfrac{k}{m} & -\dfrac{c}{m} \end{bmatrix} \begin{bmatrix} x_1 \\ x_2 \end{bmatrix} + \begin{bmatrix} 0 \\ -\dfrac{c\delta}{m} x_1 x_2 \end{bmatrix} + \begin{bmatrix} 0 \\ \dfrac{p}{m} \end{bmatrix} u(t).$$

The state vector is $x = [x_1, x_2]^T$, and u is the control input.

The control matrix is $B = [0, \frac{p}{m}]^T$, and the objective is to put the problem into the form of equation (9). This can be achieved by setting $\mathcal{N}^* = -\frac{c\delta}{p} x_1 x_2$, and to linearize the system set $u = \frac{c\delta}{p} x_1 x_2$. Additional damping can be obtained by setting $u = \frac{c\delta}{p} x_1 x_2 - \frac{\kappa}{p} x_2$.

A numerical simulation of the example is shown in Figure 17, with numerical parameters $m = 1$, $k = 5$, $c = 0.1$, $\delta = 30$ and $p = 10$. In each case the system is uncontrolled but forced sinusoidally with forcing $0.5\sin(1.2t)$ until time $t = 15$ seconds, when the feedback linearization control is switched on. In Figure 17 (a) and (b) the case of feedback linearization response on its own is shown. A clear change can be seen from the distorted non-harmonic response occurring before 15 seconds to a harmonic response afterwards. In Figure 17 (c) and (d) the case of feedback linearization response with additional viscous damping is shown. The additional viscous damping parameter in this simulation is $\kappa = 10.0$. In this case, after 15 seconds the vibrations are significantly more damped than in the previous case.

4.1 Input-output linearization

The main idea of input-output linearization is to obtain a relationship between the output and the input by repeatedly differentiating the output until the input appears. For example, in the case of the system in Figure 16, with governing equation equation (6) assume the output is the displacement, x_1, so that $y = h(x) = x_1$. Differentiating output y with respect to time t gives $\dot{y} = \dot{x}_1 = x_2$ from equation (6). But the input, u, does not appear, so

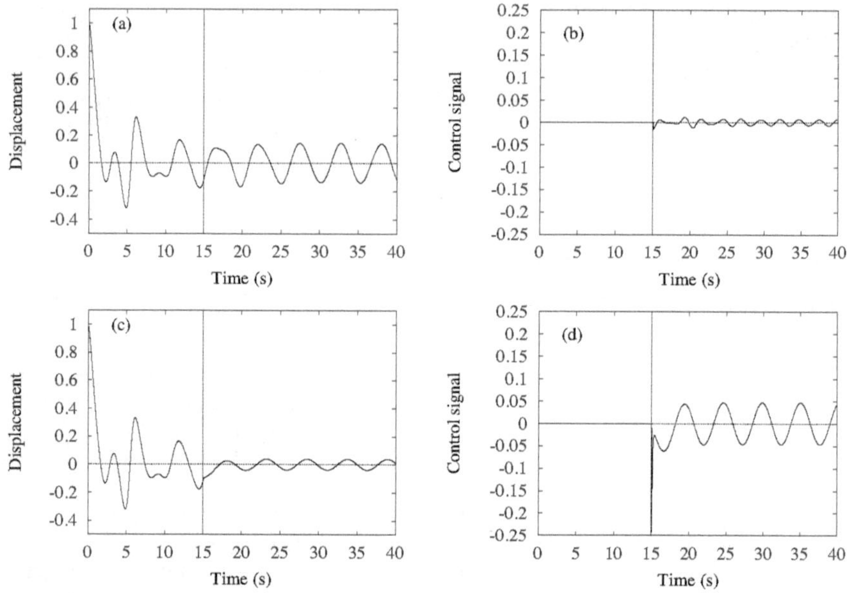

Figure 17. Feedback linearization for oscillator with nonlinear damping, with control starting at time $t = 15$ s; (a) and (b) show feedback linearization response, (c) and (d) show feedback linearization response with added damping.

differentiate again to give

$$\ddot{y} = \dot{x}_2 = -\frac{c}{m}x_2 - \frac{k_1}{m}x_1 - \frac{k_3}{m}x_1^3 + \frac{p}{m}u(t),$$

which gives a relationship between the second derivative of the output (which is the acceleration $\ddot{y}$) and the control input u. It can be seen that choosing

$$u(t) = \frac{m}{p}\left(v(t) + \frac{c}{m}x_2 + \frac{k_1}{m}x_1 + \frac{k_3}{m}x_1^3\right),$$

leads to a linear input-output relationship

$$\ddot{y} = \dot{x}_2 = v(t),$$

where $v(t)$ is an input signal which can be chosen to achieve an additional control task.

A more systematic approach is to consider how the output from the nonlinear system $y = h(x)$ varies with time. Then, differentiating the output with respect to time gives

$$\dot{y} = \frac{\partial h(x)}{\partial x}\frac{\partial x}{\partial t} \Rightarrow \frac{\partial h(x)}{\partial x}\dot{x}, \tag{10}$$

where $\frac{\partial h(x)}{\partial x}$ is an $1 \times N$ row vector and $\frac{\partial x}{\partial t}$ is a $N \times 1$ column vector. equation 10 means that the rate of change of the output with time can be expressed as the rate of change of the output with the state multiplied by the system velocity vector. Substituting for $\dot{x}$ from equation (1) gives

$$\dot{y} = \frac{\partial h(x)}{\partial x}(f(x) + g(x)u) = \frac{\partial h(x)}{\partial x}f(x) + \frac{\partial h(x)}{\partial x}g(x)u,$$

which can be rewritten as

$$\dot{y} = L_f h(x) + L_g h(x)u, \tag{11}$$

where $L_f h(x)$ and $L_g h(x)$ are the Lie derivatives of $h(x)$ with respect to $f(x)$ and $g(x)$. Effectively, the Lie derivatives are the the directional derivative of the output function, $h(x)$, along the vector fields $f(x)$ and $g(x)$.

Now, by choosing the control to remove the system dynamics and replace them with a new control signal, $v(t)$, results in a control input of the form

$$u = \frac{1}{L_g h(x)}(v(t) - L_f h(x)), \qquad L_g h(x) \neq 0, \tag{12}$$

to give $\dot{y} = v(t)$. This gives a linear relationship between the new input $v(t)$ and the derivative of the output $\dot{y}$.

The number of times the equations need to be differentiated to get an input-output relation corresponds to the *relative degree* of the system. In other words, if the condition $L_g h(x) \neq 0$ is true, the system is said to have relative degree one and no more differentiation is required. However, if the output does not appear directly in the expression $L_g h(x) = 0$, the Lie derivative process needs to be iterated until it does (Slotine and Li, 1991).

Let us now consider using input-output linearization to linearize the following nonlinear oscillator

$$m\ddot{x} + c\dot{x}(1 + \delta x^2) + kx = pu(t), \tag{13}$$

where $u(t)$ is the control input. If we assume that the output is the displacement so that $y = x$, then we first, write the governing equations of motion for the oscillator in the form $\dot{x} = f(x) + g(x)u$, to give

$$\begin{bmatrix} \dot{x}_1 \\ \dot{x}_2 \end{bmatrix} = \begin{bmatrix} x_2 \\ -\frac{c}{m}x_2(1+\delta x_1^2) + \frac{k}{m}x_1 \end{bmatrix} + \begin{bmatrix} 0 \\ \frac{p}{m} \end{bmatrix} u(t).$$

The state vector is $x = [x_1, x_2]^T$, and u is the control input. The other vectors are

$$f(x) = \begin{bmatrix} x_2 \\ -\frac{c}{m}x_2(1+\delta x_1^2) + \frac{k}{m}x_1 \end{bmatrix}, \qquad g(x) = \begin{bmatrix} 0 \\ \frac{p}{m} \end{bmatrix}.$$

To compute the Lie derivative, the first step is to compute $\frac{\partial h(x)}{\partial x}$, which in this case with $h(x) = x_1$ gives

$$\frac{\partial h(x)}{\partial x} = [1,\ 0].$$

Note that this is a row vector. Now the Lie derivatives can be computed

$$L_f h(x) = [1, 0] \begin{bmatrix} x_2 \\ -\frac{c}{m}x_2(1+\delta x_1^2) + \frac{k}{m}x_1 \end{bmatrix} = x_2, \qquad L_g h(x) = [1, 0] \begin{bmatrix} 0 \\ \frac{p}{m} \end{bmatrix} = 0.$$

As $L_g h(x) = 0$, the Lie derivative process needs to be repeated.

To do this, first compute the derivative of $L_f h(x)$ giving

$$\frac{\partial L_f h(x)}{\partial x} = [0,\ 1].$$

Then compute the second Lie derivative to give

$$L_f^2 h(x) = [0,\ 1] \begin{bmatrix} x_2 \\ -\frac{c}{m}x_2(1+\delta x_1^2) - \frac{k}{m}x_1 \end{bmatrix} = -\frac{c}{m}x_2(1+\delta x_1^2) - \frac{k}{m}x_1,$$

and

$$L_g L_f h(x) = [0,\ 1] \begin{bmatrix} 0 \\ \frac{p}{m} \end{bmatrix} = \frac{p}{m}.$$

So the system is relative degree two, and, as $L_g L_f h(x) \neq 0$, a control input can be formed using

$$u = \frac{1}{L_g L_f h(x)}(v(t) - L_f^2 h(x)), \qquad L_g L_f h(x) \neq 0. \tag{14}$$

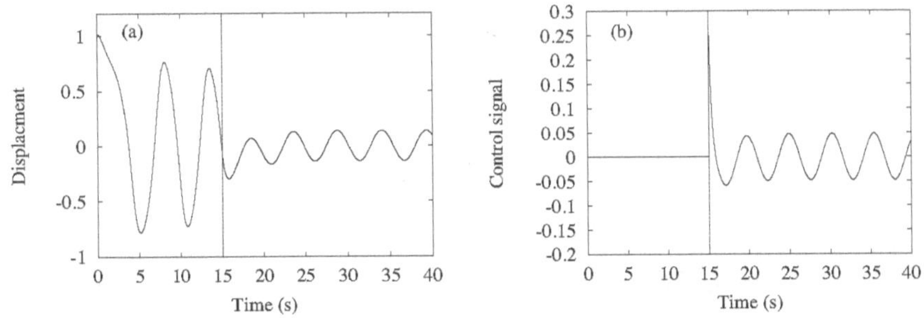

Figure 18. Feedback linearization for input-output example, with control starting at time $t = 15$ s and $v(t) = -x_1 - 3x_2$ (a) shows x_1 response and (b) shows u.

to give

$$u = \frac{m}{p}\left(v(t) + \frac{c}{m}x_2(1 + \delta x_1^2) + \frac{k}{m}x_1\right). \tag{15}$$

When this control signal is used the input-output relationship becomes $\ddot{y} = v(t)$, where $v(t)$ is a new control signal which can be defined to give the required linear behaviour.

A numerical simulation of this example is shown in Figure 18, with numerical parameters $m = k = 1$, $c = 0.1$, $\delta = 30$ and $p = 10$. In this case the system is Eqn. 13 plus sine wave forcing $0.5\sin(1.2t)$. The system is uncontrolled until time $t = 15$ seconds, when the input-output linearization control given by Eqn. 15 is switched on. The control objective is to eliminate nonlinear vibration, so $v(t)$ is chosen to be $v(t) = -x_1 - 3x_2$, which is a stable linear oscillator with large damping. The result is a very sudden reduction in displacement response, leaving just sinusoidal linear response, after the control is turned on.

Note we have not discussed the concept of zero dynamics which can cause problems for these controllers. This is not a typical situation for vibration control, but the interested reader can find a detailed treatment of this and other related issues in Slotine and Li (1991), Khalil (1992) and Sastry (1999).

4.2 Adaptive feedback linearization

It was shown in Chapter 1 how adaptive control can be applied to a linear system. Now we consider nonlinear equations in the form of equation (9),

$$\dot{x} = Ax + B(\mathcal{N}^*(x) + u),$$

where setting $u = -\mathcal{N}^*$ gives the feedback linearization control signal for the system in the case when A is a stable matrix. If the parameters in $\mathcal{N}^*$ are time varying or otherwise difficult to identify, an adaptive control law can be devised which allows variation over time, such that $\mathcal{N}^* + u = 0$ for all time, despite the uncertainty in $\mathcal{N}^*$. To do this, first redefine both $\mathcal{N}^*$ and u as

$$\mathcal{N}^* = \xi^T a^*, \quad u = -\xi^T a(t),$$

where ξ is a vector of nonlinear state terms, such as $x_1^2, x_1 x_2$, etc. a^* is a vector of parameters and $a(t)$ is a vector of time-varying control gains. Substituting these expressions into the governing equation gives

$$\dot{x} = Ax + B(\xi^T a^* - \xi^T a(t)) = Ax + B\xi^T \phi, \tag{16}$$

where $\phi = a^* - a(t)$ is the parameter error, meaning the difference between adaptive gain a_i and the uncertain parameter a_i^* which it is trying to match. Ideally $a \to a^*$ and so $\phi \to 0$, which linearizes the system.

In Chapter 1 we assumed that the control law for the adaptive control gains a were defined by using a proportional plus integral term. Here, we outline how it can be done by using a form of Lyapunov function analysis extended for multi-input, multi-output systems. First consider the following Lyapunov function for the system defined by equation 16

$$V = x^T P x + \phi^T \Gamma^{-1} \phi, \tag{17}$$

where P and Γ are to be defined as part of the control design. Differentiating with respect to time t gives the rate of change of V as

$$\dot{V} = \dot{x}^T P x + x^T P \dot{x} + \dot{\phi}^T \Gamma^{-1} \phi + \phi^T \Gamma^{-1} \dot{\phi}. \tag{18}$$

The expressions for $\dot{x}$ and $\dot{x}^T$ can be substituted from equation 16 and its transpose to give

$$\dot{V} = (x^T A^T + \phi^T \xi B^T) P x + x^T P (Ax + B\xi^T \phi) + \dot{\phi}^T \Gamma^{-1} \phi + \phi^T \Gamma^{-1} \dot{\phi}. \tag{19}$$

Expanding the brackets and gathering terms relating to the matrix A, gives

$$\dot{V} = x^T (PA + A^T P) x + \phi^T \xi B^T P x + x^T P B \xi^T \phi + \dot{\phi}^T \Gamma^{-1} \phi + \phi^T \Gamma^{-1} \dot{\phi}.$$

The matrix sum $(PA + A^T P)$ can be made negative definite by the appropriate choice of P and is usually written as $(PA + A^T P) = -Q$ so that

$$\dot{V} = -x^T Q x + \phi^T \xi B^T P x + x^T P B \xi^T \phi + \dot{\phi}^T \Gamma^{-1} \phi + \phi^T \Gamma^{-1} \dot{\phi}.$$

Now choosing $\dot{\phi} = -\Gamma\xi B^T P x$ results in

$$\dot{V} = -x^T Q x, \tag{20}$$

which is negative definite with the appropriate choice of P. Note that this result also assumes that $P^T = P$ and $\Gamma^T = \Gamma$.

This shows what choice of $\dot{\phi}$ leads to a stable system, but the adaptive control law for $a(t)$ still needs to be defined. Integrating the expression for $\dot{\phi}$ gives

$$\phi = -\int_0^t \Gamma\xi B^T P x \mathrm{d}x = a^* - a(t),$$

so the time-varying adaptive gain is given by

$$a(t) = a^* + \int_0^t \Gamma\xi B^T P x \mathrm{d}x. \tag{21}$$

In other words, the adaptive gain is the initial value a^* (or an initial estimation) plus the variation due to the changes in the parameters. The variation is an integral gain expression which involves the states, via ξ and x which are assumed to be accessible. The matrix Γ can be selected as part of the control design, and can be considered to be a control gain (sometimes also called adaptive weightings) matrix, in which the amount of adaptive effort can be selected by the control designer.

Consider the example where we have to use adaptive feedback linearization to linearize the following nonlinear oscillator

$$m\ddot{x} + c\dot{x}(1 + \delta x) + kx + \mu x_1^3 = bu(t),$$

where $u(t)$ is the control input and both δ and μ are uncertain parameters. Assume that the mass, $m = 1$ kg, stiffness, $k = 1$ N/m^2 and damping, $c = 0.1$ Ns/m. The control gain has the value $b = 10$ N/volt. Both δ and μ have some uncertainty and initial estimated values can be assumed to be $\delta = 3$ and $\mu = 7$

First, write the governing equations of motion for the oscillator in the form $\dot{x} = A(x,t) + B(\mathcal{N}^*(x) + u)$, to give

$$\begin{bmatrix} \dot{x}_1 \\ \dot{x}_2 \end{bmatrix} = \begin{bmatrix} 0 & 1 \\ -\dfrac{k}{m} & -\dfrac{c}{m} \end{bmatrix} \begin{bmatrix} x_1 \\ x_2 \end{bmatrix} + \begin{bmatrix} 0 \\ \dfrac{p}{m} \end{bmatrix} \left(-\frac{c\delta}{b} x_1 x_2 - \frac{\mu}{b} x_1^3 + u(t) \right).$$

Where $\mathcal{N}^*$ is defined as $\mathcal{N}^* = -\frac{c\delta}{b} x_1 x_2 - \frac{\mu}{b} x_1^3$. The nonlinear state terms in $\mathcal{N}^*$ are $x_1 x_2$ and x_1^3, so form a vector $\xi = [x_1 x_2, x_1^3]^T$ so that $\mathcal{N}^*$ can be written. To do this, first redefine both $\mathcal{N}^*$ and u as

$$\mathcal{N}^* = \xi^T a^* = [x_1 x_2,\ x_1^3] \begin{bmatrix} -\dfrac{c\delta}{b} \\ \\ -\dfrac{\mu}{b} \end{bmatrix},$$

whilst defining $u = -\xi^T a(t)$ gives an equation in the form of equation (16)

$$\begin{bmatrix} \dot{x}_1 \\ \dot{x}_2 \end{bmatrix} = \begin{bmatrix} 0 & 1 \\ -\dfrac{k}{m} & -\dfrac{c}{m} \end{bmatrix} \begin{bmatrix} x_1 \\ x_2 \end{bmatrix} + \begin{bmatrix} 0 \\ \dfrac{b}{m} \end{bmatrix} [x_1 x_2,\ x_1^3] \begin{bmatrix} \phi_1 \\ \phi_2 \end{bmatrix}.$$

The A matrix is stable, since $m, c, k > 0$, and the Lyapunov stability analysis can be satisfied when matrices P and Q can be defined such that $PA + A^T P = -Q$ with P and Q positive definite. In this example this leads to the relationship

$$\begin{bmatrix} p_{11} & p_{12} \\ p_{21} & p_{22} \end{bmatrix} \begin{bmatrix} 0 & 1 \\ -\dfrac{k}{m} & -\dfrac{c}{m} \end{bmatrix} + \begin{bmatrix} 0 & -\dfrac{k}{m} \\ 1 & -\dfrac{c}{m} \end{bmatrix} \begin{bmatrix} p_{11} & p_{12} \\ p_{21} & p_{22} \end{bmatrix} = - \begin{bmatrix} q_{11} & q_{12} \\ q_{21} & q_{22} \end{bmatrix},$$

which, by assuming that $p_{12} = p_{21}$ (a requirement of the Lyapunov analysis), can be written as

$$\begin{bmatrix} -2p_{21}\dfrac{k}{m} & p_{11} - p_{21}\dfrac{c}{m} - p_{22}\dfrac{k}{m} \\ p_{11} - p_{21}\dfrac{c}{m} - p_{22}\dfrac{k}{m} & 2p_{21} - 2p_{22}\dfrac{c}{m} \end{bmatrix} = - \begin{bmatrix} q_{11} & q_{12} \\ q_{21} & q_{22} \end{bmatrix}.$$

For Q to be positive definite the conditions are that $q_{11} > 0$ and $q_{11}q_{22} - q_{21}q_{12} > 0$. This means that $p_{11} < (p_{21}\frac{c}{m} - p_{22}\frac{k}{m})$ and $p_{21} < p_{22}\frac{c}{m}$. For P to be positive definite, the conditions are that $p_{11} > 0$ and $p_{11}p_{22} - p_{21}p_{12} > 0$.

So selecting $p_{21} = 0.025$ gives $q_{11} = 0.05$, then choosing $p_{22} = 0.5$ gives $q_{22} = 0.05$. Finally, selecting $p_{11} = 0.5$ gives $q_{21} = q_{12} = 0.0025$, and the Lyapunov criteria, equation (20) is satisfied.

Now the adaptive control law can be found from equation (21). This gives

$$\begin{bmatrix} a_1 \\ a_2 \end{bmatrix} = \begin{bmatrix} a_1^* \\ a_2^* \end{bmatrix} + \int_0^t \begin{bmatrix} \gamma_{11} & 0 \\ 0 & \gamma_{22} \end{bmatrix} \begin{bmatrix} x_1 x_2 \\ x_1^3 \end{bmatrix} \left[0, \frac{b}{m}\right] \begin{bmatrix} 0.5 & 0.025 \\ 0.025 & 0.5 \end{bmatrix} \begin{bmatrix} x_1 \\ x_2 \end{bmatrix} \mathrm{d}x, \tag{22}$$

where the control gain matrix has been taken as diagonal, which satisfies the condition that $\Gamma^T = \Gamma$. This expression defines the adaptive gains, and

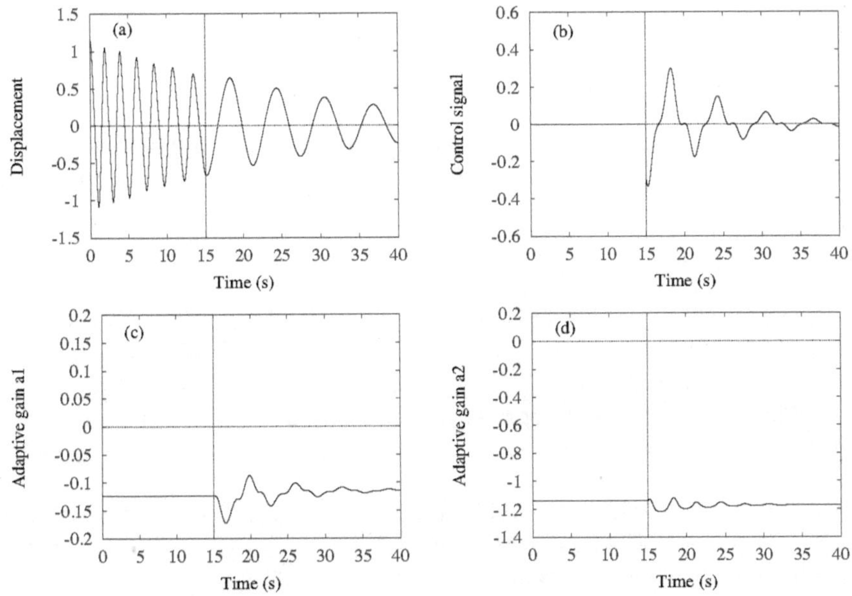

Figure 19. Adaptive feedback linearization example, with control starting at time $t = 15$ s; (a) shows the displacement response (b) shows the control signal, (c) and (d) show the adaptive gains.

the weightings γ_{11} and γ_{22} can be chosen by the control designer to give the required adaptive effort.

A numerical simulation of this is shown in Figure 19, with numerical parameters $m = k = 1$, $c = 0.1$, $\delta = 13$, $\mu = 12$ and $b = 10$. The system is uncontrolled until time $t = 15$ seconds, when the input-output linearization control is switched on. The uncertain parameters have 5% error in the initial value plus a 30% sinusoidal fluctuation. The adaptive weightings are selected as $\gamma_{11} = 0.05$ and $\gamma_{22} = 0.2$. This leads to a rapid adaption when the control is switched on at 15 seconds. As the parameter error is reduced, the adaptive gains become steady state and oscillatory to compensate for the fluctuating parameters. The system is also linearized after $t = 15$.

Note that the adaptive gain values are not unique, and they depend to some extent on the arbitrary choices of the P and Q matrices. More sophisticated adaptive control techniques (which are beyond the scope of this chapter) can be used, so that the gains can be utilised as part of a system identification process. Useful discussions on this are given in Åström and Wittenmark (1995), Slotine and Li (1991), Khalil (1992) and

Fradkov et al. (1999).

5 Control of multi-modal systems

In this section, the case of multi-modal vibrations is introduced, which is the situation most commonly encountered in the practice of vibration engineering.

5.1 Modal control

Transforming vibration problems into a modal space is a key modelling technique, and the basic concept for linear multi-degree-of-freedom systems. A similar approach can be applied to nonlinear systems. Control forces can be included in the modal representation, and, if the control objective is formulated in terms of modal quantities as well, the approach can be called *modal control.*

For example, in Wagg and Neild (2009), assuming proportional damping, it was shown that nonlinear vibrations in a particular class of beams result in modal equations of the form

$$\sum_{i=1}^{N}\sum_{j=1}^{N}\sum_{k=1}^{N} \ddot{q}_j + \zeta_j\omega_{nj}\dot{q}_j + \omega_{nj}^2 q_j + \mu_{ijk} q_i q_j q_k = \alpha_j F_c(t), \tag{23}$$

where $q(t)$ is the modal displacement, ω_{nj} the modal natural frequency, ζ_j the modal damping ratio, μ_{ijk} the nonlinear coefficient, $F_c(t)$ is the control force and α_j is the modal participation factor. The summations over i, j and k and the coefficients μ_{ijk} represent the nonlinear cubic terms, which typically include coupling between the modes.

Consider an example when taking measurements from two sensors shown in the schematic representation in Figure 20. The transverse displacement of the beam is $w(x,t)$, where x is the length along the beam. So the transverse displacement at two points a and b is $w(a,t)$ and $w(b,t)$ respectively. These physical displacements are taken as the control outputs for the system, $y_a = w(a,t)$ and $y_b = w(b,t)$. The outputs are related to the modal displacements q_1 and q_2 by a modal matrix, so that

$$\begin{bmatrix} y_a(t) \\ y_b(t) \end{bmatrix} = \begin{bmatrix} w(a,t) \\ w(b,t) \end{bmatrix} = \begin{bmatrix} \phi_1(a) & \phi_2(a) \\ \phi_1(b) & \phi_2(b) \end{bmatrix} \begin{bmatrix} q_1 \\ q_2 \end{bmatrix},$$

where $\phi(x)$ is the beam mode-shape at point x along the beam. So, assuming a negligible contribution to the response from modes $3, 4, ..., \infty$ (i.e. no observation spillover) the output vector $y = [\Phi]q$, where $[\Phi]$ is the 2×2

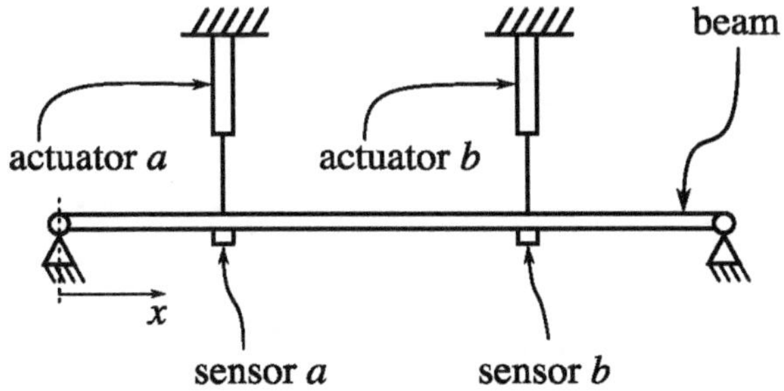

Figure 20. Vibration control of a beam with two collocated actuators and sensors.

modal matrix, and $q = [q_1, q_2]^T$. As a result the modal displacement vector can be estimated directly from $q = [\Phi]^{-1}y$. If it is possible to place the sensors so that $\phi_2(a) \approx 0$ and $\phi_1(b) \approx 0$, then a direct relationship can be obtained between the outputs and the modal displacements. The modal velocities, $\dot{q}_1$ and $\dot{q}_2$ also need to be estimated from the outputs, using the fact that $\dot{y} = [\Phi]\dot{q}$.

Now consider the beam shown in Figure 20 where we want to use feedback linearization control techniques to linearize the two mode nonlinear system defined by

$$\frac{\mathrm{d}}{\mathrm{d}t}\begin{bmatrix} q_1 \\ q_2 \\ \dot{q}_1 \\ \dot{q}_2 \end{bmatrix} = \begin{bmatrix} 0 & 0 & 1 & 0 \\ 0 & 0 & 0 & 1 \\ -\omega_{n1}^2 & 0 & -\zeta_1\omega_{n1} & 0 \\ 0 & -\omega_{n2}^2 & 0 & -\zeta_2\omega_{n2} \end{bmatrix}\begin{bmatrix} q_1 \\ q_2 \\ \dot{q}_1 \\ \dot{q}_2 \end{bmatrix} - \begin{bmatrix} 0 \\ 0 \\ \mu_1 q_1^3 + \delta_1 q_1^2 q_2 \\ \mu_2 q_2^3 + \delta_2 q_2^2 q_1 \end{bmatrix} + \begin{bmatrix} 0 \\ 0 \\ \alpha_1 p_1 \\ 0 \end{bmatrix} u_1 + \begin{bmatrix} 0 \\ 0 \\ 0 \\ \beta_2 p_2 \end{bmatrix} u_2, \tag{24}$$

where δ_1 and δ_2 are constant terms which determine the level of nonlinear cross-coupling between modes 1 and 2. Assume that both observation and control spillover are negligible and that the outputs are the modal displacements $y_1 = q_1$ and $y_2 = q_2$.

First, by inspection of equation (24), it can be seen that setting

$$u_1 = \frac{1}{\alpha_1 p_1}(\mu_1 q_1^3 + \delta_1 q_1^2 q_2) \quad \text{and} \quad u_2 = \frac{1}{\beta_2 p_2}(\mu_2 q_2^3 + \delta_2 q_2^2 q_1)$$

will linearize each mode directly. In fact, additional damping can also be included by using velocity feedback (assuming $\dot{q}_1$ and $\dot{q}_2$ can be measured), in which case the control signals become

$$u_1 = \frac{1}{\alpha_1 p_1}(\mu_1 q_1^3 + \delta_1 q_1^2 q_2 - \kappa_1 \dot{q}_1) \quad \text{and} \quad u_2 = \frac{1}{\beta_2 p_2}(\mu_2 q_2^3 + \delta_2 q_2^2 q_1 - \kappa_2 \dot{q}_2).$$

To obtain an input-output linearization, first take the outputs $y_1 = q_1$ and $y_2 = q_2$, differentiate twice to get the relationship with the control inputs given by

$$\ddot{q}_1 = -\omega_{n1}^2 q_1 - \zeta_1 \omega_{n1} \dot{q}_1 - \mu_1 q_1^3 - \delta_1 q_1^2 q_2 + \alpha_1 p_1 u_1,$$

$$\ddot{q}_2 = -\omega_{n2}^2 q_2 - \zeta_2 \omega_{n2} \dot{q}_2 - \mu_2 q_2^3 - \delta_2 q_2^2 q_1 + \beta_2 p_2 u_2,$$

then choosing

$$u_1 = \frac{1}{\alpha_1 p_1}(v_1(t) + \omega_{n1}^2 q_1 + \zeta_1 \omega_{n1} \dot{q}_1 + \mu_1 q_1^3 + \delta_1 q_1^2 q_2),$$

$$u_2 = \frac{1}{\beta_2 p_2}(v_2(t) + \omega_{n2}^2 q_2 + \zeta_2 \omega_{n2} \dot{q}_2 + \mu_2 q_2^3 + \delta_2 q_2^2 q_1),$$

will give an input-output linearization with the result that $\ddot{q}_1 = v_1(t)$ and $\ddot{q}_2 = v_2(t)$, where $v_1(t)$ and $v_2(t)$ are the new control signals, which can be chosen to give the desired linear system response.

Feedback linearization techniques can be applied to multi-modal systems when the modes are decoupled (or very weakly coupled) via the control forces. The modes themselves can be coupled, as has been shown in the last example, although, to apply this control technique, detailed knowledge of the modal equations is required, and access to all modal states needs to be assumed. Note also that the nonlinear cross-coupling terms typically give rise to nonlinear resonance phenomena, which can dominate the vibration response — see Wagg and Neild (2009) and references therein for details.

6 Morphing structures

The concept of morphing is to get a structure to change between two (or possibly more) different shapes. This shape change is required for the structure to perform a particular function. For example, many space structures are required to be packed into small containers for travel into space, and then they are required to deploy into their operating form. An example is shown in Figs, 21 and 22. Deployment is also required for structures such

Figure 21. The International Space Station (ISS) showing the solar arrays. Photo credit NASA.

as the solar sail shown in Figure 4. In this case the use of tape springs is being developed in order to deploy the sail, see for example Seffen and Pellegrino (1999) and references therein.

The other main morphing application which has been investigated to date is that of control surface morphing for aerospace structures. This type of morphing can be broadly divided into two categories. The fist type is passive morphing, where the control surface is designed such that it will morph (i.e. change shape) in response to specific aerodynamic loads. These methods have received considerable attention for use in morphing aircraft structures (Bharti et al., 2004; Lucato et al., 2004; Spadoni and Ruzzene, 2007; Wang et al., 2007; Thill et al., 2008; Baker and Friswell, 2009; Daynes et al., 2010; Bae et al., 2005). In some of these designs the nonlinear characteristics of particular structures are deliberately exploited, such as bi-stable shells (Wagg et al., 2007). This is still an active area of research, and the passive actuation method is often designed to work in conjunction with an active control element (Daynes et al., 2010). An

Figure 22. Deployment of solar arrays for the ISS. Each of the Solar Array Wings are 34 m long by 12 m and they are folded for compact delivery to space (shown in this figure). Once in orbit, they are deployed. Photo credit NASA.

example of an aircraft which uses morphing is the Boeing X-53A shown in Figure 23. This aircraft has a passive morphing system called the Active Aeroelastic Wing. The objective is to avoid an undesirable phenomena known as aileron reversal which can occur when the aircraft performs roll manoeuvres. When higher aerodynamic forces occur, the control surfaces are automatically deflected into the air stream in a manner that produces favourable wing twist instead of the reduced control generally associated with aileron reversal. This system combines passive morphing with an active control trigger.

In general, when any form of active control is present, the system is said to be an active morphing system. An example is shown in Figure 24 of a morphing aerofoil cross section developed by Continuum Dynamics Inc/Lockheed Martin. This type of aerofoil (or wing) morphing has received considerable attention in the literature (Gern et al., 2002; Bornengo et al., 2005; Coutu and Brailovski, 2009; Hubbard, 2006; Inoyama et al., 2007; Vos et al., 2007a,b; Wickenheiser and Garcia, 2007; Gandhi and Anusonti-

Figure 23. Boeing X-53A Active Aeroelastic Wing. Photo credit Jim Ross/NASA.

Inthra, 2008; Diaconu et al., 2008; Popov et al., 2008; Sofla et al., 2010), with some inspiration from nature, especially the flight of the swift — see for example Lentink et al. (2007).

There are considerable challenges in developing this type of system, not least in implementing a robust actuation system. The use of SMAs has been found to be particularly difficult and alternative methods are also being developed. One of the most popular is to use piezoelectric materials, which have been used extensively in structural control and other smart structure applications (Preumont et al., 2003; Khajepour and Golnaraghi, 1997; Preumont et al., 1992; Schultz and Hyer, 2003; Ashour and Nayfeh, 2002; Chen and Chen, 2004; Zhou and Wang, 2004; Moheimani and Vautier, 2005; Preumont et al., 2005; Song et al., 2006; Moheimani and Fleming, 2006; de Marneffe and Preumont, 2008; Harari et al., 2009). Of particular interest is the increasing development and use of Macro Fibre Composite actuators (MFC) (Song et al., 2006; Deraemaeker et al., 2009) which allow large curvature deflections to be measured and actuated, which is a

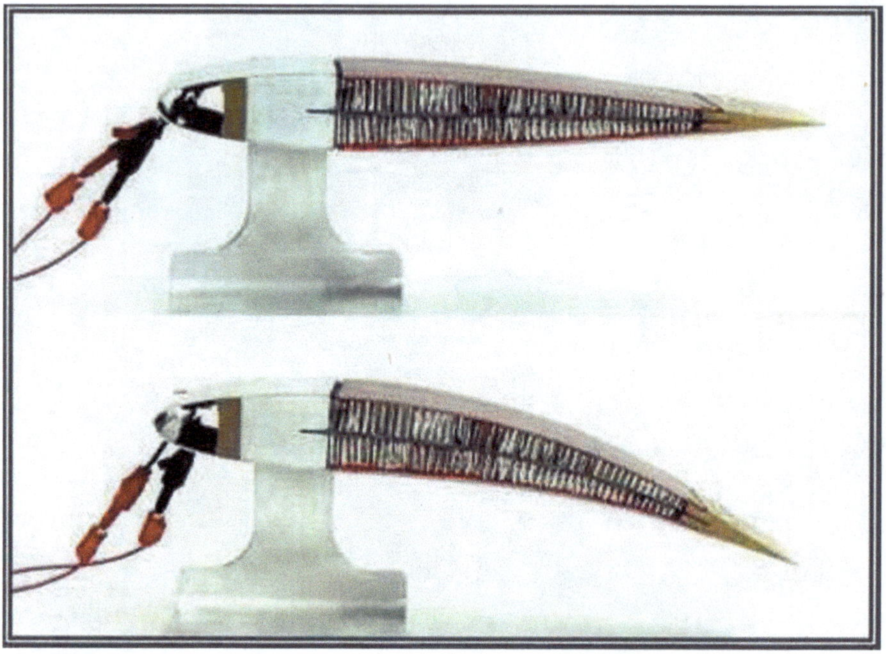

Figure 24. Continuum Dynamics Inc/Lockheed Martin Continuously Deformable shape memory alloy (SMA) wires are used to control the aerofoil cross-section. Photo credit Continuum Dynamics Inc/Lockheed Martin .

particular benefit for morphing structures.

The ultimate aim of active morphing is to create structures with continuous surfaces which can be morphed into multiple shapes for high performance flight control. An example of NASAs concept morphing aircraft is shown in Figure 25. Now we will discuss the concept of snap-through which is being considered as a method for creating morphing structures.

6.1 Snap through mechanisms

A classic engineering example of nonlinear behaviour is the buckling of an axially-loaded (planar) vertical column, as shown for example in Fig 26. In Fig 26 (a) a perfectly straight, planar column is loaded with an axial load p, and the mid-point transverse deflection is q. As the axial load reaches the critical Euler buckling load a pitchfork bifurcation occurs, which is shown in Figure 27 (a) and (b). Figure 27 (a) shows a supercritical pitchfork

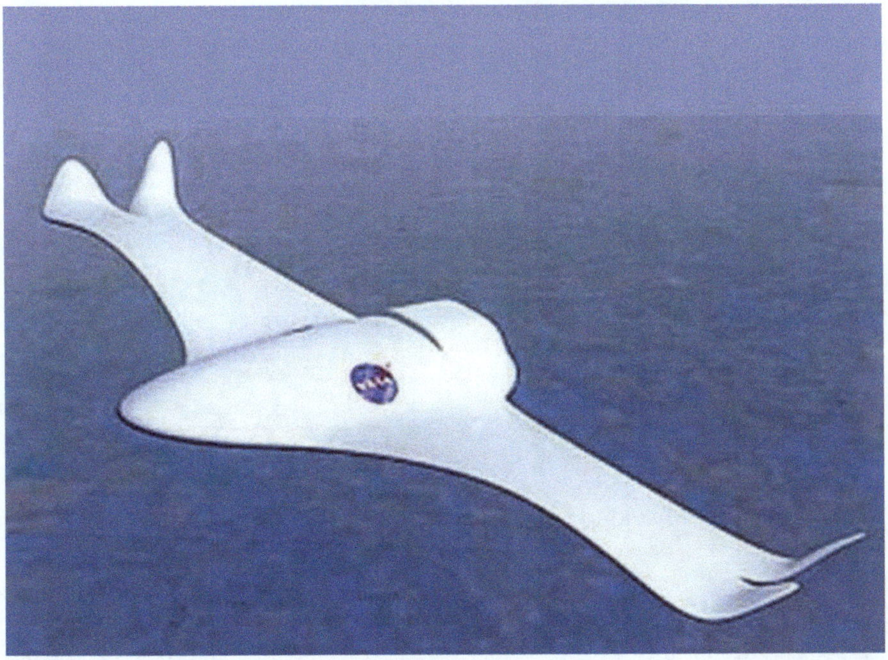

Figure 25. Future active/passive morphing. This artist's impression shows future concepts NASA anticipates for an aircraft of the future. Photo credit NASA.

bifurcation, which corresponds to the physical case when the column adopts a buckled shape but does not collapse — said to be the postbuckled state. Figure 27 (b) shows a subcritical pitchfork bifurcation, which corresponds to the physical case when the column fails catastrophically at the point of bifurcation. The two dashed curves linking the bifurcation point to zero correspond to the collapse solutions to the left or right.

In the supercritical case, Figure 27 (a), after the bifurcation point, the original straight solution becomes unstable (shown as a dashed line) and two stable solutions emerge corresponding to the column buckling either to the left or the right. As the column is perfectly straight, there is an equal chance of the column buckling in either direction. Physically, the column is never perfectly straight, and so the case shown in Fig 26 (b) is for an imperfect column, where the initial imperfection is represented by the deflection ϵ. The initial imperfection means that the column will always buckle in the same direction. The case for positive ϵ is shown in Figure

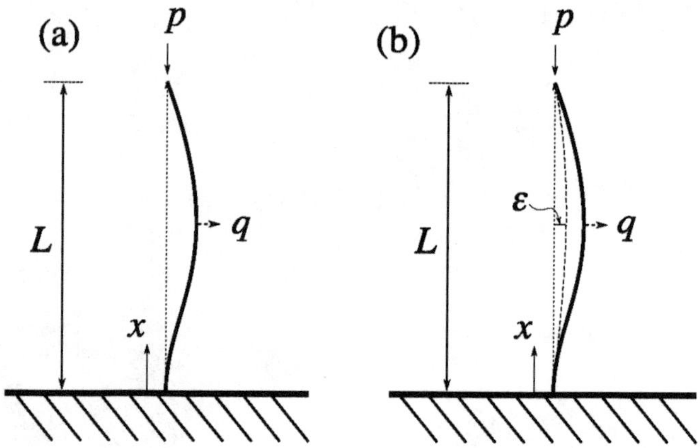

Figure 26. Beam buckling (a) perfect column, (b) column with eccentricity.

27 (c), where it can be seen that there are now two disconnected solution paths. Increasing p from zero always leads to a buckled shape to the right. If the beam is forced into the opposite (left-hand) buckled shape, and loaded above the Euler load, it can be held in this position, for example at point A in Figure 27 (c). Physically, the axial load is holding the beam in the buckled state which is opposite to its initial imperfection. Then if the axial load is decreased, at the Euler load the beam will suddenly snap-through to the other branch of solutions. The point of snap-through is a saddle-node bifurcation where the stable branch joins an unstable branch which corresponds to the original unbuckled solution.

Notice that in Figure 27 the solid lines indicate the paths of the stable equilibrium points (node/spiral) as p is varied and the dashed lines indicate the unstable equilibrium points (saddles). The unstable and stable branches join at the bifurcation point.

The physical system shown in Figure 28 (a) has a geometric nonlinearity due to the angle, θ, of the springs. This type of nonlinearity can be approximated by a Duffing-type oscillator with nonlinear stiffness shown in Figure 28 (b). The equation of motion is given by

$$m\ddot{x} + c\dot{x} - \mu x + \alpha x^3 = 0,$$

where μ and α are coefficients which depend on k, θ and L and c is (added) viscous damping. The derivation of a Duffing oscillator from the snap-

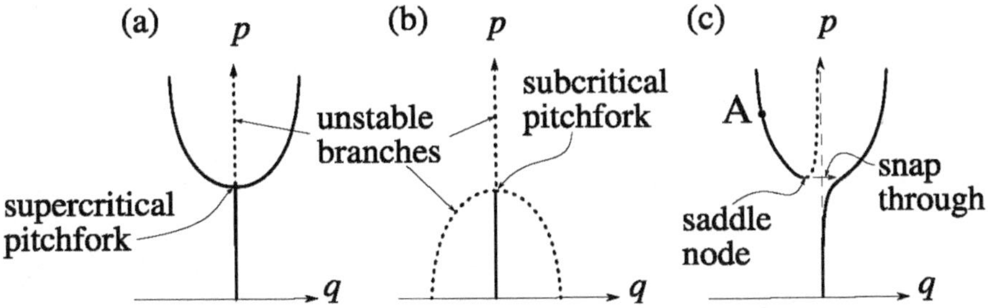

Figure 27. Pitchfork bifurcation (a) supercritical, (b) subcritical, and (c) imperfect.

through system can be found from a number of sources — see for example Virgin (2007).

Consider the case when $m = 1$ and $\alpha = 1$, and we are asked to find the change in behaviour which occurs as the linear stiffness parameter, μ, is varied and changes sign. First, rewriting the system into first-order form

$$\begin{aligned} \dot{x}_1 &= x_2 = f_1, \\ \dot{x}_2 &= \mu x_1 - x_1^3 - c x_2 = f_2. \end{aligned}$$

Then by inspection, the equilibrium points for this system can be found by equating $f_1 = f_2 = 0$ which gives

$\mu < 0$	$x_1 = 0$	$x_2 = 0,$	one equilibrium point
$\mu = 0$	$x_1 = 0$	$x_2 = 0,$	one equilibrium point
$\mu > 0$	$x_1 = 0$	$x_2 = 0$	
and	$x_1 = \pm\sqrt{\mu}$	$x_2 = 0,$	three equilibrium points.

To investigate the behaviour, the system is linearised locally close to the equilibrium points. For all μ values the equilibrium point $x_a^* = (x_1 = 0, x_2 = 0)$ exists. For $\mu > 0$ values, two additional equilibrium points exist and are labelled as $x_{b,c}^* = (x_1 = \pm\sqrt{\mu}, x_2 = 0)$. In general, the Jacobian for the system is

$$D_x f = \frac{\partial(f_1, f_2)}{\partial(x_1, x_2)} = \begin{bmatrix} \dfrac{\partial f_1}{\partial x_1} & \dfrac{\partial f_1}{\partial x_2} \\ \dfrac{\partial f_2}{\partial x_1} & \dfrac{\partial f_2}{\partial x_2} \end{bmatrix} = \begin{bmatrix} 0 & 1 \\ \mu - 3x_1^2 & -c \end{bmatrix}.$$

First for $x_a^* = (x_1 = 0, x_2 = 0)$, the Jacobian becomes

$$D_{x_a^*} f = \begin{bmatrix} 0 & 1 \\ \mu & -c \end{bmatrix}.$$

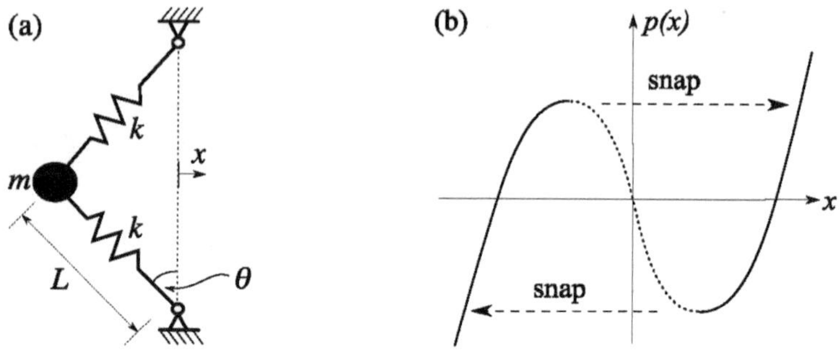

Figure 28. Snap-through system showing, (a) schematic, and (b) nonlinear stiffness function.

So for equilibrium point x_a^*, $\mathrm{tr}(A) = -c$ and $\det(A) = -\mu$.

For equilibrium points $x_{b,c}^* = (x_1 = \pm\sqrt{\mu}, x_2 = 0)$, the Jacobian becomes

$$D_{x_a^*} f = \begin{bmatrix} 0 & 1 \\ -2\mu & -c \end{bmatrix}.$$

So in this case $\mathrm{tr}(A) = -c$ and $\det(A) = 2\mu$.

Note that the expression for $\mathrm{tr}(A)$ and $\det(A)$ are computed assuming that $\mu > 0$. In the case when $\mu < 0$ the sign of μ terms *will change*. So for equilibrium point x_a^* (using Figure 5 from Chapter 1) when

$$\begin{array}{llll} \mu < 0, & \mathrm{tr}(A) = -c, & \det(A) = -(-\mu) = \mu, & \text{stable node/spiral} \\ \mu = 0, & \mathrm{tr}(A) = -c, & \det(A) = 0, & \text{degenerate case} \\ \mu > 0, & \mathrm{tr}(A) = -c, & \det(A) = -\mu, & \text{saddle} \end{array}$$

so this equilibrium point changes from a stable node/spiral to a saddle point as μ passes through zero. In general, for $\mu > 0$ the discriminant is $\Delta = \mathrm{tr}^2 - 4\det = c^2 + 4\mu$. So the μ value at which $\Delta = 0$ is $\mu = -c^2/4$, marking the degenerate node case from Figure 5 from Chapter 1. So for $-c^2/4 > \mu > 0$, x_a^* is a stable node and for $\mu < -c^2/4$ a stable spiral.

For equilibrium points $x_{b,c}^*$ when

$$\begin{array}{llll} \mu < 0, & \text{n/a}, & & \text{no equilibrium point} \\ \mu = 0, & \mathrm{tr}(A) = -c, & \det(A) = 0, & \text{degenerate case} \\ \mu > 0, & \mathrm{tr}(A) = -c, & \det(A) = 2\mu, & \text{stable node/spiral} \end{array}$$

So, for $\mu < 0$, there are no equilibrium points. In this case the discriminant is $\Delta = \mathrm{tr}^2 - 4\det = c^2 - 8\mu$. So the μ value at which $\Delta = 0$ is $\mu = 1/8$,

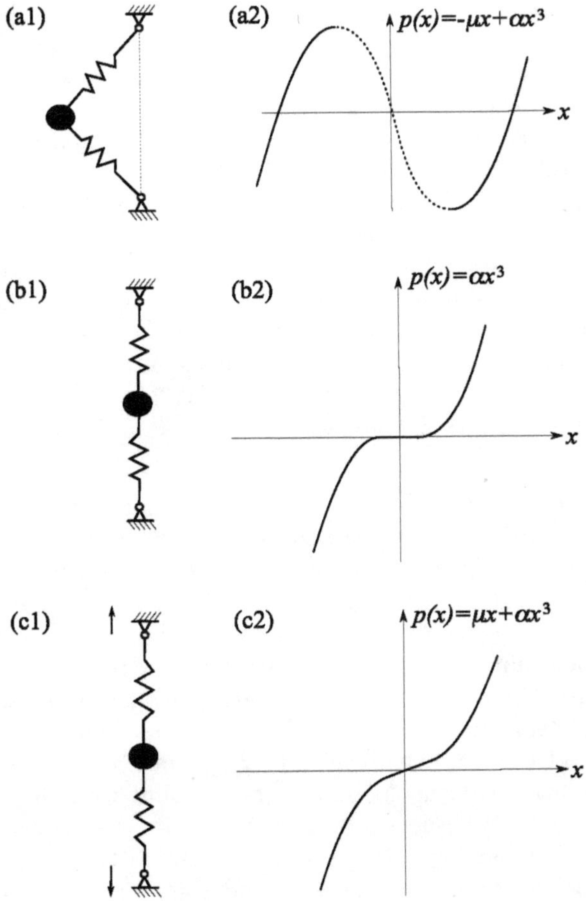

Figure 29. Change in stiffness function as μ varies for this example.

marking the degenerate node case from Figure 5 from Chapter 1. So for $0 < \mu < c^2/8$, $x^*_{b,c}$ is a stable node and for $\mu > c^2/8$ it becomes a stable spiral.

Physically changing μ from negative to positive corresponds to the system in Figure 28 having positive linear stiffness. Geometrically, this corresponds to the case when the support points are moved outwards, such that $\theta \to 0$.

The physical interpretation of the snap-through can be seen from Figure 29, where in (a1) and (a2) the linear stiffness is negative, (b1) and (b2) shows the $\mu = 0$ case and (c1) and (c2) shows the case where the linear stiffness is positive. This corresponds to moving the end supports apart from (a1) to (b1) and finally (c1). In (b1) there is no tension or compression in the springs, whereas in (c1) the springs are in tension.

This type of transition is known as a *cusp bifurcation*, because if plots (a2), (b2) and (c2) are combined into a surface plot with μ as the additional coordinate, then the surface has a cusp at $\mu = 0$. See Thompson (1982) for further details of this phenomena.

By deliberately designing a structure to have snap-through characteristics like (a1) (a2), we can create a hinge mechanism where the system can switch, via snapping, between two different stable states. One way this can be achieved in practice is using composite shells, and this is discussed next.

6.2 Multi-form shell structures

Composite shell structures made from a polymer-matrix, fibre-reinforced composite can be made to be bi-stable (or even multi-stable) by combining laminate design with a manufacturing process which involves cooling from a high temperature. Thermally induced stresses occur during cooling, and as a result of these stresses the plate is a cylindrical shell shape when fully cooled. In fact, the cooled static equilibrium shape has been shown to be the shape which minimizes the potential energy of the laminate, see Hyer (1998) (see also Chap. 7. of Wagg et al. (2007)) during cooling. This process is nonlinear, and the Kirchhoff strain assumptions are taken as an appropriate model for the shell behaviour Wagg and Neild (2009). However, the shell can also have other potential stable configurations. It should be noted that bi-stability is not unique to composite materials, for example steel arches and dome-like shells can have bi-stable behaviour. Typically, there is one other stable configuration and so the shell is said to be *bi-stable*. More generally, composite laminates can be manufactured which have *multi-stable* states.

An example is shown in Figure 30, where a bi-stable plate has been fabricated with a flat plate joined to the left-hand edge to form a rectangular plate-like structure Mattioni et al. (2008). In Figure 30 (a) the bi-stable plate (on the right of the sub-figure) is in the curvature up position, and in Figure 30 (b) the bi-stable plate is in the down (or flatter) position. This idea has been taken a stage further, and used to construct a small scale prototype of a morphing winglet, as shown in Figure 31.

The moment required to make the bi-stable composite change state has been found to be, for example, in the range 1.186–1.243 Nm for a 254 mm $\times$ 254 mm $[90_4/0_4]_T$ specimen, see Schultz and Hyer (2003) (see also Chap.

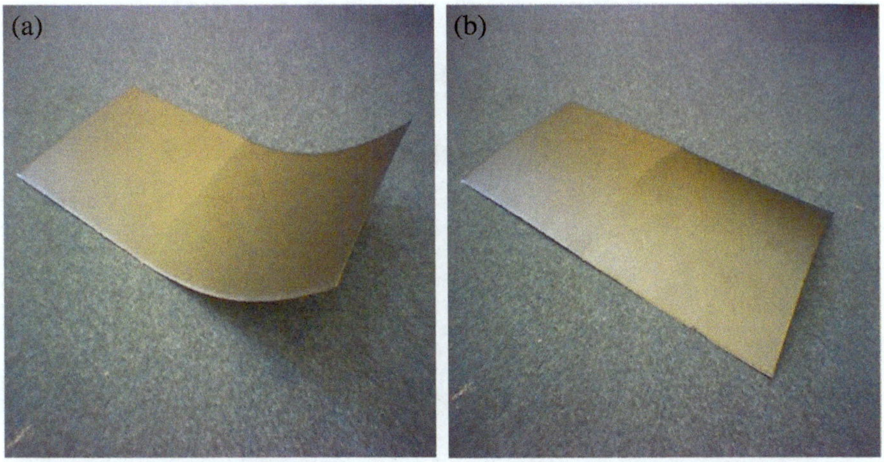

Figure 30. Flat plate and bi-stable plate manufactured together, (a) bi-stable state 1, and (2) bi-stable state 2. Reproduced with kind permission from Mattioni et al. (2006).

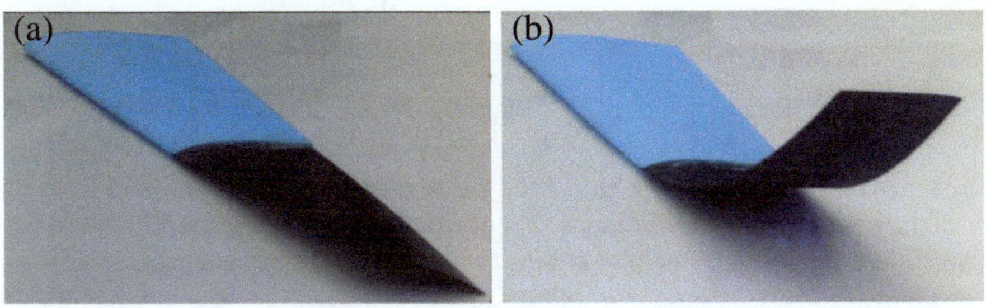

Figure 31. Bi-stable plate applied to a morphing wing concept (a) winglet lowered, and (b) winglet raised. Reproduced with kind permission from Mattioni et al. (2006).

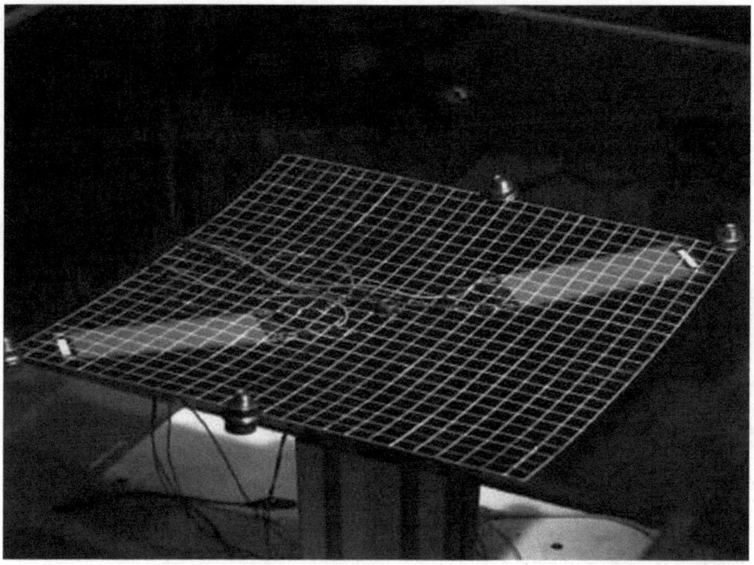

Figure 32. Change in stiffness function as μ varies for this example.

7. of Wagg et al. (2007)). When the change in state occurs, the behaviour is almost identical to the *snap-through* system already discussed, shown in Figure 28. This is because after a certain level of deflection, the sign of the stiffness term appears to suddenly reverse, propelling the system into the other state. This idea of negative linear stiffness is considered in the buckling example, where the snap-through system is modelled as a Duffing oscillator. If required, the model developed for the composite shell, can be applied to the large deflection case which occurs physically with snap-through.

These types of bi-stable materials are being used to create future adaptive structures where multiple states are required. The process of a structure changing from one shape to another is called *morphing*. In some aerospace literature this is used to describe just hinged wing aircraft, but more recently the term has become used more widely to describe any shape change in a structure.

Techniques for actuating bi-stable plates using both piezoelectric actuators and shape memory alloys have been described by in Hyer *et. al.* in Chap. 7. of Wagg et al. (2007). These techniques are designed to overcome the static moment required to change between the two stable states. An example is shown in Figure 32, which shows a bistable plate with two macrofibre composite (MFC) piezo actuators mounted along the diagonal axis of the plate. In this experiment, the MFCs where used to switch the plate from one static stable state to another.

When operating these structures in a dynamic environment, it may be possible to use some of the vibration energy to assist with the state change. It is already known that repeated dynamic snapping (non-periodic) of the laminate can be achieved by forcing it close to a resonance, see Diaz et al. (2007). Deliberately operating near a resonance would significantly reduce the moment required to actuate between states, but the high amplitudes would be disadvantageous at other times. So, in this type of scenario active vibration control would also be required. These and other similar applications of 'smart' structures offer an exciting new set of engineering challenges in which nonlinear vibrations and control will play an major part.

Bibliography

K. J. Åström and B. Wittenmark. *Adaptive Control.* Addison Wesley: Menlo Park, CA, 1995.

M Ahmadian. On the isolation properties of semiactive dampers. *Journal of Vibration & Control*, 5(2):217–232, MAR 1999. ISSN 1077-5463.

N. A. Alexander and F. Schilder. Exploring the performance of a nonlinear tuned mass damper. *Journal of Sound and Vibration*, 319(1-2):445–462, JAN 9 2009. ISSN 0022-460X. doi: 10.1016/j.jsv.2008.05.018.

O. N. Ashour and A. H. Nayfeh. Adaptive control of flexible structures using a nonlinear vibration absorber. *Nonlinear Dynamics*, 28(3-4):309–322, MAY 2002. ISSN 0924-090X.

J. S. Bae, T. M. Seigler, and D. J. Inman. Aerodynamic and static aeroelastic characteristics of a variable-span morphing wing. *Journal of Aircraft*, 42(2):528–534, 2005.

A. Bahar, F. Pozo, L. Acho, J. Rodellar, and A. Barbat. Hierarchical semi-active control of base-isolated structures using a new inverse model of magnetorheological dampers. *Computers & Structures*, 88(7-8):483–496, APR 2010. ISSN 0045-7949. doi: 10.1016/j.compstruc.2010.01.006.

D. Baker and M. I. Friswell. Determinate structures for wing camber control. *Smart Materials & Structures*, 18(3):1–13, MAR 2009. ISSN 0964-1726. doi: 10.1088/0964-1726/18/3/035014.

M. J. Balas. Stable feedback control of linear distributed parameter systems: Time and frequency domain conditions. *Journal of Mathematical Analysis and Applications*, 225(1):144–167, SEP 1 1998. ISSN 0022-247X.

M. J. Balas. Feedback control of flexible systems. *IEEE Transactions on Automatic Control*, 23(4):673–679, 1978. ISSN 0018-9286.

K. A. Bani-Hani and M. A. Sheban. Semi-active neuro-control for base-isolation system using magnetorheological (mr) dampers. *Earthquake Engineering & Structural Dynamics*, 35(9):1119–1144, July 2006.

A. H. Barbat, J. Rodellar, E. P. Ryan, and N. Molinares. Active control of nonlinear base-isolated buildings. *Journal of Engineering Mechanics—ASCE*, 121(6):676–684, JUN 1995. ISSN 0733-9399.

C. F. Beards. *Vibration analysis and control system dynamics.* Ellis Horwood: Chichester, 1981.

F. H. Besinger, D. Cebon, and D. J. Cole. An experimental investigation into the use of semi-active dampers on heavy lorries. *Vehicle System Dynamics*, 20:57–71, 1992.

S Bharti, M Frecker, G Lesieutre, and D Ramrakhyani. Active and passive material optimization in a tendon actuated morphing aircraft structure. In Flatau, AB, editor, *Smart Structures and Materials 2004: Smart Structures and Integrated Systems*, volume 5390 of *(SPIE)*, pages 247–257, 2004. ISBN 0-8194-5307-2. doi: 10.1117/12.540198. Smart Structures and Materials 2004 Conference, San Diego, CA, MAR 15-18, 2004.

D. Bornengo, F. Scarpa, and C. Remillat. Evaluation of hexagonal chiral structure for morphing airfoil concept. *Proceedings of the Institution of Mechanical Engineers Part G-Journal of Aerospace Engineering*, 219(G3):185–192, JUN 2005. ISSN 0954-4100. doi: 10.1243/095441005x30216.

F. Casciati, G. Magonette, and F. Marazzi. *Semiactive devices and applications in vibration mitigation.* Wiley: Chichester, 2006.

G. D. Chen and C. Q. Chen. Semiactive control of the 20-story benchmark building with piezoelectric friction dampers. *Journal of Engineering Mechanics—ASCE*, 130(4):393–400, APR 2004. ISSN 0733-9399.

R. L. Clark, W. R. Saunders, and G. P. Gibbs. *Adaptive structures; dynamics and control.* John Wiley: New York, 1998.

D. J. Cole and D. Cebon. Truck suspension design to minimize road damage. *Proceedings of the Institution of Mechanical Engineers Part D—Journal of Automobile Engineering*, 210(2):95–107, 1996. ISSN 0954-4070.

D. Coutu and P. Brailovski, V.and Terriault. Promising Benefits of an Active-Extrados Morphing Laminar Wing. *Journal of Aircraft*, 46(2): 730–731, MAR-APR 2009. ISSN 0021-8669. doi: 10.2514/1.40657.

S. Daynes, S. J. Nall, P. M. Weaver, K. D. Potter, P. Margaris, and P. H. Mellor. Bistable Composite Flap for an Airfoil. *Journal of Aircraft*, 47 (1):334–338, JAN-FEB 2010. ISSN 0021-8669. doi: 10.2514/1.45389. AIAA/ASME/ASCE/AHS/ASC 50th Structures, Structural Dynamics and Materials Conference, Palm Springs, CA, MAY 02-07, 2009.

B. de Marneffe and A. Preumont. Vibration damping with negative capacitance shunts: theory and experiment. *Smart Materials & Structures*, 17 (3), JUN 2008. ISSN 0964-1726. doi: 10.1088/0964-1726/17/3/035015.

J. P. Den Hartog. *Mechanical Vibrations*. McGraw-Hill: New York, 1934.

A. Deraemaeker, H. Nasser, A. Benjeddou, and A. Preumont. Mixing Rules for the Piezoelectric Properties of Macro Fiber Composites. *Journal of Intelligent Material Systems & Structures*, 20(12):1475–1482, AUG 2009. ISSN 1045-389X. doi: 10.1177/1045389X09335615.

C. G. Diaconu, P. M. Weaver, and F. Mattioni. Concepts for morphing airfoil sections using bi-stable laminated composite structures. *Thin-Walled Structures*, 46(6):689–701, JUN 2008. ISSN 0263-8231. doi: 10.1016/j.tws.2007.11.002.

A. F. Arrieta Diaz, F. Mattioni, S. A. Neild, P. M. Weaver, D. J. Wagg, and K. Potter. Nonlinear dynamics of a bi-stable composite laminate plate with applications to adaptive structures. In M. R. Vetrano and G. Degrez, editors, *Proceedings of the Second European Conference for Aerospace Sciences*, number 4.03.01, Brussels, July 2007. 2nd European Conference for Aerospace Sciences.

I. J. Fialho and G. J. Balas. Design of nonlinear controllers for active vehicle suspensions using parameter-varying control synthesis. *Vehicle System Dynamics*, 33(5):351–370, 2000. ISSN 0042-3114.

M. Fliess, J. L. Lévine, P. Martin, and P. Rouchon. Flatness and defect of non-linear systems: introductory theory and examples. *International Journal of Control*, 61(6):1327–1361, 1995.

A. L. Fradkov, I. M. Miroshnik, and V. O. Nikiforov. *Nonlinear and adaptive control of complex systems*. Kluwer: Dordrecht, 1999.

C. R. Fuller, S. J. Elliot, and P. A. Nelson. *Active control of vibration*. Academic Press, 1996.

F. Gandhi and P. Anusonti-Inthra. Skin design studies for variable camber morphing airfoils. *Smart Materials & Strucures*, 17(1), FEB 2008. ISSN 0964-1726. doi: 10.1088/0964-1726/17/01/015025.

W. K. Gawronski. *Advanced Structural Dynamics and Active Control of Structures*. Springer-Verlag: New York, 2004.

F. H. Gern, D. J. Inman, and R. K. Kapania. Structural and aeroelastic modeling of general planform wings with morphing airfoils. *AIAA Journal*, 40(4):628–637, APR 2002. ISSN 0001-1452.

Apama Ghosh and Biswajit Basu. Seismic vibration control of nonlinear structures using the liquid column damper. *Journal of Structural Engineering—ASCE*, 134(1):146–153, JAN 2008. ISSN 0733-9445. doi: 10.1061/(ASCE)0733-9445(2008)134:1(146).

N Giorgetti, A Bemporad, HE Tseng, and D Hrovat. Hybrid model predictive control application towards optimal semi-active suspension. *International Journal of Control*, 79(5):521–533, MAY 2006. ISSN 0020-7179. doi: 10.1080/00207170600593901.

G. C. Goodwin, S. F. Graebe, and M. E. Salgado. *Control System Design*. Pearson, 2000.

S. Harari, C. Richard, and L. Gaudiller. New Semi-active Multi-modal Vibration Control Using Piezoceramic Components. *Journal of Intelligent Material Systems & Structures*, 20(13):1603–1613, SEP 2009. ISSN 1045-389X. doi: 10.1177/1045389X09102561.

K. S. Hong, H. C. Sohn, and J. K. Hedrick. Modified skyhook control of semi-active suspensions: A new model, gain scheduling, and hardware-in-the-loop tuning. *Journal of Dynamic Systems Measurement and Control-Transactions of the ASME*, 124(1):158–167, March 2002.

G. W. Housner, L. A. Bergman, T. K. Caughey, A. G. Chassiakos, R. O. Claus, S. F. Masri, R. E. Skelton, T. T. Soong, B. F. Spencer, and J. T. P. Yao. Structural control: Past, present, and future. *Journal of Engineering Mechanics—ASCE*, 123(9):897–971, SEP 1997. ISSN 0733-9399.

D Hrovat. Survey of advanced suspension developments and related optimal control applications. *Automatica*, 33(10):1781–1817, OCT 1997. ISSN 0005-1098.

J. E. Hubbard. Dynamic shape control of a morphing airfoil using spatially distributed transducers. *Journal of Guidance Control and Dynamics*, 29 (3):612–616, MAY-JUN 2006. ISSN 0731-5090.

M. W. Hyer. *Stress analysis of fibre-reinforced composite materials*. McGraw Hill, 1998.

D. J. Inman. *Vibration with control*. Wiley: Chichester, 2006.

D. Inoyama, B. P. Sanders, and J. J. Joo. Topology synthesis of distributed actuation systems for morphing wing structures. *Journal of Aircraft*, 44 (4):1205–1213, JUL-AUG 2007. ISSN 0021-8669. doi: 10.2514/1.25535.

A. Isidori. *Nonlinear Control Systems*. Springer, 1995.

N Jalili. A comparative study and analysis of semi-active vibration-control systems. *Journal of Vibration & Acoustics-Transactions of the ASME*, 124(4):593–605, OCT 2002. ISSN 1048-9002. doi: 10.1115/1.1500336.

D. Karnopp. Active and semiactive vibration isolation. *Journal of Mechanical Design*, 117(Sp. Iss. B):177–185, JUN 1995. ISSN 1050-0472.

K. Kasai, Y. M. Fu, and A. Watanabe. Passive control systems for seismic damage mitigation. *Journal of Structural Engineering—ASCE*, 124(5): 501–512, MAY 1998. ISSN 0733-9445.

A Khajepour and MF Golnaraghi. Experimental control of flexible-structures using nonlinear modal coupling: Forced and free vibration. *Journal of Intelligent Materials Systems & Structures*, 8(8):697–710, AUG 1997. ISSN 1045-389X.

H. K. Khalil. *Nonlinear Systems.* Macmillan: New York, 1992.

K. J. Kitching, D. J. Cole, and D. Cebon. Performance of a semi-active damper for heavy vehicles. *Journal of Dynamic Systems Measurement & Control—Transactions of the ASME*, 122(3):498–506, SEP 2000. ISSN 0022-0434.

Miroslav Krstić, Ioannis Kanellakopoulos, and Petar Kokotović. *Nonlinear and adaptive control design.* John Wiley, 1995.

D. Lentink, U. K. Muller, E. J. Stamhuis, R. de Kat, W. van Gestel, L. L. M. Veldhuis, P. Henningsson, A. Hedenstrom, J. J. Videler, and J. L. van Leeuwen. How swifts control their glide performance with morphing wings. *Nature*, 446(7139):1082–1085, APR 26 2007. ISSN 0028-0836. doi: 10.1038/nature05733.

D. J. Leo. *Smart material systems.* Wiley: New York, 2007.

Y Liu, TP Waters, and MJ Brennan. A comparison of semi-active damping control strategies for vibration isolation of harmonic disturbances. *Journal of Sound & Vibration*, 280(1-2):21–39, FEB 7 2005. ISSN 0022-460X. doi: 10.1016/j.jsv.2003.11.048.

SLDSE Lucato, J Wang, P Maxwell, RM McMeeking, and AG Evans. Design and demonstration of a high authority shape morphing structure. *International Journal of Solids & Structures*, 41(13):3521–3543, JUN 2004. ISSN 0020-7683.

F. Mattioni, P. M. Weaver, K. Potter, and M. I. Friswell. Multi-stable composites application concept for morphing aircraft. In M. Bernadou, J. Cagnol, and R. Ohayon, editors, *Proceedings of the 16th International Conference on Adaptive Structures and Technologies*, pages 45–52, 2006. ISBN 978-1-932078-57-2. 16th International Conference on Adaptive Structures and Technologies, Paris, FRANCE, OCT 09-12, 2005.

F. Mattioni, P. M. Weaver, K. D. Potter, and M. I. Friswell. Analysis of thermally induced multistable composites. *International Journal of Solids & Structures*, 45(2):657–675, JAN 15 2008. ISSN 0020-7683. doi: 10.1016/j.ijsolstr.2007.08.031.

Y. V. Mikhlin and S. N. Reshetnikova. Dynamical interaction of an elastic system and a vibro-impact absorber. *Mathematical Problems in Engineering*, 2006:1–15, 2006. Article ID 37980.

Z. Milovanovic, I. Kovacic, and M. J. Brennan. On the Displacement Transmissibility of a Base Excited Viscously Damped Nonlinear Vibration Isolator. *Journal of Vibration and Acoustics-Transactions of the ASME*, 131 (5), OCT 2009. ISSN 1048-9002. doi: 10.1115/1.3147140.

S. O. R. Moheimani and A. J. Fleming. *Piezoelectric transducers for vibration control and damping.* Springer-Verlag: London, 2006.

S. O. R. Moheimani and B. J. G. Vautier. Resonant control of structural vibration using charge-driven piezoelectric actuators. *IEEE Transactions on Control Systems Technology*, 13(6):1021–1035, NOV 2005. ISSN 1063-6536. doi: 10.1109/TCST.2005.857407.

S. O. R. Moheimani, D. Halim, and A. J. Fleming. *Spatial control of vibration.* World Scientific, 2003.

J. E. Mottershead and Y. M. Ram. Inverse eigenvalue problems in vibration absorption: Passive modification and active control. *Mechanical Systems and Signal Processing*, 20(1):5–44, JAN 2006. ISSN 0888-3270. doi: 10.1016/j.ymssp.2005.05.006.

S. Nagarajaiah and S. Narasimhan. Seismic control of smart base isolated buildings with new semiactive variable damper. *Earthquake Engineering & Structural Dynamics*, 36(6):729–749, MAY 2007. ISSN 0098-8847. doi: 10.1002/eqe.650.

P. B. Nair and A. J. Keane. Passive vibration suppression of flexible space structures via optimal geometric redesign. *AIAA Journal*, 39(7):1338–1346, July 2001.

H. Nijmeijer and A. van der Schaft. *Nonlinear Dynamical Control Systems.* Springer-Verlag: New York, 1990.

B. Panda, E. Mychalowycz, and F. J. Tarzanin. Application of passive dampers to modern helicopters. *Smart Materials & Structures*, 5(5): 509–516, October 1996.

K. D. Papoulia and J. M. Kelly. Visco-hyperelastic model for filled rubbers used in vibration isolation. *Journal of Engineering Materials and Technology-Transactions of the ASME*, 119(3):292–297, July 1997.

A. V. Popov, M. Labib, J. Fays, and R. M. Botez. Closed-Loop Control Simulations on a Morphing Wing. *Journal of Aircraft*, 45(5):1794–1803, SEP-OCT 2008. ISSN 0021-8669. doi: 10.2514/1.37073.

J. N. Potter, S. A. Neild, and D. J. Wagg. Generalisation and optimisation of semi-active, onoff switching controllers for single degree-of-freedom systems. *Journal of Sound & Vibration*, 329:2450–2462, 2010.

A. Preumont. *Vibration control of active structures.* Kluwer Academic Publishers: Dordrecht, 1997.

A. Preumont and K. Seto. *Active Control of Structures.* WileyBlackwell, 2008.

A. Preumont, J. P. Dufour, and C. Malekian. Active damping by a local force feedback with piezoelectric actuators. *Journal of Guidance Control and Dynamics*, 15(2):390–395, MAR-APR 1992. ISSN 0731-5090.

A Preumont, A Francois, F Bossens, and A Abu-Hanieh. Force feedback versus acceleration feedback in active vibration isolation. *Journal of Sound & Vibration*, 257(4):605–613, OCT 31 2002. ISSN 0022-460X. doi: 10.1006/jsvi.5047.

A. Preumont, A. Francois, P. De Man, and V. Piefort. Spatial filters in structural control. *Journal of Sound & Vibration*, 265(1):61–79, JUL 31 2003. ISSN 0022-460X. doi: 10.1016/S0022-460X(02)01440-2.

A. Preumont, A. Francois, P. De Man, N. Loix, and K. Henrioulle. Distributed sensors with piezoelectric films in design of spatial filters for structural control. *Journal of Sound & Vibration*, 282(3-5):701–712, APR 22 2005. ISSN 0022-460X. doi: 10.1016/j.jsv.2004.03.056.

J. C. Ramallo, E. A. Johnson, and B. F. Spencer. "smart" base isolation systems. *JOURNAL OF ENGINEERING MECHANICS-ASCE*, 128(10): 1088–1099, October 2002.

D. Sammier, O. Sename, and L. Dugard. Skyhook and H-infinity control of semi-active suspensions: Some practical aspects. *Vehicle System Dynamics*, 39(4):279–308, APR 2003. ISSN 0042-3114.

S. Sastry. *Nonlinear systems:Analysis, stability and control.* Springer-Verlag: New York, 1999.

M. R. Schultz and M. W. Hyer. Snap-through of unsymmetric cross-ply laminates using piezoceramic actuators. *Journal of Intelligent Material Systems and Structures*, 14(12):795–814, 2003.

K. A. Seffen and S. Pellegrino. Deployment dynamics of tape springs. *Proceedings of the Royal Society of London A-Mathematical Physical and Engineering Sciences*, 455(1983):1003–1048, MAR 8 1999. ISSN 1364-5021.

S. E. Semercigil, F. Collette, and D Huyni. Experiments with tuned absorber — impact damper combination. *Journal of Sound and Vibration*, 256(1): 179–188, 2002.

Y. Shen, M. F. Golnaraghi, and G. R. Heppler. Semi-active vibration control schemes for suspension systems using magnetorheological dampers. *Journal of Vibration & Control*, 12(1):3–24, January 2006.

Mohammad Shoeybi and Mehrdaad Ghorashi. Nonlinear vibration control of a system with dry friction and viscous damping using the saturation phenomenon. *Nonlinear Dynamics*, 45(3-4):249–272, AUG 2006. ISSN 0924-090X. doi: 10.1007/s11071-006-1438-2.

J-J. E. Slotine and W Li. *Applied nonlinear control.* Prentice Hall: Englewood Cliffs, NJ, 1991.

A. Y. N. Sofla, S. A. Meguid, K. T. Tan, and W. K. Yeo. Shape morphing of aircraft wing: Status and challenges. *Materials & Design*, 31(3):1284–1292, MAR 2010. ISSN 0261-3069. doi: 10.1016/j.matdes.2009.09.011.

G. Song, V. Sethi, and H. N. Li. Vibration control of civil structures using piezoceramic smart materials: A review. *Engieering Structures*, 28(11):1513–1524, SEP 2006. ISSN 0141-0296. doi: 10.1016/j.engstruct.2006.02.002.

T. T. Soong and G. F. Dargush. *Passive energy dissipation systems in structural engineering*. Wiley: Chichester, 1997.

A. Spadoni and M. Ruzzene. Static aeroelastic response of chiral-core airfoils. *Journal of Intelligent Materials Systems & Structures*, 18(10):1067–1075, Oct 2007. ISSN 1045-389X. doi: 10.1177/1045389XO6072361. 16th International Conference on Adaptive Structures and Technologies, Paris, France, Oct 9-12, 2005.

B. F. Spencer and S. Nagarajaiah. State of the art of structural control. *Journal of Structural Engineering-ASCE*, 129(7):845–856, July 2003.

A. V. Srinivasan and D. M. McFarland. *Smart structures*. Cambridge: New York, 2001.

J. A. Tamboli and S. G. Joshi. Optimum design of a passive suspension system of a vehicle subjected to actual random road excitations. *Journal of Sound & Vibration*, 219(2):193–205, JAN 14 1999. ISSN 0022-460X.

C. Thill, J. Etches, I. Bond, K. Potter, and P. Weaver. Morphing skins. *Aeronautical Journal*, 112(1129):117–139, MAR 2008. ISSN 0001-9240.

J. M. T. Thompson. *Instabilities and catastrophes in science and engineering*. John Wiley & Sons, 1982.

R. Vepa. *Dynamics of Smart Structures*. Wiley, 2010.

G. Verros, S. Natsiavas, and C. Papadimitriou. Design optimization of quarter-car models with passive and semi-active suspensions under random road excitation. *Journal of Vibration & Control*, 11(5):581–606, MAY 2005. ISSN 1077-5463. doi: 10.1177/1077546305052315.

M. Vidyasagar. *Nonlinear Systems Analysis*. Society for Industrial and Applied Mathematics: SIAM, 2002.

L. N. Virgin. *Vibration of Axially-Loaded Structures*. Cambridge, 2007.

R. Vos, R. Barrett, R. de Breuker, and P. Tiso. Post-buckled precompressed elements: a new class of control actuators for morphing wing UAVs. *Smart Materials & Structures*, 16(3):919–926, JUN 2007a. ISSN 0964-1726. doi: 10.1088/0964-1726/16/3/042.

R. Vos, R. De Breuker, R. Barrett, and P. Tiso. Morphing wing flight control via postbuckled precompressed piezoelectric actuators. *Journal of Aircraft*, 44(4):1060–1068, JUL-AUG 2007b. ISSN 0021-8669. doi: 10.2514/1.21292.

Ashwin Vyas, Anil K. Bajaj, and Arvind Raman. Dynamics of structures with wideband autoparametric vibration absorbers: experiment. *Proceedings of the Royal Soceity of London A.*, 460:1857–1880, 2003.

D. Wagg, I. Bond, P Weaver, and M. Friswell, editors. *Adaptive Structures: Engineering Applications.* Wiley: Chichester, 2007.

D. J Wagg and S. A. Neild. *Nonlinear Vibration with Control.* Springer-Verlag, 2009.

J. Wang, A. Nausieda, S. L. dos Santos E. Lucato, and A. G. Evans. Twisting of a high authority morphing structure. *International Journal of Solids & Structures*, 44(9):3076–3099, MAY 1 2007. ISSN 0020-7683. doi: 10.1016/j.ijsolstr.2006.09.008.

JF Wang, CC Lin, and BL Chen. Vibration suppression for high-speed raitway bridges using tuned mass dampers. *International Journal of Solids & Structures*, 40(2):465–491, JAN 2003. ISSN 0020-7683.

A. M. Wickenheiser and E. Garcia. Aerodynamic modeling of morphing wings using an extended lifthig-line analysis. *Journal of Aircraft*, 44(1): 10–16, JAN-FEB 2007. ISSN 0021-8669. doi: 10.2514/1.18323.

K. Worden, W. A. Bullough, and J. Haywood. *Smart Technologies.* World Scientific: Singapore, 2003.

H. Yoshioka, J. C. Ramallo, and B. F. Spencer. "smart" base isolation strategies employing magnetorheological dampers. *Journal of Engineering Mechanics—ASCE*, 128(5):540–551, May 2002.

A Zaremba, R Hampo, and D Hrovat. Optimal active suspension design using constrained optimization. *Journal of Sound & Vibration*, 207(3): 351–364, OCT 30 1997. ISSN 0022-460X.

Y. H. Zhou and J. Z. Wang. Vibration control of piezoelectric beam-type plates with geometrically nonlinear deformation. *International Journal of Non-linear Mechanics*, 39(6):909–920, AUG 2004. ISSN 0020-7462. doi: 10.1016/S0020-7462(03)00074-X.

FSC
www.fsc.org
MIX
Papier | Fördert
gute Waldnutzung
FSC® C083411

Kolibri 360 GmbH
Ferdinand-Jühlke-Straße 7
99095 Erfurt, Deutschland
produktsicherheit@kolibri360.de